W9-CRL-550

BIORELATED POLYMERS

BIORELATED POLYMERS
Sustainable Polymer Science and Technology

Edited by

Emo Chiellini
University of Pisa
Pisa, Italy

Helena Gil
University of Coimbra
Coimbra, Portugal

Gerhart Braunegg
Technical University of Graz
Graz, Austria

Johanna Buchert
VTT Biotechnology
Espoo, Finland

Paul Gatenholm
Chalmers University of Technology
Göteborg, Sweden

and

Maarten van der Zee
ATO B.V.
Wageningen, The Netherlands

Kluwer Academic / Plenum Publishers
New York, Boston, Dordrecht, London, Moscow

Library of Congress Cataloging-in-Publication Data

Biomedical polymers: sustainable polymer science and technology/edited by Emo
Chiellini ... [et al.].
    p.   cm.
    Includes bibliographical references and index.
    ISBN 0-306-46652-X
      1. Biopolymers—Biotechnology. 2. Polymers—Biodegradation. I. Chiellini, Emo. II.
International Conference on Biopolymer Technology (1st: 1999: Coimbra, Portugal) III.
International Conference on Biopolymer Technology (2nd: 2000: Ischia, Italy)

TP248.65.P62 B556 2001
668.9—dc21

2001038597

This publication was made possible by the financial support from the European
Commission through the FAIR programme; FAIR CT97-3132

BIOPOLYMER-NET

Combined Proceedings of the First and Second International Conference on Biopolymer Technology, organised
by the International Centre of Biopolymer Technology, held in Coimbra, Portugal on September 29–October 1,
1999 and in Ischia (Naples), Italy on October 25–27, 2000

ISBN 0-306-46652-X

©2001 Kluwer Academic / Plenum Publishers, New York
233 Spring Street, New York, New York 10013

http://www.wkap.nl/

10 9 8 7 6 5 4 3 2 1

A C.I.P. record for this book is available from the Library of Congress

Acknowledgements

We thank all the authors that have contributed to this document. Furthermore, we greatly acknowledge the financial support from the European Commission through the FAIR programme (FAIR-CT97-3132) which made it possible to organise the conferences and publish its results. And last but not least our deepest thanks go to Maria G. Viola who managed to transform all contributions into a camera-ready manuscript.

v

Contributors

JORGE ABURTO, Laboratoire de Chimie Agro-Industrielle, UMR INRA, Ecole Nationale Supérieure de Chimie de Toulouse, INP Toulouse, 118 route de Narbonne, 31077 Toulouse Cedex 04, France

GRAZYNA ADAMUS, Polish Academy of Sciences, Centre of Polymer Chemistry, ul. Marii Curie Sklodowskiej 34, 410819 Zabrze, Poland

ISABAELLE ALRIC, Laboratoire de Chimie Agro-Industrielle, UMR INRA, Ecole Nationale Supérieure de Chimie de Toulouse, INP Toulouse, 118 route de Narbonne, 31077 Toulouse Cedex 04, France

FABIOLA AYHLLON-MEIXUEIRO, Laboratoire de Chimie Agro-industrielle, UMR 1010 INRA/INP-ENSCT - 118, route de narbonne, 31077 Toulouse Cedex 04, France

JACKY BARBOT, Institut National de la Recherche Agronomique, Unité de Biochimie et Technologie des Protéines, B.P. 717627, 44316 Nantes Cedex 3, France

MAGNUS BENGTSSON, Department of Polymer Technology, Chalmers University of Technology, S-41296 Göteborg, Sweden

RODOLFO BONA, Institut für Biotechnologie, TU-Graz, Petersgasse 12, A-8010 Graz, Austria

ELIZABETH BORREDON, Laboratoire de Chimie Agro-Industrielle, UMR INRA, Ecole Nationale Supérieure de Chimie de Toulouse, INP Toulouse, 118 route de Narbonne, 31077 Toulouse Cedex 04, France

GERHART BRAUNEGG, Institut für Biotechnologie, TU-Graz, Petersgasse 12, A-8010 Graz, Austria

PAULO BRITO, Departamento de Engenharia Quimica da Faculdade de Ciècias e Tecnologia da Universidade de Coimbra, Pólo II - Pinhal de Marrocos, 3030 Goimbra, Portugal

FERNANDO CALDEIRA JORGE, Bresfor, Indústria do Formol, S.A., Apartado 13, 3830 Gafanha da Nazaré, Portugal

SERGIO CASELLA, Dipartimento di Biotechnologie Agrarie, Università di Padova, Agripolis, Padova, Italy

JOSÉ A.A.M. CASTRO, Department of Chemical Engineering, University of Coimbra, Coimbra, Portugal

EMO CHIELLINI, Department of Chemistry and Industrial Chemistry, University of Pisa, via Risorgimento 35, 56126 Pisa, Italy

PATRIZIA CINELLI, Department of Chemistry and Industrial Chemistry, University of Pisa, via Risorgimento 35, 56126 Pisa, Italy

FRANCESCA COLOMBO, Politecnico di Milano, Facoltà di Ingegneria, Milano, Italy

ANDREA CORTI, Department of Chemistry and Industrial Chemistry, University of Pisa, via Risorgimento 35, 56126 Pisa, Italy

JOÃO G. CRESPO, Departamento de Quimica - CQFB, Faculdade de Ciências e Tecnologia, Universidade Nova de Lisboa, 2825-114 Caparica, Portugal

OLOF DAHLMAN, Swedish Pulp and Paper Research Institue, Box 5604, S-11486 Stockholm, Sweden

WOLF-DIETER DECKWER, Gesellschaft für Biotechnologische Forschung mbH, Mascheroder Weg 1, D-38124, Braunschweig, Germany

CLAUDE DESSERME, Institut National de la Recherche Agronomique, Unité de Biochimie et Technologie des Protéines, B.P. 717627, 44316 Nantes Cedex 3, France

PIETER J. DIJKSTRA, Department of Chemical Technology and Institute of Biomedical Technology, University of Twente, P.O. Box 217, 7500 AE Enschede, The Netherlands

MARIA G. DUARTE, Department of Biochemistry, University of Coimbra, Coimbra, Portugal

RENÉ ESTERMANN, Composto+, Geheidweg 24, 4600 Olten, Switzerland

JAN FEIJEN, Department of Chemical Technology and Institute of Biomedical Technology, University of Twente, P.O. Box 217, 7500 AE Enschede, The Netherlands

JORGE M.B. FERNANDES DINIZ, Escola Secundária de Jaime Cortesão, Coimbra, Protugal

ANTOINE GASET, Laboratoire de Chimie Agro-Industrielle, UMR INRA, Ecole Nationale Supérieure de Chimie de Toulouse, INP Toulouse, 118 route de Narbonne, 31077 Toulouse Cedex 04, France

PAUL GATENHOLM, Department of Polymer Technology, Chalmers University of Technology, S-41296 Göteborg, Sweden

CARLOS F.G.C. GERALDES, Department of Biochemistry, University of Coimbra, Coimbra, Portugal

M. HELENA GIL, Departamento de Engenharia Quimica da Faculdade de Ciêcias e Tecnologia da Universidade de Coimbra, Pólo II - Pinhal de Marrocos, 3030 Goimbra, Portugal

SAMUEL GIRARDEAU, Laboratoire de Chimie Agro-Industrielle, UMR INRA, Ecole Nationale Supérieure de Chimie de Toulouse, INP Toulouse, 118 route de Narbonne, 31077 Toulouse Cedex 04, France

WOLFGANG GLASSER, Department of Wood Sci and Forest Production, Virginia Tech. Blacksburg, USA

ELIZABETH GRILLO FERNANDES, Department of Chemistry and Industrial Chemistry, University of Pisa, via Risorgimento 35, 56126 Pisa, Italy

JACQUES GUÉGUEN, Institut National de la Recherche Agronomique, Unité de Biochimie et Technologie des Protéines, B.P. 717627, 44316 Nantes Cedex 3, France

MARTIN GUSTAVSSON, Department of Polymer Technology, Chalmers University of Technology, S-41296 Göteborg, Sweden

VERA HAACK, Institute of Organic Chemistry and Macromolecular Chemistry, Friedrich Schiller University of Jena, Humboldstraße 10, D-07743 Jena, Germany

GUDRUN HAAGE, Institut für Biotechnologie, TU-Graz, Petersgasse 12, A-8010 Graz, Austria

STEFAN HAUSMANNS, Axiva GmbH, Industriepark Hoechst, G864, Frankfurt/Main, Germany

JOERN HEERENKLAGE, Technical University of Hamburg-Harburg, Department of Waste Management, Harburger Schloßstraße 37, 21079 Hamburg, Germany

ALEKSANDRA HEIMOWSKA, Gdynia maritime Academy, 81-225 Gdynia, Poland

THOMAS HEINZE, Institute of Organic Chemistry and Macromolecular Chemistry, Friedrich Schiller University of Jena, Humboldstraße 10, D-07743 Jena, Germany

UTE HEINZE, Institute of Organic Chemistry and Macromolecular Chemistry, Friedrich Schiller University of Jena, Humboldstraße 10, D-07743 Jena, Germany

SYED H. IMAM, Plant Polymer Research Unit, National Center for Agricultural Utilization Research, Agricultural Research Service, USDA, 1815 North University Street, Peoria, Illinois 61604, USA

HELENA JANIK, Gdynia Maritime Academy, Morska 83, 81-225 Gdynia, Poland

ZBIGNIEW JEDLINSKI, Polish Academy of Science, Centre of Polymer Chemistry, 41-819 Zabrze, Poland

MARIA JUZWA, Polish Academy of Science, Centre of Polymer Chemistry, 41-819 Zabrze, Poland

EL-REFAIE KENAWY, Department of Chemistry, Faculty of Science, University of Tanta, Tanta, Egypt

MAREK KOWALCZUK, Polish Academy of Sciences, Centre of Polymer Chemistry, ul. Marii Curie Sklodowskiej 34, 410819 Zabrze, Poland

KATARZYNA KRASOWSKA, Gdynia maritime Academy, 81-225 Gdynia, Poland

KRISTIINA KRUUS, VTT Biotechnology, Tietotie 2, Espoo, P.O. Box 1500, FIN-02044 VTT, Finland

COLETTE LARRÉ, Institut National de la Recherche Agronomique, Unité de Biochimie et Technologie des Protéines, B.P. 717627, 44316 Nantes Cedex 3, France

ANDREA LAZZERI, Department of Chemical Engineering, Industrial Chemistry and Material Science, University of Pisa, via Diotisalvi 2, 56126 Pisa, Italy

PAULO C. LEMOS, Departamento de Quimica - CQFB, Faculdade de Ciências e Tecnologia, Universidade Nova de Lisboa, 2825-114 Caparica, Portugal

JAN-PLEUN LENS, Agrotechnological Research Institute ATO, Subdivision Industrial Proteins, P.O.Box 17, 6700 AA Wageningen, The Netherlands

CÉCILE MANGAVEL, Institut National de la Recherche Agronomique, Unité de Biochimie et Technologie des Protéines, B.P. 717627, 44316 Nantes Cedex 3, France

LIJUN MAO, Plant Polymer Research Unit, National Center for Agricultural Utilization Research, Agricultural Research Service, USDA, 1815 North University Street, Peoria, Illinois 61604, USA

LUIGI MARINI, Novamont SpA, via Fauser, 28100 Novara, Italy

ELKE MARTEN, Gesellschaft für Biotechnologische Forschung mbH, Mascheroder Weg 1, D-38124, Braunschweig, Germany

PETER MERTINS, Aventis Research and Technologies GmbH & Co. KG, Industriepark Hoechst, G-864 Frankfurt/Main, Germany

HANNA MILLER, Technical University of Gdańsk, Chemical Faculty, Polymer Technology Department, Narutowicza 11/13, 80-925 Gdańsk, Poland

WIM J. MULDER, Agrotechnological Research Institute ATO, Subdivision Industrial Proteins, P.O.Box 17, 6700 AA Wageningen, The Netherlands

ROLF MÜLLER, Federal Institute of Technology, Universitätstr. 41, Zürich, Switzerland

ROLF-JOACHIM MÜLLER, Gesellschaft für Biotechnologische Forschung mbH, Mascheroder Weg 1, D-38124, Braunschweig, Germany

MYRIAM NAESSENS, Department of Biochemical and Microbial Technology, Faculty of Agricultural and Applied Biological Sciences, University of Gent, Coupure links 653, B-9000 Gent, Belgium

JÖRG NICKEL, German Aerospace Center, Institute of Structural Mechanics, Lilienthalplatz 7, D-38108 Braunschweig, Germany

MARJA-LEENA NIKU-PAAVOLA, VTT Biotechnology, Tietotie 2, Espoo, P.O. Box 1500, FIN-02044 VTT, Finland

LINA PEPINO, Departamento de Engenharia Quimica da Faculdade de Ciêcias e Tecnologia da Universidade de Coimbra, Pólo II - Pinhal de Marrocos, 3030 Goimbra, Portugal

RUI PEREIRA DA COSTA, Bresfor, Indústria do Formol, S.A., Apartado 13, 3830 Gafanha da Nazaré, Portugal

JOOP A. PETERS, Laboratory of Applied Organic Chemistry and Catalysis, Delft University of Technology, The Netherlands

ANTÓNIO PORTUGAL, Departamento de Engenharia Quimica da Faculdade de Ciêcias e Tecnologia da Universidade de Coimbra, Pólo II - Pinhal de Marrocos, 3030 Goimbra, Portugal

SILVANA POVOLO, Dipartimento di Biotechnologie Agrarie, Università di Padova, Agripolis, Padova, Italy

ANA M. RAMOS, Departamento de Quimica - CQFB, Faculdade de Ciências e Tecnologia, Universidade Nova de Lisboa, 2825-114 Caparica, Portugal

MARIA A. M. REIS, Departamento de Quimica - CQFB, Faculdade de Ciências e Tecnologia, Universidade Nova de Lisboa, 2825-114 Caparica, Portugal

ULRICH RIEDEL, German Aerrospace Center, Institute of Structural Mechanics, Lilienthalplatz 7, D-38108 Braunschweig, Germany

MARIA RUTKOWSKA, Gdynia maritime Academy, 81-225 Gdynia, Poland

FLORIAN SCHELLAUF, Institut für Biotechnologie, TU-Graz, Petersgasse 12, A-8010 Graz, Austria

GERALD SCHENNINK, ATO, Department of Polymers, Composites and Additives, P.O.Box 17, 6700 AA, Wageningen, The Netherlands

BEA SCHWARZWÄLDER, Composto+, Geheidweg 24, 4600 Olten, Switzerland

LUÍSA S. SERAFIM, Departamento de Quimica - CQFB, Faculdade de Ciências e Tecnologia, Universidade Nova de Lisboa, 2825-114 Caparica, Portugal

FRANÇOISE SILVESTRE, Laboratoire de Chimie Agro-industrielle, UMR 1010 INRA/INP-ENSCT - 118, route de narbonne, 31077 Toulouse Cedex 04, France

ROBERTO SOLARO, Department of Chemistry and Industrial Chemistry, University of Pisa, via Risorgimento 35, 56126 Pisa, Italy

RAINER STEGMANN, Technical University of Hamburg-Harburg, Department of Waste Management, Harburger Schloßstraße 37, 21079 Hamburg, Germany

WIM M. STEVELS, Department of Chemical Technology and Institute of Biomedical Technology, University of Twente, P.O. Box 217, 7500 AE Enschede, The Netherlands

ANITA TELEMAN, Swedish Pulp and Paper Research Institute, Box 5604, S-11486 Stockholm, Sweden

IVAN TOMKA, Federal Institute of Technology, Universitätstr. 41, Zürich, Switzerland

VÉRONIQUE TROPINI, Laboratoire de Chimie Agro-industrielle, UMR 1010 INRA/INP-ENSCT - 118, route de narbonne, 31077 Toulouse Cedex 04, France

JOWITA TWARDOWSKA, Gdynia Maritime Academy, Morska 83, 81-225 Gdynia, Poland

CARLOS VACA-GARCIA, Laboratoire de Chimie Agro-Industrielle, UMR INRA, Ecole Nationale Supérieure de Chimie de Toulouse, INP Toulouse, 118 route de Narbonne, 31077 Toulouse Cedex 04, France

MAARTEN VAN DER ZEE, ATO, BU Renewable Resources, Department Polymers, Composites and Addtives, P.O. Box 17, NL-6700 AA Wageningen, The Netherlands

JAAP VAN HEEMST, ATO, Department of Polymers, Composites and Additives, P.O.Box 17, 6700 AA, Wageningen, The Netherlands

ROBERT VAN TUIL, ATO, Department of Polymers, Composites and Additives, P.O.Box 17, 6700 AA, Wageningen, The Netherlands

ERICK J. VANDAMME, Department of Biochemical and Microbial Technology, Faculty of Agricultural and Applied Biological Sciences, University of Gent, Coupure links 653, B-9000 Gent, Belgium

MICHEL VERT, CRBA - UMR CNRS 5473, University of Montpellier 1, Faculty of Pharmacy, 15 Ave. Charles Flahault, 34060 Montpellier, France

LIISA VIIKARI, VTT Biotechnology, Tietotie 2, Espoo, P.O. Box 1500, FIN-02044 VTT, Finland

ELISABETH WALLNER, Institut für Biotechnologie, TU-Graz, Petersgasse 12, A-8010 Graz, Austria

ZHIYUAN ZHONG, Department of Chemical Technology and Institute of Biomedical Technology, University of Twente, P.O. Box 217, 7500 AE Enschede, The Netherlands

Preface

Application of polymers from renewable resources - also identified as biopolymers - has a large potential market due to the current emphasis on sustainable technology. For optimal R&D achievements and hence benefits from these market opportunities, it is essential to combine the expertise available in the vast range of different disciplines in biopolymer science and technology.

The International Centre of Biopolymer Technology - ICBT - has been created with support from the European Commission to facilitate co-operation and the exchange of scientific knowledge between industries, universities and other research groups. One of the activities to reach these objectives, is the organisation of a conference on Biopolymer Technology.

In September 1999, the first international conference on Biopolymer Technology was held in Coimbra, Portugal. Because of its success - both scientifically and socially - and because of the many contacts that resulted in exchange missions or other ICBT activities, it was concluded that a second conference on Biopolymer Technology was justified. This second conference was held in Ischia, Italy in October 2000. And again, the scientific programme contained a broad spectrum of presentations in a range of fields such as biopolymer synthesis, modification, technology, applications, material testing and analytical methods.

The originality and the high scientific quality of the presented work have convinced us to publish selected papers from both conferences. We regard the result as an excellent overview of the current "state of the art" of the European activities in the field of fundamental and applied research on biorelated polymeric materials and relevant bioplastic items.

The Editors
Emo Chiellini, Helena Gil, Gerhart Braunegg,
Johanna Buchert, Paul Gatenholm, and Maarten van der Zee

Contents

Chapter 4
**ISOLATION, CHARACTERISATION AND MATERIAL
PROPERTIES OF 4-*O*-METHYLGLUCURONOXYLAN FROM
ASPEN**
*Martin Gustavsson, Magnus Bengtsson, Paul Gatenholm, Wolfgang Glasser,
Anita Teleman and Olof Dahlman*

Chapter 5
**AN ORIGINAL METHOD OF ESTERIFICATION OF CELLULOSE
AND STARCH**
*Samuel Girardeau, Jorge Aburto, Carlos Vaca-Garcia, Isabaelle Alric,
Elizabeth Borredon and Antoine Gaset*

PART 2. BIOPOLYMER TECHNOLOGY AND APPLICATIONS

Chapter 6
**BIOPOLYMERS AND ARTIFICIAL BIOPOLYMERS IN
BIOMEDICAL APPLICATIONS, AN OVERVIEW**
Michel Vert

PART 3. (BIO)SYNTHESIS AND MODIFICATIONS

PART 4. MATERIAL TESTING AND ANALYTICAL METHODS

Chapter 24
BIODEGRADATION OF POLYMERIC MATERIALS: An Overview of Available Testing Methods
Maarten van der Zee

Chapter 25
COMPARISON OF TEST SYSTEMS FOR THE EXAMINATION OF THE FERMENTABILITY OF BIODEGRADABLE MATERIALS
Joern Heerenklage, Francesca Colombo and Rainer Stegmann

Chapter 26
STRUCTURE-BIODEGRADABILITY RELATIONSHIP OF POLYESTERS
Rolf-Joachim Müller, Elke Marten and Wolf-Dieter Deckwer

Chapter 27
BIODEGRADATION OF THE BLENDS OF ATACTIC POLY[(R,S)-3-HYDROXYBUTANOIC ACID] IN NATURAL ENVIRONMENTS

Chapter 32
COMPARISON OF QUANTIFICATION METHODS FOR THE
CONDENSED TANNIN CONTENT OF EXTRACTS OF *PINUS
PINASTER* BARK
*Lina Pepino[1], Paulo Brito, Fernando Caldeira Jorge, Rui Pereira Da Costa,
M. Helena Gil and António Portugal*

Chapter 33
THE ROLE OF LIFE-CYCLE-ASSESSMENT FOR
BIODEGRADABLE PRODUCTS: BAGS AND LOOSE FILLS
Bea Schwarzwälder, René Estermann and Luigi Marini

PART 1

BIOPOLYMERS AND RENEWABLE RESOURCES

Potato Starch Based Resilient Thermoplastic Foams

ROBERT VAN TUIL, JAAP VAN HEEMST and GERALD SCHENNINK
ATO, Department of polymers, composites and additives, P.O. Box 17, 6700 AA, Wageningen, The Netherlands

Abstract: Assuming a starch based material can be found able to replace extruded polystyrene foam, a renewable alternative would be available that strongly reduces the amount of foamed plastics in waste streams. In this study it is shown that by the foaming of potato starch based expandable beads such a material can be produced. Expandable beads out of pure potato starch were produced by extrusion compounding. Extrusion conditions and material composition were chosen such as to enable full destructurization of the starch while minimizing degradation. Extrusion yielded totally amorphous expandable beads with a glass transition temperature ranging from 70 to 120°C, depending on the water concentration. In a successive step, the expandable beads were foamed on an injection molding machine. The resulting foam properties depended strongly on the actual processing conditions and on the plasticizer content. A material composition and processing conditions were found that facilitate the processing of potato starch into resilient thermoplastic foams. At a density of 35 kg/m^3 and an average cell size of 85 μm, the properties of these foams are comparable to those of extruded polystyrene foam.

1. INTRODUCTION

As a result of the increasing amount of raw materials that is used to manufacture foamed products and their environmental impact in waste streams the awareness to reduce the use of these materials and to stimulate re-use and recycling is growing. However, these processes are limited by the complexity of the waste streams and high costs involved in re-use and recycling of foams. A possible solution could be found in materials from

Biorelated Polymers: Sustainable Polymer Science and Technology
Edited by Chiellini *et al.*, Kluwer Academic/Plenum Publishers, 2001

renewable resources that are either soluble in water or can be composted in waste streams.

Starch is one of these renewable resources. Starch is widely available and is especially suited for the production of foamed thermoplastic materials due to the intrinsic presence of a blowing agent. Destructurization of starch by means of extrusion compounding will facilitate the formation of expandable beads or foams of thermoplastic starch.

The properties of foams made from pure starch are limited however: mechanical properties are poor and starch foam is very sensitive to changes in relative humidity. To improve these characteristics, starch is chemically and physically modified or blended with other additives or polymers. This approach has proven to be very effective. Various starch-based foams have been introduced in the market, mainly in loose fill applications[1, 2.]

As an alternative to these materials the objective in this research will be to produce starch based resilient thermoplastic foams based totally on native potato starch. This is done in two steps: the production of expandable thermoplastic starch beads by extrusion and foaming of the expandable beads after conditioning in a successive foaming step. The properties of these foams will be studied as a function of processing parameters and material composition. A further objective is to determine the ultimate properties of pure starch based foams.

2.   MATERIALS AND METHODS

2.1   Materials

The starch used in this research was native potato starch, supplied by Avebe BV, The Netherlands. Additives used in this study are water, glycerol and microtalc particles. Water is used as plasticizer and blowing agent. Glycerol has a dual function: it is used as plasticizer and lubricant. The microtalc is added as nucleating agent. The grade used is MicroTalc Extra AT where 80% of the particles are smaller than 15 μm, obtained from Norwegian Talc.

2.2   Extrusion

The objective of extrusion compounding is to produce amorphous expandable starch beads with a defined glass transition temperature. Extrusion is done on a Clextral BC45 co-rotating twin screw extruder

(L/D=23) under mild processing conditions using different screw configurations (maximum temperature 110°C, maximum screw speed 50 rpm). To destructurize the starch grains properly, water is added as plasticizer (25-35% b.w. based on dry starch content). Glycerol, 3% b.w. based on dry starch content, acts as plasticizer and lubricant. As nucleating agent microtalc particles are added to the starch/glycerol mixture (1% b.w.). The extruded material was granulated, dried and conditioned to obtain beads with various moisture contents.

2.3   Foaming

The thermoplastic starch beads are foamed on a Demag D60 injection molding machine equipped with a standard PE screw (compression ratio 1:2). Beads with the desired moisture content (10-15% b.w.) are fed into the injection unit, melted and injected in the same fashion as in a regular injection molding process. The exception, however, is that the melt is injected into free air instead of a mould. The temperature used for foaming ranges from 150 to 200°C.

2.4   Analysis

To characterize the materials after extrusion, the beads were conditioned for one week at 20°C at different relative humidity values to obtain various moisture contents. Characteristics were obtained using differential scanning calorimetry (DSC) and X-ray measurements. DSC measurements are performed to determine the glass transition temperature of the beads. X-ray measurements provide an indication about the degree and, if present, the type of crystallinity in the extruded materials.

The resulting foam is conditioned at 60% R.H. at 20°C for one week. Mechanical tests, i.e. compressive testing and resilience testing, are performed according to the method described by Tatarka and Cunningham[3]. The foam density was calculated by sand replacement volumetric measurement. The cell structure was analyzed by Scanning Electron Microscopy.

3.   RESULTS AND DISCUSSION

3.1   Extrusion compounding

To produce amorphous thermoplastic beads out of native starch the granular and crystalline structure of the starch grains must be broken down. Destructurization can be achieved by the combined effect of shear and increased temperature involved in extrusion under the addition of sufficient plasticizer. However, subjecting starch to shear forces and high temperatures in this fashion also causes the unwanted effect of degradation. The final properties of the extruded beads and the resulting foam will therefore depend on the degree of destructurization and degradation.

The degree of destructurization strongly depends on the material composition and the processing history. Aichholzer and Fritz[4] found that for water plasticized starches at a maximum extrusion temperature of 120°C full destructurization was observed using cross-polarized light microscopy and that at these conditions the degree of destructurization was independent of screw speed and screw design when using only positive kneader blocks.

Fig 1 shows the two screw configurations that were used for the extrusion of starch. Design A has different zones with positive kneader elements while design B uses reverse screw elements (RSE). Using design B will result in an increase of the filled region and in a change of the pressure profile along the extruder axis.

Figure 1. Screw designs used in extrusion (right hand side is exit from barrel). Design A has positive kneader elements; design B has reverse screw elements.

To determine the degree of destructurization X-ray diffraction measurements were performed on the extruded beads. Fig 2 shows the results of these measurements. In the figure the X-ray diagram of native potato starch is added as a reference. The curves clearly show that the initial peaks that can be attributed to native starch disappear upon processing.

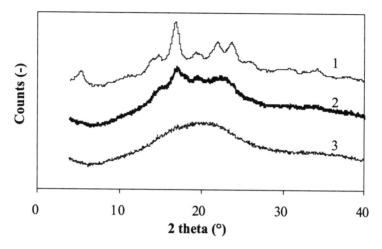

Figure 2. X-ray diagrams of 1. native potato starch; 2. extruded starch using screw design A and 3. extruded starch using screw design B.

In the case of design B a fully amorphous hump is observed, while the curve resulting from design A shows some residual B-type crystallinity[5]. This crystallinity is most likely not a result of re-crystallization due to the fast cooling after extrusion and the relatively high T_g's of the starch based materials. Furthermore, the figure shows no peaks that Van Soest[5] attributes to V-type crystallinity resulting from the processing history. Using design B for extrusion yields a truly amorphous material. This can most likely be allocated to the increase in residence time when compared to design A and to a higher pressure in the extruder, facilitating destructurization. The results of the X-ray measurements demonstrate that full destructurization of the native starch is achieved. The unwanted effect of degradation that is incorporated in the extrusion process has been minimized by optimizing the process conditions and material composition during extrusion. The mechanical and thermal history of the material plays an important role in the degree of destructurization and degradation. The addition of water and glycerol will lower the viscosity and minimizes wall slip, which in consequence lowers the applied shear. Extruding at a relatively low temperature (at a high water concentration), a low screw speed and a simple screw design minimizes degradation. In the present study, the presence of degradation was accessed in terms of die pressure, torque and material color. This has resulted in the optimal extrusion conditions as mentioned in Table 1.

Table 1. Extrusion conditions for the extrusion of potato starch based expandable beads.

Temperature [°C]	Screw speed [rpm]	Throughput [kg/h]	Added water [%]	Die pressure [bar]	Torque [A]
110 (max)	50 (max)	9-10	25-35	45-55	38-43

Since water is a plasticizer for starch, the glass transition temperature depends on the water content. DSC measurements were used to determine the glass transition temperature as a function of the water content. Fig 3 shows the results of these measurements. Della Valle et al.[6] have collected data on the dependency of the T_g on the water content from several sources. The area between the dashed lines in the figure represents the band in which this data can be found. From the figure it is obvious that the present measurements, represented by the solid line, fully agree with this data.

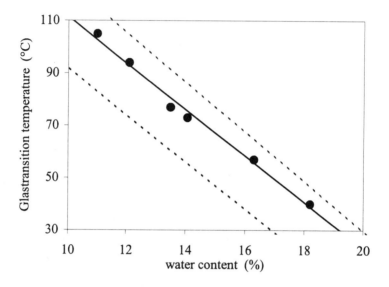

Figure 3. Glass transition temperature of potato starch beads as function of the water content. Dashed lines interpretation of Della Valle et al.[6]

3.2   Foam processing-property relationships

The expandable beads that have been produced by extrusion compounding are conditioned to a certain water content and are foamed by means of injection molding into free air. In this process the water in the beads has a dual function; it acts as plasticizer and as blowing agent. Water as plasticizer, in combination with the elevated process temperature, has to provide sufficient softening of the starch to allow for the large material deformations involved in bubble expansion. However, the amount of plasticizer is limited. Too much plasticizer will lower the melt viscosity too much and the melt will not be able to sustain gas bubbles finally leading to cell collapse. Water as blowing agent, again in combination with the elevated temperature, has to develop sufficient vapor pressure for expansion of the foam, i.e. overcoming the opposing internal forces in the material when stretching the cell walls. In this respect a high water content is favorable for a higher 'driving force' enabling expansion to low foam densities.

Therefore there is a strong interaction between the plasticizer and the blowing agent during foaming. Continuing loss of plasticizer (diffusion of water to the bubbles) from the starch leads to a continuing increase of melt viscosity. The water gained in the bubbles is favorable in a further pressure build-up needed for continuing foam expansion. At some point the material has stiffened that much due to loss of plasticizer that the gas pressure becomes insufficient to stretch the cell walls further and the expansion stops. At this point a freeze up of the foam structure can be observed. Water furthermore diffuses to the outside environment. This loss of water obviously competes with pressure build-up within the cells.

3.2.1   The influence of the processing temperature

The influence of the temperature during foaming on the properties of the resulting foam was studied at temperatures ranging from 150°C to 200°C. It was found that at temperatures below 150°C the material did not fully melt. Temperatures above 190°C cause the onset of degradation. Furthermore, the driving force of the water vapor at temperatures below 150°C is insufficient to expand the very high viscous starch melt. Within the range suited for foaming, variations in process temperature show their influence on resulting foam properties. Foam density and cell size strongly depend on the temperature. This is shown in Figs 4 and 5. Apart from the water content, the material composition is the same for all foaming experiments.

The foam density was found to decrease with increasing processing temperature (Fig 4). This can be attributed to the fact that at higher temperatures the water vapor has a larger driving force and facilitates a larger expansion. Furthermore, at higher temperatures the viscosity of the material decreases allowing for larger deformations. The diffusion of water to the outside environment leads to an increase of the glass transition temperature and to freeze up of the starch. This effect is more pronounced at higher temperatures. This whole foaming process comprising nucleation and growth, expansion, T_g and viscosity increase and cell structure freeze up takes place in the order of 1 second and accounts for the fact that coalescence of bubbles hardly occurs.

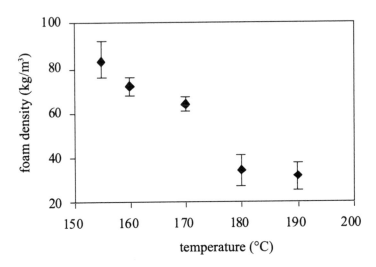

Figure 4. Foam density as function of the processing temperature for potato starch expandable beads with a water content of 10% at constant injection speed.

The result is that with increasing temperatures the cell diameter decreases, as shown in Fig 5. At higher temperatures a larger number of microbubbles are nucleated[7] that reach full expansion (also see section 3.2.3), which explains why at higher temperatures both the cell diameter and density decrease.

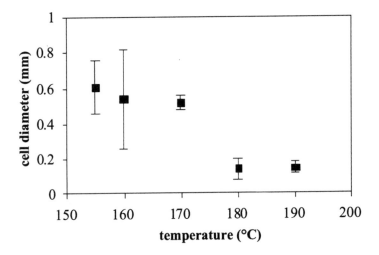

Figure 5. Cell diameter as function of the processing temperature for potato starch expandable beads with a water content of 10% at constant injection speed.

3.2.2   The influence of the water content

The influence of the water content during foaming on the properties of the foam was studied in the range of 10% to 15%. The lower limit in water content represents a boundary where it is still possible to properly melt the beads. Below this limit, a temperature above the onset of degradation would be needed to enable the melting of the material. Increasing the moisture content has an increasing plasticizing effect and decreases the T_g. This facilitates bubble expansion and is a wanted effect up to a certain limit. Beyond this limit the plasticizing effect of water in the starch continues to exist after expansion to maximum volume because not all the water has diffused from the material. This remaining water counters the freeze up of the existing cell structure and invokes cell collapse leading to high densities. In the high moisture content region, however still below the 'collapse limit', foam shrinkage will occur as the foam cools down. This is reflected in the foam density as shown in Fig 6.

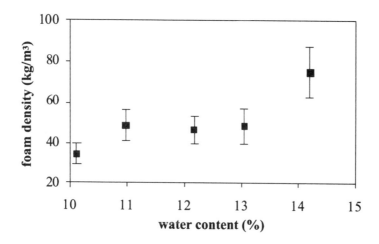

Figure 6. Foam density as function of the water content for potato starch expandable beads at 180°C and at constant injection speed.

Fig 6 also clearly shows the duality in blowing agent and plasticizer activity of the water. If the water would only have a plasticizing function one would expect a decrease in the density as a function of the water content as can be seen in Fig 4 for the dependence of the density on temperature. If the water would solely be a blowing agent not more than 2% water would be needed to reach a density in the range of 30 kg/m^3. Therefore, even lower densities might be expected at water contents lower than 10% if the plasticizing function is taken over by another additive. Increasing the water content above 10% facilitates cell coalescence and collapse.

3.2.3   The influence of the flow rate and nucleating agent

At a constant water content (10%) and a constant foaming temperature (180°C), the influence on the flow rate and addition of microtalc was studied. The flow rate is controlled by setting the injection speed on the injection molding machine. Obviously, setting the injection speed to a certain value at a constant die opening also influences the material pressure at the die. Increasing the injection speed also increases this pressure. This has been confirmed by experiments using an injection unit equipped with a pressure sensor.

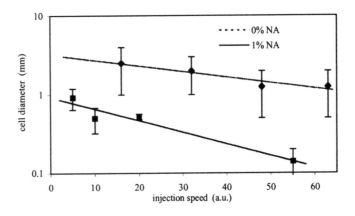

Figure 7. The cell diameter as function of the injection speed for potato starch expandable beads with a water content of 10% at 180°C. NA denotes nucleating agent.

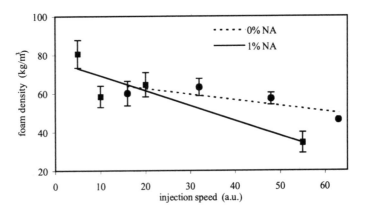

Figure 8. The foam density as function of the injection speed for potato starch expandable beads with a water content of 10% at 180°C. NA denotes nucleating agent.

The influence of the injection speed on the cell diameter and the foam density is shown in Figs 7 and 8. The figures further show the effect of the addition of microtalc. Increasing the injection speed influences the magnitude and the speed of the pressure drop. The pressure drop becomes larger and the pressure drop occurs faster. This is favourable in several ways.

First, it facilitates nucleation[8]; the number of nucleated microbubbles increases[9] and the number of microbubbles with a pressure exceeding the critical pressure for expansion also increases[10]. Effectively, the number of cells that coexist and expand without coalescence increases which results in a smaller cell diameter. From Fig 7, this effect is not that strong for starch without a nucleating agent, but the presence of microtalc certainly intensifies this effect. In general, the presence of a nucleating agent is favourable because it provides a larger number of so-called 'hot spots' for bubble nucleation and the nucleation of satellite bubbles[11, 12] than material without a nucleating agent. Furthermore, the presence of a nucleating agent contributes to a more uniform foam structure. The differences in cell sizes are smaller which helps in limiting the coalescence of cells.

Secondly, the initial driving force (i.e. the water vapour pressure) increases with increasing injection speed. This enables a larger expansion of the foam resulting in a lower foam density. This effect is only stronger for starch with a nucleating agent at higher flow rates.

3.3   Cell structure and mechanical properties

From the experiments described in the previous sections foaming conditions have been found that produce a starch foam with small cells and low density. The cell structure and mechanical properties of the resulting foam have been analyzed. All foam samples are conditioned at 20°C at 60% R.H. for one week prior to measurements.

When using the appropriate processing conditions it is possible to obtain a high expansion factor resulting in a low foam density. The combination of a low water content, high injection speed, high process temperature and material composition produces foam densities ranging from 30 to 40 kg/m^3. This indeed represents a high expansion factor, since the density of the expandable starch beads typically is about 1400 kg/m^3.

Figs 9 and 10 show the cell structure of the potato starch foam. From Fig 9 it is evident that the foaming process produces closed cells. The number of cell faces varies slightly but on average is 12 to 14, which is a realistic value for foams according to Gibson and Ashby[13]. Though this figure shows that there are small differences in cell size, Fig 10 shows that the cell structure has a high degree of uniformity in size and distribution. The cell size is very small, up to 150 μm, which is in the same range as cross-linked PE.

Figure 9. SEM photo of potato starch foam (magnification 400x).

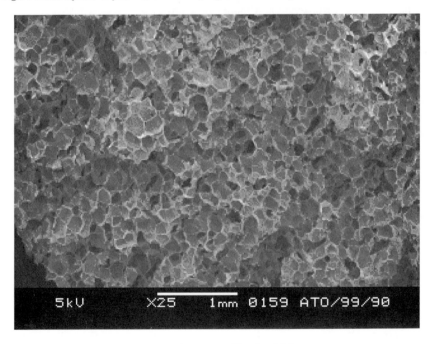

Figure 10. SEM photo of potato starch foam (magnification 25x)

Mechanical testing on the foam is performed to measure the compressive stress and the resiliency. The results from these tests are shown in table 2, in which the properties of extruded polystyrene foam (XPS) and commercial starch-based loose-fill foams (Eco-foam and Mater-Bi) together with EPS loose-fill foam (Pelaspan Pac) are added for comparison. The values of XPS are obtained from tests on typical XPS retail packaging trays. The table shows that the compressive stress reached with potato starch foam is comparable with that of XPS. Through the cell structure of the potato starch foam (high cell density, very small cells) a good resiliency can be obtained, although pure starch plastics exhibit brittle fracture behavior. This brittle fracture still is present on the microscopic scale of the individual cells but due to the cell density, the foam exhibits resiliency on macroscopic scale.

Table 2 shows also that while Eco-Foam and Mater-Bi foams have comparable properties to PS-based loose-fill materials, the properties of potato starch foams are comparable with XPS. This indicates that potato starch foams could be suitable for XPS-like applications.

Table 2. Comparison of properties of XPS, EPS, potato starch foam, Eco-foam and Mater-Bi. Data for XPS, potato starch foam and Mater-Bi from tests in this study. Data of Eco-foam and Pelaspan Pac from Tatarka and Cunningham[3] (conditioned at 23°C at 50% R.H.)

Material	Foam density [kg/m^3]	Cell diameter [mm]	Compressive stress [MPa]	Resiliency [%]
XPS	45.6	0.55	0.14	72.8
Potato starch foam	34.5	0.14	0.17	55.3
Pelaspan Pac (EPS)	9.6	-	0.082	79.3
Mater-Bi	16.8	-	0.040	62.3
Eco-foam	19.1	-	0.055	70.7

Potato starch foam samples have been conditioned at several relative humidity values. In these experiments, the commercially available Eco-Foam and Mater-Bi have been tested in the same fashion for comparison. From these experiments it was found that the potato starch foam has a good humidity stability up to values of about 75 to 80%. Only at high humidity the foam absorbs that much moisture from the environment that the T_g decreases so the foam shrinks noticeable or collapses. The results are shown in Fig 11 and demonstrate that the humidity stability of the potato starch foam competes very well with these commercial foams up to 78% R.H.

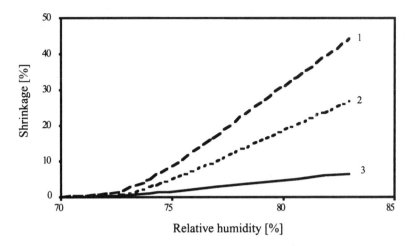

Figure 11. Foam shrinkage as function of the relative humidity for 1. Eco-foam; 2. Potato starch foam and 3. Mater-Bi

4.   CONCLUSION

In this research study, the possibility of producing expandable beads out of pure potato starch by extrusion compounding has been studied. By choosing the parameters governing the extrusion process adequately, it is possible to destructurize starch completely while minimizing degradation. Extrusion yields a totally amorphous material with a glass transition temperature ranging from 70 to 120°C.

Within a certain processing window it is possible to foam the expandable beads. The resulting foam properties depend strongly on the actual processing conditions during foaming. The correct parameters enable the production of a resilient thermoplastic foam from pure potato starch. Material freeze up due to a fast drop in T_g is the predominant mechanism for cell growth and stabilization of these foams.

The characteristics of the developed material demonstrate that pure potato starch can be used for the production of foams without the need of modified starches or blends with other additives or polymers. This will make new applications for thermoplastic starch based foams possible.

REFERENCES

1. Belloti, V., Bastioli, C., Rallis, A. and Del Tredici, G., 1995, Expanded articles of biodegradable plastic material and a process for the preparation thereof, EP0667369.

2.   Lacourse, N. and Altieri, 1989 P. Biodegradable packaging material and the method of preparation thereof, US4,863,655
3.   Tatarka, P.D. and Cunningham, R.L., 1998, Properties of protective loose-fill foams, *J. Appl. Pol. Sci.* **67** (7), pp. 1157-1176
4.   Aichholzer, W. and Fritz, H-G, 1996, Charakterisierung der Stärkedestrukturierung bei der Aufbereitung von bioabbaubaren Polymerwerkstoffen. In *Stärke* **48** (11/12), pp. 434-444
5.   Van Soest, J. ,1996, PhD thesis *Starch plastics: structure-property relationships*, University of Utrecht
6.   Della Valle, G., Vergnes, B. Colonna, P. and Patria, A., 1997, Relations between rheological properties of molten starches and their expansion behaviour in extrusion, *J. Food Eng.* **31** (3), pp. 277-295
7.   Van Heur B., 1991, PhD Thesis *In-situ foaming and its application in the production of a leading edge*, Delft University of Technology
8.   Throne, J. 1996, *Thermoplastic foams*, Sherwood Publishers, Hinckly
9.   Hobbs S.Y., 1975, Bubble growth in thermoplastic structural foams, *G.E.P. technical information series*, nr. 75CRD210
10. Gent, A.N. and Tompkins, D.A., 1969, Nucleation and growth of gas bubbles in elastomers, *J. Appl. Physics* **40** (6), p. 2520
11. Han, C.D. and Villamizar, C.A., 1978, Studies on structural foam processing I, the rheology of foam extrusion., *Pol. Eng. Sci.* **18** (9), pp. 687-698
12. Villamizar, C.A. and Han, C.D., 1978, Studies on structural foam processing II, bubble dynamics in foam injection molding, *Pol. Eng. Sci.* **18** (9), pp. 699-710
13. Gibson, L. and Ashby, M, 1988, *Cellular solids; structures and properties*, Pergamon Press, Oxford

The New Starch

[1]IVAN TOMKA, [1]ROLF MÜLLER, [2]STEFAN HAUSMANNS and [3]PETER MERTINS

[1]*Federal Institute of Technology, Universitätstr. 41, Zürich, Switzerland; [2]Axiva GnbH, Industriepark Hoechst, G 864, Frankfurt/Main, Germany, [3]Aventis Research and Technologies GmbH & Co KG, Industriepark Hoechst, G-864, Frankfurt/Main, Germany*

Abstract: A polymer - analogous to starch - was previously prepared by biotransformation from sucrose and the enzyme amylosucrase. The reaction yields a crystalline precipitate of a molecule of degree of polymerisation 60-100 glucose units and fructose monomer. The molecular characterisation of the polymer by ^{13}C-NMR revealed a 1,4-coupling of the anhydroglucose units via the alpha glycosidic bond in a strictly linear manner. The new polymer is soluble in 1n KOH and DMSO at 20°C. The crystalline precipitate, as synthesised, shows in X-ray wide angle powder diffraction the B-structure of native potato starch. Upon heating in the presence of 18% by weight of water the crystallites transform to the A-structure of native wheat starch. The phase diagram of the new polymer, starch and the solvent/non-solvent shows - above a critical concentration - the formation of a gel phase that comprises an elastically active network. In this network the cross-links are the crystallites of the two polymers with an entirely new structure and the tie molecules are the long branches of the amylopectine. Such gels can be made by an extrusion process and shaped into oriented, elastic fibers and films.

1. INTRODUCTION

Several attempts were reported in recent years[1-3] for the production of approximately linear chains of poly(glucanes) (PG). The selected general route to target the linear compounds was by using native branched PGs and modify them in biotransformation using selected debranching enzymes. In such enzyme essays the enzyme-substrate reaction can't be fully completed due to the fact that the substrate is a macromolecule. The resulting

Biorelated Polymers: Sustainable Polymer Science and Technology
Edited by Chiellini *et al.*, Kluwer Academic/Plenum Publishers, 2001

poly(glucanes)[1-3] still contain residues of branched molecules of an extremely broad molar mass distribution.

In native starches, long-branched amylopectine molecules form the crystalline phase and the eventually present short-branched amylose constitutes the amorphous state. Pairs of the branches - with each of its constituents strictly linear and a degree of polymerisation of 26 - fold into the native, parallel oriented double-helices, which again build a planar-hexagonal packed crystalline phase. In the metastable B-structure, one of three sites of this packing are filled by water and in the stable A-structure the hexagonal closed packing comprises solely the double-helices. As usual only strictly linear poly(glucanes) are able to attain regular conformations and ready to crystallise.

Especially in the case of amylose phosphoric acid ester residues contaminate these native starch molecules. The enzyme essay doesn't allow to get rid of such contaminations. The resulting poly(glucane) may show some enhancement of crystallinity but a high degree of it could not achieved along this route.

The basic motivation to gain the linear poly(glucanes) is in the attempt to raise the crystallisation tendency of the native poly(glucane).

The target of the present work was to produce poly(glucanes) in the crystallinity range of 95-100 % by weight. For this reason a polymer - analogous to starch - was sythesised from sucrose and the enzyme amylosucrase.

2.   EXPERIMENTAL

The following procedure to prepare the specific poly(1,4-α-D-glucane) – which we will call as "neoamylose" in the present paper - is described[4]: in a batch reactor 20 % by weight of sucrose was dissolved in sodium citrate buffer at pH 6.5. The temperature was set to 37 °C and the reaction was started by adding 20 ST units/ml of amylosucrase to the sucrose solution under stirring at 200 rpm. The duration of the reaction was 72 hours. The reaction yields a crystalline precipitate and fructose monomer.

The methods for the molecular characterisation by ^{13}C NMR-spectroscopy, the detection of crystallinity by wide angle X-ray diffraction and the registration of thermal transitions in the heat flow calorimeter were described[5].

3.   RESULTS

The crystalline precipitate is a new polymer – "neoamylose" - soluble in 1n KOH and DMSO at 20°C and in water above 140 °C in a confined volume.

The molecular characterisation of the "neoamylose" by ^{13}C-NMR revealed a 1,4-coupling of the anhydroglucose units via the alpha glycosidic bond in a strictly linear manner[5] and a chain length of degree of polymerisation 60-100 glucose units

The crystalline precipitate, as synthesised, shows in X-ray wide angle powder diffraction which is equivalent to the B-structure of native potato starch. Upon heating in the presence of 18% by weight of water the crystallites transform to the A-structure of native wheat starch.

The new polymer can be purified by recrystallisation from water. Water dissolves at 140 °C 30 % by weight of the crystalline substance (Fig 1).

By cooling the solution to a temperature < 80 °C the polymer precipitates (Fig 2) forming spheres (Fig 4) of 100 % crystallinity (Fig 3) of the native crystalline A structure of starch. The crystalline spheres vary in size – in the range of 50 nm to 5 000 nm - depending on the temperature and concentration during precipitation as well as on the duration of the crystal growth.

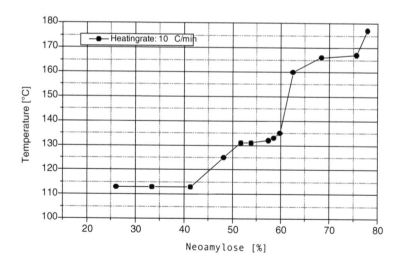

Figure 1. Peak melting temperature of the "Neoamylose" plotted against its concentration.

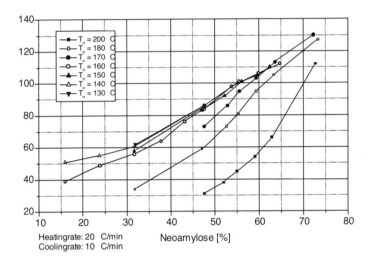

Figure 2. Temperature of crystallization Tr of the "neoamylose" as a function of its concentration and of the melting temperature Tm. The parameter Th is the highest temperature the sample was exposed during dissolution.

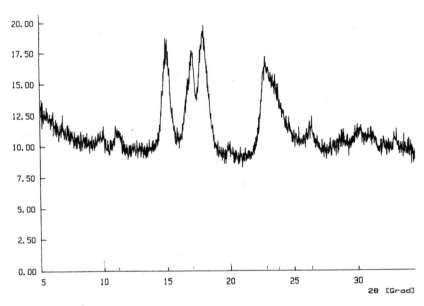

Figure 3. Wide angel X-ray diffraction pattern of the "neoamylose" Scattered intensity plotted against scattering angel 2θ.

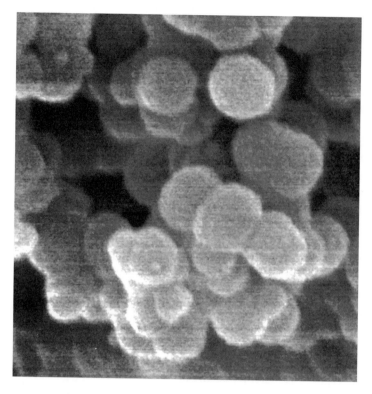

Figure 4. Spherical particles are crystals of "neoamylose"-; magnification 18'000x

The new polymer co-crystallises with amylopectine from solution. Nucleation and growth of the crystallites, of course, are influenced by the degree of over saturation. The phase diagram of the new polymer, starch and the solvent/non-solvent shows - above a critical concentration - the formation of a gel phase that comprises an elastically active network. The network shows limited swelling in water below 140 °C and it dissolves above this temperature.

4. DISCUSSION

In native starches, branched amylopectine molecules form the crystalline phase and the eventually present amylose constitutes the amorphous state. Pairs of the branches with a degree of polymerisation of 26 fold into the native, parallel oriented double-helices, which build a planar-hexagonal packed crystalline phase. In the metastable B-structure, one of three sites of this packing are filled by water and in the stable A-structure the hexagonal closed packing comprises solely the double-helices. In the course of the enzymatic synthesis the new polymer folds to the thermally more stable double-helix, as soon as the necessary degree of polymerisation of 52 is reached and these precipitate by forming the planar lattice in which the axes of the helices are perpendicular to the plane of hexagonal order. Folding and packing leading to the precipitation of the crystalline phase are concurring synthesis, which explains the resulting somewhat broader molar mass distribution compared with the minimal length of the stable double-helical regular conformer. Native amylose is not capable to form these conformers due to its much larger overall molecular length and therewith associated steric hindrance in the condensed solution phase. Nucleation and growth of the crystallites, of course, are influenced by the degree of over saturation.

The "neoamylose" co-crystallises with amylopectine from solution. In the network formed by cocrystallisation of the two polymers the tie molecules are the long branches of the amylopectin and the network-branches are the crystallites. Such gels can be made by an extrusion process, melting the starch and the new polymer and shaped into oriented, elastic fibers and films.

The "Neoamylose" forms with iodine and analogous compounds molecular complexes. X-ray diffractograms show the same intensity distribution with scattering angle as complexes of amylose from native potato starch. In these molecular complexes amylose formes V-helices. The binding capacity for iodine was higher for the new polymer.

5.   CONCLUSION

By biotransformation of sucrose with amylosucrase a strictly linear polysaccharide a poly(1,4-α-D-glucane), the "neoamylose" was prepared. The new polymer is fully crystalline free of amorphous content and resists enzymatic attack by α-amylase. It forms via cocrystalisation with starch or amylopectine from solution elastically active gels.

The two-fold capability of the "new starch" to bind iodine in a single helical V-type regular conformer and the double helical regular folding similar to that of the amylopectin branches in native starch *distinguish the new polymer as the starch per se.*

REFERENCES

1. Kossmann, J., Büttcher, V., Welsh, T. Patent: Dna Sequences Coding For Enzymes Capable Of Facilitating The Synthesis Of Linear 'Alpha'-1,4 Glucans In Plants, Fungi And Microorganisms, WO-A-95/31 553

2. Harada, T., Misaki, A., Akai, H., Yokobayashi, K. and Sugimoto, K., 1972, Characterisation of *Pseudomonas* isoamylase by its actions on amylopectin and glycogen: Comparison with *Aerobacter* pullulanase, *Biochim. Biophys. Acta*, **268**: 497.
3. Hijka, H., Joshida, M. Patent: Process For The Production Of Amylose Films, US-A-3734760
4. Gallert, K-CH., Bengs, H. and Simandi, C. Patent: Polyglucan And Polyglucan Derivatives Which Can Be Obtained From Amylosucrase By Biocatalytic Production In The Presence Of Biogenic Substances, WO-A-00/22155.
5. Willenbücher, R. W., Tomka, I., and Müller, R., 1992, Thermally Induced Structural-Transitions in the Starch-Water System. In *Prog. Plant. Polym. Carbohydr. Res. (II part: Proc. Int. Symo. Plant. Polym. Carbohydr.)* 7th volume, Behr's Verlag, Hamburg, Germany, pp. 219-225.

Structural Materials Made Of Renewable Resources (Biocomposites)

JÖRG NICKEL and ULRICH RIEDEL
German Aerospace Center, Institute of Structural Mechanics, Lilienthalplatz 7, D-38108 Braunschweig, Germany

Abstract: In view of the increasing shortage of resources as well as growing ecological damage, the aspects of the exploitation of raw materials and the recovery after the end of the lifetime of products have to increasingly be taken into consideration. In addition, the aspect of saving energy by means of lightweight constructions must also be regarded. The use of conventional, i.e. petrochemically-based plastics and fibre-reinforced polymers, the production process, as well as usage and recovery are often very difficult and demand considerable technical resources. An answer to solve all these problems may be provided by natural fibre-reinforced biopolymers based upon renewable resources, called biocomposites in the following. By embedding plant fibres, e.g. from flax, hemp, or ramie (cellulose fibres) into biopolymeric matrices, e.g. derivatives from cellulose, starch, shellac, or plant oils, fibre-reinforced polymers are obtained that can be integrated into natural cycles in an environmentally-friendly manner, e.g. by classic recycling, by CO2-neutral incineration (including recovery of energy), and possibly by composting.

1.   INTRODUCTION

The original area of application for fibre composite materials has been in the field of aerospace. In the meantime, however, these structural materials are also being used for a number of other technical applications. The use of fibre composite materials makes particular sense when the goal is to achieve a high degree of stiffness and strength at a low weight. The low density of the matrix resins being used (unsaturated polyester, phenol resins, epoxy resins) and the imbedded, very strong and stiff fibres (glass, aramid, and

Biorelated Polymers: Sustainable Polymer Science and Technology
Edited by Chiellini *et al.*, Kluwer Academic/Plenum Publishers, 2001

carbon fibres) are responsible for the good specific, i.e. weight-related properties. In addition, the possibility to tailor each component to specific requirements by arranging the load-bearing (long) reinforcement fibres in the direction of the applied forces is taken advantage of during fabrication. Thus the actual composite material is not developed until the production of the component. As a result, various production technologies have been created for this process.

Petrochemically-based fibre composite materials often pose considerable problems with regard to disposal at the end of their lifetime. Such a material's combination of various and usually very resistant fibres and matrices makes recycling very difficult. Due to increasing environmental problems, merely disposing of the product is becoming less of an option. For this reason, environmentally-friendly alternatives such as the recycling of raw materials, CO_2-neutral thermal utilisation or, in certain cases, biodegradation are being looked into. The development of a structural material based on renewable resources, made of natural fibres, and embedded in so-called biopolymers as well as the availability of economically and ecologically sound component production methods are the subject of current research work at the DLR's Institute of Structural Mechanics[33].

2.   NATURAL FIBRE-REINFORCED STRUCTURAL MATERIAL

2.1   Natural Fibres as Reinforcement

The fibres in a fibre composite material serve to reinforce the material and therefore must provide a high degree of tensile strength and stiffness, whereas the matrix usually shows a time- and temperature-dependent behaviour, considerably lower tensile strength as well as a comparably greater amount of elasticity. This means that the mechanical properties of the fibres have considerable influence on the strength and stiffness behaviour of the composite. Very thin fibres are usually used since they have a large surface/volume ratio which, in turn, has a positive effect on the adhesion between the fibres and matrix. The required quality of the fibre material depends on the desired stiffness and strength behaviour of the composite[1-5]. Further selection criteria for the determination of the appropriate reinforcement fibres are the:
–   Maximum elongation
–   Heat resistance

- Adhesion between fibres and matrix
- Dynamic behaviour
- Long-term behaviour
- Price and processing costs

Natural fibres can be subdivided into plant, animal, and mineral fibres. All plant fibres (cotton, jute, flax, hemp, etc.) are made of cellulose; animal fibres are made of protein (wool, silk, hair). Based upon their origin, plant fibres are subdivided into bast and hard fibres. Bast fibres are derived from the stems or stalks of plants; hemp, jute, ramie, and flax, for example, belong to this category. Hard fibres, on the other hand, are derived from leaves, leaf sheaths, or fruit; sisal and coconut belong to this category[6-16].

As already mentioned, specific mechanical properties are decisive when using natural fibres in composite materials. The fact that hemp, flax, and ramie natural fibres can compete with technical fibres is demonstrated in Fig 1.

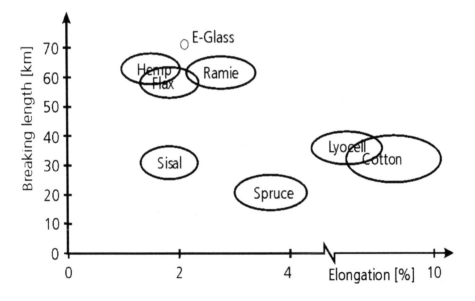

Figure 1. Comparison of the properties of different natural reinforcement fibres.

As seen in diagram 1, it is clear that the ramie, flax, and hemp fibres have the greatest breaking lengths compared to sisal, spruce, and cotton fibres. Because of a considerably greater maximum elongation of up to 10%, cotton fibre is not suitable as a reinforcement fibre for composite materials. E-glass (e.g. Al-B-silicate glass[17]) was included in the diagram as a reference fibre because it is of great importance in fibre composite technology[6, 18].

2.2   Biopolymers as Matrix Systems

The described biopolymers are embedded in a biopolymer matrix whose task is to guarantee the stability of the structure, to protect the mechanically high-quality fibres from radiation and aggressive influences, and to transmit shear forces. The polymers are usually subdivided into thermosets and thermoplastics which are equally suitable as matrices for the structural material.

Biopolymers are the type of polymer whose basic constituents consist mainly of renewable resources. These basic constituents can be formed either from the polymer chain or side chain, or from the monomers which are to be cross-linked to the polymers. A number of formation possibilities for biopolymers can be created as result.

Naturally synthesised polymers like, e.g., starch and cellulose must be physically or chemically transformed in order to be processed as a thermoplastic material. Starch, for example, can be thermoplastified by means of de-structuring with the aid of other elements (e.g. glycerine and water)[19, 20]. Afterwards, an attempt is often made to improve the properties by adding copolymers, which also can have a petrochemical origin (one product is, e.g., Mater-Bi)[21]. But this effect can also be attained by means of partial or complete esterification of the side chain hydroxyl groups with short-chained organic acids like e.g. acetic acids and possibly the addition of softening agents (one product is, eg., Sconacell A)[22]. In addition, other physical, chemical, mechanical, and thermal properties of the biopolymer are influenced by this. In order to thermoplastify the cellulose while maintaining the cellulose chain structure, primarily the hydroxyl group on the side chain is esterified (one product, e.g., is Bioceta)[23-25].

During the process of biotechnical synthesis – usually a fermentative process - polymers primarily composed of micro-organisms are derived (one product, e.g. is Biopol)[26]. These polymers serve as energy storage for the micro-organisms. In comparison, starch fulfills the task of energy storage in plants. The most important example of fermented biopolymers to be mentioned here is polyhydroxy butyric acids and their copolyesters (one product is, e.g., Biopol)[26].

Primarily small components are linked to polymers by means of chemical synthesis. These monomers, in turn, are either entirely naturally synthesised – as in the case of lactic acid – or are slightly changed by chemical modification as in the case of different epoxydated sunflower, rapeseed, flax, or soy oils. The latter basic components are still reticulated with hardeners obtained with petrochemicals (some products are, e.g., Tribest of the 3000th row, PTP, or Elastoflex)[27]. But also other natural raw materials such as cellulose and lignin can be reticulated since they have the corresponding functions and because it is possible to add additional ones.

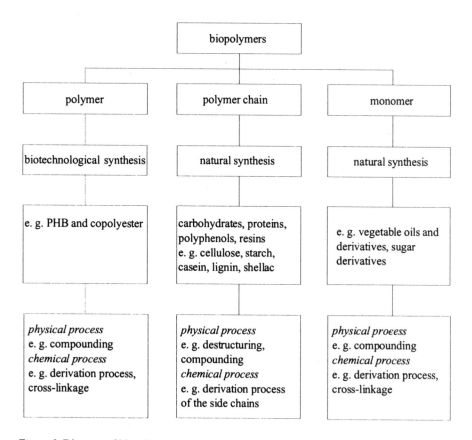

Figure 2. Diagram of biopolymers

The presented possibilities of the available matrices are diverse[28-33] which makes it necessary to select a material that suits the requirements. The choice of a matrix for a high-performance structural material is made according to the boundary conditions of application temperature, mechanical stress, and processing technology. An important requirement for the matrix is a sufficiently low viscosity so that a good impregnation of the fibres can be achieved. However, important characteristics such as the maximum elongation, which preferably should be in the natural fibre range, and a good adhesion to the natural fibres should be guaranteed. These characteristics, in addition to other matrix properties, are the basic prerequisites for the construction of an ideal fibre composite material.

At the DLR, the available biopolymers are examined to determine if they are suitable to be used as matrix material in composites. In addition to being able to fulfil the above-mentioned tasks, the possibility of using these materials for new methods is determined (e.g., resin injection methods such as the differential pressure resin transfer moulding (DP-RTM) method which was developed at the institute).

3.   MATERIAL PROPERTIES

In Fig. 3, the tensile properties of selected biocomposites and glass fibre-reinforced polymers (GFRP) are compared with each other. It is clearly shown that the GFRP properties already are almost reached while having the same fibre volume content. If the fact is taken into consideration that the density of the natural fibres at approx. 1500 kg/m^3 is clearly below that of the glass fibres at approx. 2500 kg/m^3, a greater share of fibre volume at the same component weight would be possible for the biocomposites which, in turn, would have a greater strengthening effect.

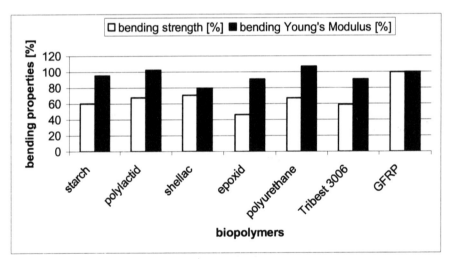

Figure 3. Bending properties of selected natural fibre fleece-reinforced biocomposites

Manufacturing technologies based on the tried-and-tested methods in the fibre plastics technology are being developed, researched, and optimised at the DLR for the production of biocomposites. In particular, the pressing technique, the hand-lay-up and the filament winding technology as well as pultrusion can be tested with only slight modifications during the manufacturing of the component. Unidirectional (UD) reinforced laminates or fabric layers are advantageous for the constructive use of the anisotropy of the fibre composite material. Nonwovens are sufficient for components or parts of the component which are not stressed as much.

4.   APPLICATION RESEARCH

In the following, an overview of the developmental work which began in 1989 is presented in which the emphasis lies on the selected industrial cooperations that produced the first mature market products or where new products are being manufactured.

4.1   Designer Office Chairs

One of the first research co-operations in the area of biocomposites was made with the Wilkhahn company (subsidised by the Ministry of Economics, Technology, and Transportation of Lower Saxony) and dealt with the development of designer office chairs whose seat pan was to be made of a biodegradable composite. In addition to meeting the requirements for the seat listed in the specifications sheet, the goal was to enable a consistent concept for a means of disposal, i.e. composting, and the same time to guarantee the greatest possible amount of creative freedom of the material by using a material that was ecologically sound. The first measurement results were used to make mechanical predictions on the feasibility of this concept by applying the Finite Element Method. In addition to an aptitude test of all biopolymers available on the market, the processing technique was thoroughly examined and necessary developments were carried out in order for the new material to be applied industrially. In summary, it was shown that the mechanical properties could be approximately fulfilled and that a consistent concept for a means of disposal was feasible. Production problems were the result of the biopolymers which were used, making it necessary for further research in this area in order to use the "bio-composite" development concept in further industrial applications.

4.2   Door Panelling Elements

Another step in putting the concept into practice could be made in a development project with the Johnson Controls Interiors company (subsidised by the Ministry of Food, Agriculture, and Forestry of Lower Saxony). The goal of this project was to create a door panelling element with a material made of 100% renewable resources since the goal was an environmentally-friendly means of disposal. In addition to the priority of economic efficiency, the established manufacturing technique was to be used in order to meet the required component specifications. A biopolymer which by then had been developed made it possible for the natural fibre composite material to meet the requirements determined in the specifications sheet.

However, this development has not yet been brought out into the industry for economical reasons.

Figure 4. Seat pan made of biocomposites

Figure 5. Door panelling element made of biocomposites

4.3   Pultruded Support Slats

According to a market analysis, high-quality beach wood for the production of support slats will no longer be available in the future at the current market price. The reason for this is overexploitation in Eastern Europe and also in other countries. In order to still be able to use a renewable resource but to avoid using components based on petrochemicals like, e.g., phenol resin glues, and to also be able to manufacture products without machine processing and without limiting artistic freedom, the answer proves to be the pultrusion of natural fibres embedded in polymers. This complex research topic was worked on together with the Thomas Technik & Innovation Company (subsidised by the Ministry of Economics, Technology, and Transportation of Lower Saxony).

In addition to research on different types of fibres and their composition, the pultrusion process was optimised with regard to the use of renewable resources without neglecting economical efficiency. Thus the required mechanical properties could be attained. Although special steps had to be taken for the latter point, it was possible to guarantee an environmentally-friendly means of disposal. These type of bed frames should be available on the market at the beginning of 2001 [34].

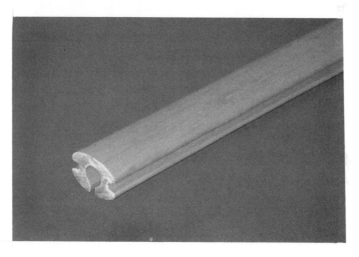

Figure 6. Biocomposite profile

4.4   New thermosets based on vegetable oils

The problems with the matrix systems which came up in several projects are being worked on in a current research project with the Cognis company (sponsored by the Specialised Agency for Renewable Resources). New types of vegetable oil-based thermosets are being developed which are to be applied in the area of fibre composites (natural fibres as a reinforcement system). In addition to their processing properties, these biopolymers are to be compatible to their petrochemical counterparts with regard to their other, e.g., mechanical features and durability. The use of the highest possible amount of renewable resources plays an important role here. Initial results are already being included in currently running projects.

Figure 7. Tube and box-type carrier on tray

4.5   Safety helmet

The concept of structural materials made of renewable resources shows
the enormous potential of this new class of materials. In this research project
together with the Schuberth Helme GmbH company (subsidised by the
Ministry of Food, Agriculture, and Forestry of Lower Saxony), the goal is to
develop an industrial safety helmet made of a minimum of 85% of
renewable resources. An appropriate fibre/matrix system based on the
concept of the biocomposite materials was developed which fulfils the DIN
EN 397 German Industrial Standard requirements for industrial safety
helmets. While working on this pure material basis, a manufacturing process
with appropriate productivity and quality had to be made available for the
helmet bowls in view of a series production later on.

The test results show that the DIN EN 397 German Industrial Standard
requirements for the helmet bowls have definitely been met or have even
been surpassed. Due to an optimised lay-up sequence as well as making use
of the light-weight construction potential of the natural fibres, it was possible
to achieve a reduction in the weight of the helmets of 5 – 10% which also
means a considerable improvement in wearing comfort.

The goal of future work will be to optimise the material concept of the
biocomposites to the extent that this product will also be able to be used in
areas with higher security standards.

Figure 8. Industrial safety helmet

4.6 Interior panelling for track vehicles

Panelling for air columns (Fig 9) were developed for the „Lirex" (Light innovative regional express which was presented to the public for the first time at the Innotrans in Berlin) concept study of the Alstom LHB company with funds from the Ministry of Food, Agriculture, and Forestry of Lower Saxony.

Figure 9. Air column panelling element

These columns are connection elements between two side windows in the train's field of vision. This was able to visibly demonstrate how efficient biocomposites are.

The guidelines for DB AG vehicles according to TL918413 were successfully maintained. In this context, the definitive and considerable requirements of the DIN 5510 - 1 German Industrial Standard for fire protection should be *primarily* mentioned here. The biocomposite was equipped with halogen-free fire retardants and was able to attain fire-protection class S4, smoke development class SR2, and fluidity class ST2.

5.   CONCLUSION

The experiences made have shown that biocomposites can be excellently processed to make structural material. The weight-related mechanical properties make it possible to strive for application areas that are still dominated by glass fibre-reinforced plastics. At this time, limitations must be accepted in areas with extreme environmental conditions. Main target groups therefore are, for example, panelling elements in automobile and freight car manufacturing, the furniture industry and the entire market of the sports and leisure industry.

In conclusion, it has been determined that long-standing developments while retaining the ambitious goal of using renewable resources with a share of a minimum of 80% do not necessarily have to lead to differences in the material. First products have already come out on the market and additional ones will follow. In order to be able to service diverse industrial areas with this new material, further research activities must be carried out in the area of manufacturing technology and in the optimisation of the components. It is expected that in the near future the market acceptance will considerably increase due to the introduction of the first biocomposite products and that new markets will be opened up.

REFERENCES

1. Michaeli, W., and Wegener M., 1990, *Einführung in die Technologie der Faserverbundwerkstoffe*. Hanser Verlag publishing company, Munich.
2. Carlsson, L.A., and Byron Pipes, R., 1989, *Hochleistungsfaserverbundwerkstoffe - Herstellung und experimentielle Charakterisierung*. B.G. Teubner-Verlag publishing company, Stuttgart.
3. Moser, K., 1992, *Faser-Kunststoff-Verbund*. VDI-Verlag GmbH publishing company, Düsseldorf.
4. Ehrenstein, G.W., 1992, *Faserverbund-Kunststoffe*. Carl Hanser Verlag publishing company, Munich.

5. DIN 60 001, 1990, *Textile Faserstoffe – Naturfasern*. Deutsches Institut für Normung (German Institute for Standardisation), Berlin.

6. Wagner, E., 1961, Die textilen Rohstoffe (Natur- und Chemiefaserstoffe). In *Teil T12 Handbuch für Textilingenieure und Textilpraktiker*, Dr. Spohr-Verlag publishing company Wuppertal-Elberfeld.

7. Flemming, M., Ziegmann, G., and Roth, S., 1995, *Faserverbundbauweisen, Fasern und Matrices*. Springer-Verlag publishing company Berlin Heidelberg, pp. 155-179.

8. Kromer, K.-H., Gottschalk, H., and Beckmann, A., 1995, Technisch nutzbare Leinfaser. In *Landtechnik* 6: 340-341.

9. Schliefer, K., 1975, Zellulosefaser, natürliche. In *Ullmanns Encyklopädie der technischen Chemie*, 4th edition, Publishing company: Verlag Chemie, Weinheim, Vol. 9: pp. 247-253.

10. Gessner, W., 1955, *Naturfasern Chemiefasern*. Fachbuchverlag, specialised books publishing company, Leipzig.

11. N.N., 1955, Volume IV. *Technik 1.Teil*. Landolt-Börnstein, 6th edition, pp. 158-295.

12. N.N., 1955, Volume IV: *Technik 1.Teil*. Landolt-Börnstein, 6th edition, pp. 322-420.

13. N.N., 1994, *Flachs sowie andere Bast- und Hartfasern, Faserstoff-Tabellen nach P.-A. Koch*. Special print from the chemical fibre/textile industry, 44th/96th year, Deutscher Fachverlag GmbH publishing company, Frankfurt/Main.

14. Haudek, H. W., and Viti, E., 1980, *Textilfasern*. Verlag Johann L. Bondi & Sohn publishing company, Wien-Perchtoldsdorf, Melliand Textilberichte KG, Heidelberg, pp. 15-73.

15. Haudek, H. W., and Viti, E., 1980, *Textilfasern*. Verlag Johann L. Bondi & Sohn publishing company, Wien-Perchtoldsdorf, Melliand Textilberichte KG, Heidelberg, pp. 122-141.

16. Haudek, H. W., and Viti, E., 1980, *Textilfasern*. Verlag Johann L. Bondi & Sohn publishing company, Wien-Perchtoldsdorf, Melliand Textilberichte KG, Heidelberg, pp. 156-161.

17. Ruge, J., 1989, *Technologie der Werkstoffe*, Vieweg-Verlag publishing company, Braunschweig.

18. Seifert, H.-J., 1969, Fallschirmwerkstoffe. In *Fallschirmtechnik und Bergungssysteme*. (Publishers H.-D. Melzig), lecture series from DGLR-DFVLR-AGARD.

19. Fritz, H.-G., 3rd-4th Sept. 1997, Innovative Polymerwerkstoffe unter Einbeziehung nachwachsender Rohstoffe. In *Werkstoffe aus nachwachsenden Rohstoffen*, Conference proceedings of the 1st International Symposium, Rudolstadt. Publishers Thüringisches Institut für Textil- und Kunststoff-Forschung e.V., Rudolstadt.

20. Aichholzer, W., 3rd-4th Sept. 1997, Bioabbaubare Faserverbundwerkstoffe auf der Basis nachwachsender Rohstoffe. In *Werkstoffe aus nachwachsenden Rohstoffen*, Conference proceedings of the 1st International Symposium, Rudolstadt. Publishers Thüringisches Institut für Textil- und Kunststoff-Forschung e.V., Rudolstadt.

21. Bastioli, C., 11th-12th Feb. 1998, Starch based bioplastics: Properties, applications and future perspectives. In *Biologisch abbaubare Werkstoffe (BAW)*, Conference Proceedings of Symposium, Würzburg.

22. Rapthel, I., and Kakuschke, R., 3rd-4th Sept. 1997, Entwicklungen und Eigenschaften von Sconacell A als vollständig abbaubare Kunststoffe auf Basis teilacetylierter Naturstärke. In *Werkstoffe aus nachwachsenden Rohstoffen*. Conference proceedings of the 1st International Symposium, Rudolstadt. Publishers Thüringisches Institut für Textil- und Kunststoff-Forschung e.V., Rudolstadt.

23. N.N., 1997, Bioceta – Biologisch abbaubares Zellulosediacetat. *Kunststoff + Kautschuk Produkte 97/98*. Yearly Handbook of the Processors, Recycling Companies, Deliverers and Furnishers, Publishing company Darmstadt, pp. 176-177.

24. Eicher, T., and Fischer, W., 1975, Zelluloseester. *Ullmanns Enzyklopädie der technischen Chemie*. 4th Edition, Volume 9, Publishing company Verlag Chemie, Weinheim, pp. 227-246.

25. Kuhne, K., 29th-30th June 1998, Neue Thermoplaste auf Zellulosebasis. In *International Wood and Natural Fibre Composites Symposium*. Proceedings of the symposium, Kassel, Publishers Universität Gesamthochschule Kassel, lecture 13.

26. Schack, D., 11th-12th Feb. 1998, Polyhydroxybutyrat/valerat Copolymere. In *Biologisch abbaubare Werkstoffe (BAW)*. Conference proceedings of the symposium, Würzburg.

27. Scherzer, D., June 1997, Pflanzenfaserverstärkte Polyurethanschäume in industriellen Anwendungsbereichen. In *Neue Produkte aus pflanzlichen Fasern*. Branchendialog, Mainz, Publishers Bildungsseminar für die Agrarverwaltung Rheinland-Pfalz.

28. Utz, H., 19th-20th Nov. 1992, Bioabbaubare Kunststoffe im Verpackungsbereich. In *Biologisch abbaubare Kunststoffe*. Conference Proceedings of Symposium No. 16105/56.253, Esslingen.

29. Fritz, H.-G., Seidenstücker, T., Bölz, U., Juza, M., Schroeter, J., and Endres, H.-J, 1994, *Study on production of thermoplastics and fibres based on mainly biological materials*. EUR 16102, Directorate-General XII Science, Research and Development.

30. Witt, U., Müller, R.-J., and Klein, J., 1997, *Biologisch abbaubare Polymere*. Study of the Franz-Patat-Zentrums für Polymerforschung e. V., Braunschweig.

31. Herrmann, A. S., Nickel, J., and Riedel, U., 1998, Construction materials based upon biologically renewable resources – from components to finished parts. In *Polymer Degradation and Stability 59*, pp. 251–261.

32. Riedel, U., and Nickel, J., 1999, Natural fibre-reinforced biopolymers as construction materials - new discoveries. In *Die Angewandte Makromolekulare Chemie*, **272**: 34-40.

33. Riedel, U., 1999, PhD Thesis *Entwicklung und Charakterisierung von Faserverbundwerkstoffen auf Basis nachwachsender Rohstoffe*. Fortschritt-Berichte VDI, 5, 575.

34. Riedel, U., and Gensewich, C., 1999, Pultrusion von Konstruktionswerkstoffen aus nachwachsenden Rohstoffen. In *Die Angewandte Makromolekulare Chemie*, **272**: 11-16.

Isolation, Characterisation and Material Properties of 4-O-Methylglucuronoxylan from Aspen

[1]MARTIN GUSTAVSSON, [1]MAGNUS BENGTSSON,
[1]PAUL GATENHOLM, [2]WOLFGANG GLASSER, [3]ANITA TELEMAN ,
and [3]OLOF DAHLMAN
[1]*Dept of Polymer Technology, Chalmers Univeristy of Technology, S-412 96 Göteborg, Sweden*
[2]*Dept of Wood Sci and Forest Prod, Virginia Tech, Blacksburg, USA*
[3]*Swedish Pulp and Paper Research Institute, Box 5604, S-11486 Stockholm, Sweden*

Abstract: Hemicelluloses are among the most abundant biopolymers although they are not yet being used as materials. This study focused on isolating polymeric 4-O-methylglucuronoxylan, which is the most abundant heteropolysaccharide (hemicellulose) in aspen wood (*Populus tremula*). The wood chips were extracted sequentially with different liquids. The isolated xylan was analysed for carbohydratate composition, molecular weight and crystallinity. The alkali extraction procedure used is selective and yields a high xylose containing polymeric xylan from aspen wood. The xylan showed fair film forming properties, which were enhanced when the xylan was carboxymethylated in homogenous alkali solution. The carboxymethylated xylan was water soluble and formed excellent films which were partially crystalline.

1   INTRODUCTION

Hemicelluloses are among the world's most abundant biopolymers. They are heteropolysaccharides present in large quantities in wood and the majority of plant tissues. The total global production of lignocellulosic materials in forests is estimated at 140 billion tons per year [1]. The assumption that about 25 % of the lignocellulosic materials are hemicelluloses gives an estimation of the annual hemicellulose biosynthesis of 35 billion metric tons.

Biorelated Polymers: Sustainable Polymer Science and Technology
Edited by Chiellini *et al.*, Kluwer Academic/Plenum Publishers, 2001

Although hemicelluloses are biosynthesised to this large extent and are believed to play an important role in the formation of the hierarchical structure of wood, these polysaccharides are not yet being used as polymeric materials to any great extent in modern society because the material is difficult to isolate and is severely degraded in the present pulp process[2,3].

The linear backbone of xylan consists of xylopyranose units linked by β-1,4-glycosidic bonds. Hardwood xylans are partially acetylated and branched with randomly distributed 4-*O*-methylglucuronic acid (MeGlcA) groups[4].

The degree of polymerisation of the methylglucuronoxylan is higher when the xylan is isolated by alkali extraction than when it is isolated by steam explosion[5]. The xylan is partially depolymerised during steam explosion due to high temperature and pressure, while alkali extraction hydrolyses the naturally occurring acetyl groups and results in a deacetylated material.

The aim of the study was to isolate the polymeric xylan from aspen and investigate the material properties of this material.

2   MATERIALS AND METHODS

Swedish aspen wood chips (AC) were obtained from Rockhammar's Bruk, Frövi, Sweden. The carbohydrate composition of the wood chips was analysed according to section 2.9. The starting material composition is summarised in Table 1.

2.1   Refining

A refiner (8-inch Sprout Waldron, 105-A) was used for grinding the chips mechanically through rotating disks. Water was used as the cooling medium during refining. Aspen wood chips (17.1 kg) with a solid content (sc) of 64%, i.e. a dry weight (dw) of 10.9 kg, were collected in about 2000 L of water-fiber suspension. To separate water from the wood fibres, the suspension was run through a filter centrifuge (Bock, 305TX, rpm = 1740).

2.2   Prehydrolysis

The 10.9 kg (dry weight) of aspen wood was divided after refining in three batches. The batches were treated equally in a Pfaudler, E47186, reactor with a volume of 60 L. A liquid : fibre weight ratio of 12:1 was used in the prehydrolysis step. The fibres were soaked in water for 2 h, after which HCl solution was added to a final concentration of 0.05 N [pH 1.4]. The reactor was heated to 70 °C for 2 h and the content was then cooled to room temperature. The pH of the content was adjusted to 10.2 with concentrated ammonium hydroxide and the mixture was stirring overnight. The reactor content was filtered and washed twice with water and then with ethanol. The filtrate was discarded.

2.3   Alkaline ethanol extraction

80% ethanol was used in a liquid : fibre ratio of 12:1 for this extraction. NaOH was added to a final concentration of 1% w/w. The reactor was sealed and heated to 78 °C for 2 h. The internal pressure was 0.33 bar above atmospheric. The content was filtered and washed with ethanol and hot water (50 °C). The filtrate was discarded.

2.4   Room temperature alkaline water extraction

The fibres were stirred under nitrogen in 4% w/w NaOH at room temperature for 14 h. The liquid: fibre ratio was 10:1. The fibres were filtrated and washed once with hot water (50 °C). The filtrates from the extraction and the wash were collected and labelled X1.

2.5   Hot alkaline water extraction

The fibres were stirred under nitrogen in 4% wt NaOH at 70 °C for 2 h. The liquid : fibre ratio was 10:1. The fibres were filtered and washed once with hot water (50 °C). The filtrates from the extraction and the wash were collected and labelled X2.

2.6   Bleaching

Filtrates X1 and X2 (~360L) from the extraction steps were mixed and pH adjusted with concentrated HCl to pH 11. NaEDTA was added to a final concentration of 1% wt. Two L of 50 % hydrogen peroxide solution were added in two halves 5 hours apart. The reaction temperature was 40 °C. The reactor was left stirring for 2 days, with subsequent additions of NaOH to maintain the pH at about 11.

2.7   Ultrafiltration

Ultrafiltration equipment (DC 30 P, Amicon, Grace) with an MWCO of 3000 was used to concentrate the 360 L of X1+X2 solutions to 80 L. This solution was dia-filtered under steady state conditions for 2 days to exchange 98% of the water. The 80 L were reduced to 22 L after two days.

2.8   Spray drying

The ultrafiltrated solution was pumped (33.3 ml/min) through an atomizer. The fine droplets vaporised in the drying air sprayed through the heating vessel, and a dry powder was collected in a cyclone.

Settings:

Drying air: T = 250 °C,

Air flow rate = 0.6 m³/min

Heater: 255 °C

Pump: 33.3 mL/min

2.9   Carbohydrate and lignin analysis during isolation

The carbohydrate composition (neutral carbohydrates only) was monitored during the isolation process with a High Performance Liquid Chromatography (HPLC) method[6]. The lignin content was determined gravimetrically and spectrometrically in the usual manner [7,8].

2.10 Carbohydrate analysis by Capillary Zone Electrophoresis

The carbohydrate composition (neutral carbohydrates and uronic acid residues) of the isolated xylan was determined by enzymatic hydrolysis and subsequent capillary zone electrophoresis analysis (ENZ/CZE) according to a method described elsewhere [9].

2.11 NMR spectroscopy

For ^{1}H NMR analysis, 2 mg methylglucuronoxylan was dissolved in 0.7 ml D_2O or 1 M NaOD. ^{1}H NMR spectra were obtained at 400.13 MHz using a Bruker DPX400 spectrometer and a probe temperature of 70 °C. The typical acquisition parameters employed were a 90° pulse, a spectral width of 4000 Hz and a repetition time of 9 s.

For ^{13}C-NMR analysis, 200 mg carboxymethylated methylglucurono-xylan was dissolved in 3 ml D_2O. ^{13}C NMR spectra were obtained at 100.57 MHz using a Varian 400 MHz spectrometer and a probe temperature of 25 °C. The typical acquisition parameters employed were a 90° pulse, a spectral width of 25000 Hz and a repetition time of 19 s.

2.12 Size exclusion chromatography

For size exclusion chromatography (SEC) analysis, 3 mg xylan was dissolved in 0.15 ml DMSO (75 °C, 2h) and 2.85 ml 0.5% LiCl/DMAc was then added. The molecular weight of the xylan isolated was analyzed using SEC[10,11].

2.13 Carboxymethylation

The isolated glucuronoxylan was carboxymethylated in homogenous 1 N alkali solution. 10 g of glucuronoxylan was dissolved in 500 ml water, containing 20 g NaOH and 17.9 g chloroacetic acid, sodium salt. The reaction vessel was heated to 70 °C. After 3 h the reaction vessel was cooled to room temperature and the pH of the content was adjusted to 10 with 4 N HCl. The content was precipitated in 4 l of ethanol, filtered and air dried overnight. The air dried product was dissolved in 250 ml and reprecipitated in 1.2 l of ethanol. After yet another air drying over night, the product was dissolved in 150 ml water and freeze dried for 5 days (<10 μbar). 9.1 g of the freeze dried product was recovered.

2.14 Wide angle x-ray scattering

The samples were also characterised by wide-angle x-ray diffraction (WAXD). The xylan was solvent cast from water onto petri dishes and allowed to air dry. The film was placed in the diffractometer in 4 layers. Diffractograms were recorded in the reflection geometry on a Siemens D5000 diffractometer using CuK_a radiation. Diffractograms were taken between 5° and 30°(2θ) at a rate of 1°(2θ) per minute and a step size of 0.1°(2θ).

3   RESULTS AND DISCUSSION

The neutral carbohydrate composition of the starting material (aspen wood) is shown in Table 1. If it is assumed that the native xylan contains one 4-*O*-methylglucuronic acid group and seven *O*-acetyl groups out of ten xylose residues, approximately 7 weight% has to be added. The consequence of this assumption is that the *O*-acetyl-(4-*O*-methylglucurono)xylan accounts for approximately 25% of the weight of this wood. The glucose:mannose ratio in aspen glucomannan has been reported to be 1:2[4], which means a glucomannan content of approximately 4%. Both the xylose and mannose contents of the starting material correspond well to the amounts of *O*-acetyl-

(4-*O*-methylglucurono)xylan   and   glucomannan   usually   found   in
hardwood[4,12]. The lignin content of aspen has been reported to be 21%[4]. If
the rest is assumed to be extractives, starch, pectin and ash, a diagram of the
chemical composition of the starting material can be proposed as shown in
Fig. 1.

Table 1. Neutral carbohydrate composition of aspen wood

	Glc	Xyl	Gal	Man	Total sugar
Weight %	43	18	1.1	3.0	65.1

3.1   Mass balance

The xylose/glucose ratio of the fibres was used to monitor the course of
extraction. The fibre residue was weighed after each step and analysed for
solid contents. This made it possible to calculate how much material was
solubilised in each extraction medium. Table 2 shows the mass balance and
the xylose to glucose ratio for selected samples.

The weight given for the filtrates is calculated on the basis of solid's
weight difference before and after the extraction step.

Table 2. Mass balance and carbohydrate ratios

	Weight Kg	Weight % of AC	Mol Xyl/Glc	Xylose wt %
Aspen chips, AC	10.9	100	0.49	18
Prehydrolysis				
- Filtrate	1.2	11	∞	ND
Alkaline EtOH	9.7	89	ND	ND
- Filtrate	1.4	13	ND	ND
Alkaline water 1	8.3	76	ND	ND
- Filtrate X1	0.68	6.2	ND	ND
Alkaline water 2	7.6	70	ND	ND
- Filtrate X2	0.35	3.2	ND	ND
- Filtrate X1+ X2	1.0	9.4	21	ND
Residue	7.2	67	0.17	7.3
Spray dried xylan	1.0	9.2	480	63

ND = Not determined

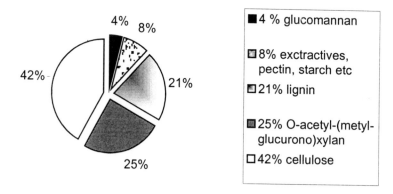

Figure 1. Chemical composition of aspen wood

The solubles' weights are summarised in Figure 2. It can be seen that the water/ammonium solubles from the prehydrolysis correspond well to the glucomannan and extractives fraction of wood.

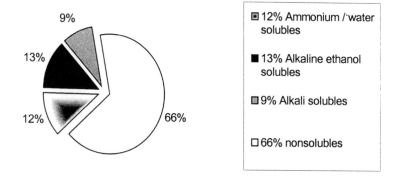

Figure 2. Extraction of aspen wood (%-wt)

The alkaline ethanol extraction, which is the delignification step, seems quite ineffective since only about 60% of the weight corresponding to the total lignin was removed. This was still higher than previously reported[13]. The content of the nonsolubles is significantly larger than the cellulose content of the wood chips, indicating that the residue contains significant amounts of lignin and glucuronoxylan. The latter is confirmed in the composition diagram in Table 2.

3.2   Glucuronoxylan yield

Table 3 summarises the yield of the isolation process.

Table 3. Glucuronoxylan yield

	Weight Kg	Weight % of AC	Xylose kg	Xylose % of xyl in AC
Aspen chips, AC	10.9	100	2.0	100
Spray dried xylan	1.0	9.2	0.64	32

The results correspond well to previously performed alkali extraction of aspen[13]. The final lignin content of the bleached and spray dried product was 4.8% (Klason + acid soluble lignin), analysed according to a method described elsewhere[7].

3.3   ENZ/CZE and SEC

Carbohydrate analysis using enzymatic hydrolysis and subsequent capillary zone electrophoresis (ENZ/CZE) revealed the number of 4-*O*-methylglucuronic acid units to be approximately one unit per nine D-xylose residues (Table 4). No glucuronic acid units were detected. The carbohydrate composition of the polysaccharide isolated here from aspen wood is quite similar to that previously reported for typical hardwood xylans[4,14]. SEC analysis indicated the molecular weight (M_w) to be 15000 g/mol.

Table 4. Relative amounts of acidic and neutral monosaccharide residues in the 4-*O*-methylglucuronoxylan isolated by alkali extraction from aspen

	Xyl	MeGlcA	Man	Ara, Glc, Gal
Weight %[a]	83	14	2	<1
Mol %[b]	87	10	2	<1

[a] Relative weight % of the carbohydrates in the sample
[b] Relative mol % of the carbohydrates in the sample

3.4   NMR spectroscopy

The xylan obtained was studied by NMR spectroscopy. One major and two minor signals were observed in the anomeric proton region between 4.4 and 5.5 ppm in the ¹H NMR spectrum (Fig. 3). The chemical shifts and three-bond coupling constants in the ¹H NMR spectrum indicate that the major structural element in the xylan was a β-(1→4)-D-xylopyranosyl residue and that a 4-*O*-methylglucuronic acid side-group and a disubstituted xylopyranosyl residue were minor structural elements[13,15]. The spectrum resembles the spectrum for alkali-extracted 4-*O*-methylglucuronoxylan from aspen described earlier[13]. It can thus be concluded that the isolated carbohydrate consists of an almost pure 4-*O*-methylglucuronoxylan. The relative intensities of the signals indicate that the relative content ofMeGlcA is 8 mol%.

Acetyl signals at 2.2 ppm, as recently reported for *O*-acetyl-(4-*O*-methylglucurono)xylan isolated from aspen[16], were not detected. Such *O*-acetyl groups are most probably removed under the alkaline conditions used here during the extraction of the xylan.

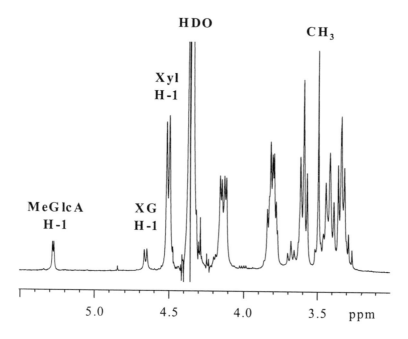

Figure 3. ^{1}H NMR spectrum of alkali-extracted 4-*O*-methylglucuronoxylan from aspen (D$_2$O, 70 °C). XG denotes a disubstituted xylopyranosyl residue

The ^{13}C NMR investigation of the carboxymethylated xylan focused on the anomeric and carbonyl carbons at 98-103 ppm and 173-175 ppm, respectively. The carbonyl group at the reducing end could not be detected because of a poor signal to noise ratio. The DS of carboxymethyl groups was estimated at 0.1.

3.5 Material Properties

The isolated glucuronoxylan was evaluated with respect to material properties. The solubility in water was limited, but it was fully soluble in 0.5 N NaOH solution. Heat treatment of xylan in water at 95 °C for 15 minutes gave an improvement in solubility. This solution was cast onto petri dishes and the film formation was visually evaluted. The crystallinity of films was determined by X-ray diffraction and was found to be 40%, determined by integration of the peak 17°-21° in Fig. 4, which corresponds well to previous studies[13]. The carboxymethylated xylan was fully soluble in water and formamide. A water solution was film cast and visually evaluated. The carboxymethylated xylan showed an improvement in film forming properties. The crystallinity was found to be 27% (Fig. 4). The reduction in crystallinity is believed to be caused by the reduction of straight unsubstituted xylose segments in the xylan backbone.

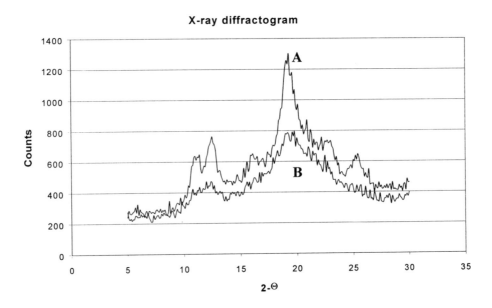

Figure 4. X-ray diffraction pattern of xylan (A) and carboxymethylated xylan (B)

4   CONCLUSIONS

The extraction of polymeric xylan with alkali following mild hydrolytic pretreatment and delignification in accordance with O'Dwyer[17] and as adopted by Glasser et al[18]. is selective in isolating high xylose containing heteropolysaccharides in high yield directly from aspen wood. The xylan is deacetylated as confirmed by ^{1}H-NMR spectroscopy. The isolated xylan is substituted with 4-O-methylglucuronic acid groups (about 1 per 9 anhydroxylose units as determined by ENZ/CZE). Mild carboxymethylation (to a DS of about 0.1 according to ^{13}C- NMR estimates) results in better film forming properties, increases the solubility, and lowers the crystallinity.

ACKNOWLEDGEMENTS

The financial support of NUTEK (PROFYT program), Stiftelsen Gunnar Sundbladhs Forskningsfond and Stiftelsen Bengt Lundqvists Minne is gratefully acknowledged. We wish to thank Robert Wright for his assistance during the isolation process and the Division of Wood Chemistry at the Royal Institute of Technology, Stockholm, Sweden, for allowing us to use their NMR spectrometer. We also thank Tom Glass for ^{13}C NMR analysis and Rickard Berggren for performing the SEC analysis.

REFERENCES

1. Hon, D.N.-S., 1996, *Chemical Modification of Lignocellulosic Materials*, Chap 1, Marcel Dekker Inc., New York, pp 2-3
2. Kleppe, P.J., 1970, Kraft pulping, *Tappi*, **53**: 35-47
3. Roberts, J.C., and El-Karim, S.A., 1983, The behavior of surface adsorbed xylans during the beating of a bleached kraft pine pulp, *Cellulose Chemistry and Technology*, **17**: 379-386
4. Timell, T.E., 1967, Recent progress in the chemistry of wood hemicelluloses *Wood Science Technology*, **1**: 45-70
5. Glasser, W.G., Kaar, W.E., Jain, K.R., and Sealey, J.E., 2000, Isolation options for non-cellulosics heteropolysaccharides (HetPS), *Cellulose* **41**: 299-317
6. Kaar, W.E., Cool, L.G., Merriman M.M., and Brink, D.L., 1991, The complete analysis of wood polysaccharides using HPLC, *J. Wood Chemistry and Technology*, **11**: 447-463
7. Kaar, W.E., Brink, D.L., 1991, Simplified analysis of acid soluble lignin, *J. Wood Chemistry and Technology*, **11**: 465-477
8. Kaar, W.E., Brink, D.L., 1991, Summative analysis of nine common north American woods, *J. Wood Chemistry and Technology*, **11**: 479-494
9. Dahlman, O., Jacobs, A., Liljenberg, A. and Ismail Olsson, A., 2000, Analysis of carbohydrates in wood and pulps employing enzymatic hydrolysis and subsequent capillary zone electrophoresis, *J. Chromatogr.* A **891**: 157-174
10. Berggren, R., Berthold, F., Sjöholm, E. and Lindström, M., 2000, Degradation patterns and strength development of fibres subjected to ozone degradation or acid hydrolysis, *6th European Workshop on Lignocellulosics and Pulp*, Bordeaux, pp. 205-208.
11. Sjöholm, E., Gustafsson, K., Berthold, F. and Colmsjö, A., 2000, Influence of carbohydrate composition on the molecular weight distribution of kraft pulps, *Carbohydrate Polymers* **41**: 1-7.
12. Sjöström, E., 1981, *Wood Chemistry, Fundamentals and Applications*, San Diego, CA, Academic Press
13. Gabrielli, I., Gatenholm, P., Glasser, W.G, Jain, R.K., and Kenne, L., 2000, Separation, characterization and hydrogel-formation of hemicellulose from aspen wood, *Carbohydrate Polymers* **43**: 367-374.
14. Shimizu, K., Sudo, K., Ono, H., and Fujii, T., (1989), *Wood Processing and Utilization*, (J. F. Kennedy, G.O. Phillips, P. A. Williams, eds.) Ellis Horwood, Chicester, pp 407-412

15. Teleman, A., Harjunpää, V., Tenkanen, M., Buchert, J., Hausalo, T., Drakenberg, T. and Vuorinen, T., 1995, Characterisation of 4-deoxy-b-L-*threo*-hex-4-enopyranosyluronoc acid attached to xylan in pine kraft pulp and pulping liquor by ^{1}H and ^{13}C NMR spectroscopy, *Carbohydrate Research* **272**: 55-71.

16. Teleman, A., Lundqvist, J., Tjerneld, F., Stålbrand, H. and Dahlman, O., 2000, Characterization of acetylated 4-*O*-methylglucoronoxylan isolated from aspen employing ^{1}H and ^{13}C NMR spectroscopy, *Carbohydrate Research* **329**: 807-815.

17. O'Dwyer, M.H. (1926) LXXXVI. The hemicelluloses. Part IV. *Biochem. J.*

18. Glasser, W.G., Jain, R.K. and Sjöstedt, M.A., (1995) Thermoplastic pentosan-rich polysaccharides from biomass, US patent #5,430,142

An Original Method of Esterification of Cellulose and Starch

SAMUEL GIRARDEAU, JORGE ABURTO, CARLOS VACA-GARCIA,
ISABELLE ALRIC, ELIZABETH BORREDON and ANTOINE GASET
Laboratoire de Chimie Agro-Industrielle, UMR INRA, Ecole Nationale Supérieure de Chimie de Toulouse, INP Toulouse, 118 route de Narbonne, 31077 Toulouse Cedex 04, France

Abstract: New acylation techniques were developed for the fabrication of fatty esters of cellulose and starch. They exclude the use of organic solvents and are readily achieved. Emulsification of the fatty acid in water allowed the intimate contact between the fatty reagent and the polysaccharide. Fatty acid salts (soap) were used as both catalyst and emulsifying agent. Reaction conditions were optimized using an experimental design. Starch and cellulose octanoates were obtained having a DS of 0.52 and 0.23 with a recuperation yield of 70 and 85 %, respectively. Both polysaccharide esters showed a marked hydrophobic character.

1.   INTRODUCTION

In the last years, an increasing interest has been observed on the synthesis and the non-food applications of fatty acid esters (FAE) of cellulose and starchy materials. These materials present improved and/or new properties as water repellency, thermal stability and thermoplasticity.

Conventional synthesis of such materials employs fatty-acid chlorides or anhydrides in organic solvents[1]. Recently, a free-solvent method to esterify starch was developed in our laboratory using formic acid and octanoyl chloride as the gelatinizing and acylating agents respectively. After optimization of reaction conditions, starch octanoate with DS of 1.7 was obtained with a recuperation yield of 89 %[3].

Acylation reactions of cellulose with fatty acids have been accomplished in the absence of solvent with the help of a co-reagent and the solvent exchange technique as a pretreatment for cellulose. The latter included soaking of cellulose with water, followed by washing with ethanol and finally with the fatty acid[2]. In the present work, we propose a new technique for the synthesis of cellulose and starch fatty esters by esterification with fatty acids without the use of co-reagent or organic solvent. This technique passes through an emulsion to accomplish an intimate contact between the polysaccharide and the fatty acid.

2.   EXPERIMENTAL PROCEDURES

Materials. Alpha cellulose (4% pentosans) and amylose (70%) were obtained from Sigma France. Starches from all other sources were kindly supplied from INRA Nantes, France. Vegetal oils were obtained from Sidobre-Sinova (France). Other chemicals were of reagent grade and were purchased from usual providers and were used without further purification or treatment.

Esterification reaction. A homogeneous mixture of polysaccharide (cellulose or amylose), water or ethanol, soap (added or created *in situ*) and fatty (octanoic) acid was obtained by emulsification at 2000 rpm using a high speed stirrer (homogenizer). Water or ethanol were then distilled off at 130°C for 30 min followed by the esterification reaction at 195°C for 2-6 hr.

Degree of substitution (DS). 0.5 g of purified sample was stirred for 30 min in 40 ml of aqueous ethanol (70%). After addition of 20 ml of a 0.5 N NaOH aqueous solution, the stirring was continued for 48 h at 50°C. The unreacted NaOH was back-titrated with 0.5 N aqueous HCl. The solid was recovered by filtration and thoroughly washed with deionized water and ethanol, then oven-dried at 50°C for 48 h. The absence of ester functions in the saponified solid was confirmed by FTIR spectroscopy. Ester content was calculated as:

$$EC (\%) = [(A\text{-}B) \times N_B - (D\text{-}C) \times N_A] \times M/(10 \times w),$$

where A and B are respective volumes of NaOH solution added to sample and blank (ml); N_B and N_A are respective normality of NaOH and HCl solutions; D and C are respective volumes of HCl added to sample and blank (ml); M is the molecular weight of the grafted acyl residue and w is the weight of sample (dry basis in g).

Degree of substitution was then calculated as:

$$DS = 162 \times EC / [M \times 100 - ((M\text{-}1) \times EC)],$$

where 162 is the molecular weight of anhydroglucose monomer.

Recovery yield (RY). The RY is defined as the ratio of the precipitated fraction to the theoretical total cellulose ester. It is assumed that the DS of the non-precipitated part is the same as that of the precipitated fraction. The RY is calculated as :

$$RY\ (\%) = 100 \times W_p\ /\ (W_0 \times (162 + DS(M\text{-}1))/162),$$

where W_0 is the weight of initial cellulose; W_p is the weight of purified product and M is the molecular weight of the fatty acyl substituent (127 for the octanoyl).

Optimization. The influence of the reaction parameters on the yield of the reaction were studied using response surface methodology (RSM). A central composite design was used and both canonical analysis of the second-order equation and discussion of the isoresponse curves were used for the interpretation of the results.

3.   RESULTS AND DISCUSSION

3.1   Starch fatty esters

A central composite design permitted us to evaluate the effect of catalyst (NaOH) concentration and reaction time on the degree of substitution (DS) and the recovery yield (RR) (Figs 1a and 1b).

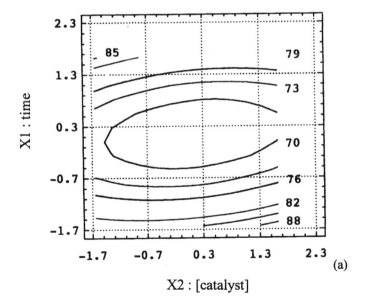

(a)

X2 : [catalyst]

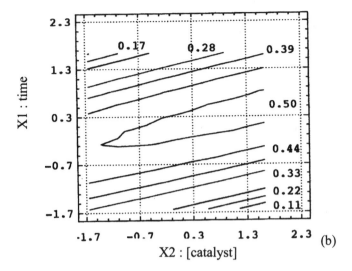

Figure 1. Effect of the reaction time and NaOH catalyst concentration on the (a) degree of substitution and (b) recuperation yield of amylose ester obtained from reaction between amylose and octanoic acid at 195°C. X1 (Time) : 2 to 6 h; X2 ([Catalyst]) : 1.5 to 23 meq/eq OH

The ethanol concentration in the emulsions was fixed at 2.5 eq/eq OH. The response surface analysis showed that it is necessary to set the reaction conditions at [NaOH] = 23 meq/eqOH and time = 6 hr to synthesize an amylose ester with a high DS and RR, that is DS_8 = 0.50 and RR = 70%. These conditions were used to study the effect of the catalyst (Table 1) and biopolymer type (Table 2) on the DS and RR at the fatty esters.

Table 1. Amylose esterification at 195°C with octanoic acid in the presence of different catalysts.

Catalyst (23 meq/eqOH)	DS_8	RR (%)
KOH	0.50	70
NaOH	0.52	68
LiOH	0.40	72
Sodium acetate	0.36	76
CH_3ONa	0.36	79
K_2CO_3	0.02	100

Strong alkalis were revealed to be better catalysts for this reaction than weak bases. Indeed, KOH or NaOH allowed to obtain a high DS around 0.5 with a RR of 70%. On the contrary, a DS of only 0.02 was obtained with K_2CO_3. Thus, in the following experiments, the NaOH was chosen as the catalyst to evaluate the effect of different starch sources on the grafting extent of octanoic acid (Table 2).

Table 2. Esterification of different starchy materials at 195°C with octanoic acid in the presence of NaOH catalyst (23 meq/eqOH).

Polysaccharide	DS_8	RR (%)
Amylose	0.52	68
Amylopectine	0.17	82
Potato starch	0.05	95
Gelatinized potato starch	0.17	85
Maize starch	0.34	77
Wheat starch	0.25	81
Rice starch	0.30	79

Amylose or amylopectine led to the highest DS because its linear structure and absence of granular envelope. The swelling and disruption of granules during the thermal gelatinization of potato starch causes the release of amylose and amylopectine in the reaction medium. Thus, gelatinized potato starch allowed to triplicate the DS compared to native potato starch.

3.2 Cellulose fatty esters

The optimization of the reaction conditions was done by using a central composite experimental design. In this case, the variables studied were the catalyst, reagent and water concentrations. The maximum DS (0.23) was obtained with a recuperation yield (RR) of 85% at the following conditions: [potassium laurate] = 0,01 eq/eq OH; [octanoic acid] = 10 eq/eq OH and [H_2O] = 3 eq/eq OH.

The kinetics of the reaction (Fig 2) was studied at the optimized conditions by following the evolution of DS, the RR and the DP (degree of polymerization).

The DS increased with the reaction time and seems to attain a plateau after 5 hours of reaction. The RR and DP values diminish both throughout the reaction to drop since the first hour of reaction because of an important degradation of the polymer. The cellulose is presumably hydrolyzed under these reactions conditions. The oligomers cannot precipitate at the end of reaction if their DP is too low and therefore they remained in the liquid phase.

Furthermore, we studied the performance of other catalysts, using the above optimized conditions. The results are shown in Table 3.

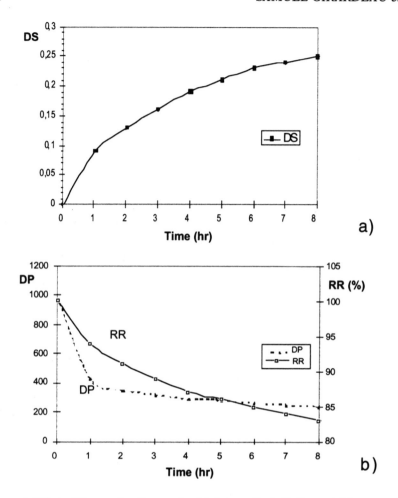

Figure 2. Effect of the reaction time on the (a) degree of substitution and (b) recuperation yield and degree of polymerization of cellulose ester obtained from reaction between cellulose and octanoic acid at 195°C. [potassium laurate] = 0,01 eq/eq OH; [octanoic acid] = 10 eq/eq OH and [H₂O] = 3 eq/eq OH.

Table 3. Cellulose esterification at 195°C with octanoic acid in the presence of different catalysts.

Catalyst	DS	RR (%)
C_{12}, K^+	0.23	85
C_{18}, K^+	0.20	85
C_{18}, Na^+	0.14	92
C_{12}, Na^+	0.17	85
C_2, Na^+	0.16	88
NaOH	0.19	85
K_2CO_3	0.19	86

Slight differences were observed when the cation of the salts was varied. Potassium salts were the most efficient catalysts. On the other hand, the chain length of the catalyst showed no significant influence on the grafting of fatty chains. The potassium laurate allowed to obtain the highest DS, that is 0.23. This catalyst was therefore used to react different fatty acids under the same optimized conditions as shown in Table 4.

Table 4. Cellulose esterification at 195°C with different fatty acids in the presence of potassium laurate as catalyst.

Fatty acid (C_n)	DS	RR (%)
C_8	0.23	85
C_{10}	0.17	82
C_{12}	0.12	84
C_{14}	0.11	82
C_{16}	0.10	83
C_{18}	0.10	80

It was observed that DS decreases as the length of the fatty acid increases. Longer chains renders difficult the diffusion of the fatty acid into cellulose fibers and consequently the overall reactivity decreases. However, all of the ester samples showed a remarkable hydrophobic character.

4.   CONCLUSION

Emulsification of cellulose or starch with water or ethanol, soap and a fatty acid is a performing method to allow the grafting of fatty chains onto the biopolymer backbone. An experimental design allowed to determine the best conditions to synthesize starch fatty esters with DS ≤ 0,5 using natural raw materials as reagents.

DS ≤ 0,23 were obtained for cellulose fatty esters because of its highly crystalline structure, decreasing its reactivity compared to starch. Despite their relatively low DS values, cellulose and starch fatty esters showed a marked hydrophobic character.

REFERENCES

1. Malm, C., Mench, J., Kendall, D., and Hiatt, G., 1951. Aliphatic acid esters of cellulose. Preparation by acid chloride-pyridine procedure. *Ind. Eng. Chem.,* **43**: 684-688.
2. Vaca-Garcia, C., and Borredon, M. E., 1999. Solvent-free fatty acylation of cellulose and lignocellulosic wastes II. Reactions with fatty acids. *Bioresource Technol.* **70**: 135-142.
3 Aburto, J., Alric, I., and Borredon, E., 1999. Preparation of long-chain esters of starch using fatty acid chlorides in the absence of an organic solvent. *Starch/Stärke*, **51**: 132-135.

PART 2

BIOPOLYMER TECHNOLOGY AND APPLICATIONS

Biopolymers and Artificial Biopolymers in Biomedical Applications, an Overview

MICHEL VERT
CRBA – UMR CNRS 5473, University Montpellier 1, Faculty of Pharmacy, 15 Ave. Charles Flahault, 34060 Montpellier, France

Abstract: Nowadays, the domains of life-respecting, therapeutic polymeric systems and materials are among the most attractive areas in polymer science. Increasing attention is being paid to polymeric compounds that can be bioassimilated, especially in the field of time-limited therapeutic applications. Basically biopolymers are of interest because of their inherent biodegradability. However, a close look at the requirements to be fulfilled show that only a few of them can be used in the human body. The interest and the strategy to make artificial biopolymers, i.e. polymers of non-natural origin that are made of prometabolite building blocks and that can serve as components of biomedical or pharmacological therapeutic systems, are recalled. A few examples of artificial biopolymers for biomedical applications are presented.

1.   INTRODUCTION

Human beings started using available materials for therapeutic purposes several thousands years ago. Since then, the strategy of the therapist has always been the same: trying to take advantage of any novel materials to treat diseases or trauma. It is the reason why metals, alloys, ceramics and also biopolymers such as wood, cellulose or proteins, have been used as therapeutic materials long before the concept of biomaterial be defined. Actually, the first scientific societies of the field were founded in the 60's. During the last three decades, attention was increasingly paid to the confrontation of the properties of a biomaterial to the list of specifications of a given application in attempts to achieve better matching between the properties of the biomaterial and those required by a given application. From a general viewpoint, the list of specifications of a biomaterial includes criteria related to biocompatibility and biofunctionality, the former reflecting a proper reaction of the host organism, the latter reflecting the ability of the material to fulfil the task for which it is designed (Table 1).

Table 1. Criteria to be fulfilled by a biomaterial to be of practical interest

Biocompatibility	Biofunctionality
Non-toxic	Material-related properties
Non-immunogenic	Chemical
Non-carcinogenic	Physical
Non-thrombogenic	Physico-chemical
	Thermal
	Mechanical
	Biological
	Application-related properties
	Easy to manipulate
	Sterilisable
	Storable

Considering the various applications of biomaterials in therapy, it is easy to see that some of these applications are aimed at replacing a lost function or organ and request a therapeutic device made of biomaterials (prosthesis) for the rest of the patient's lifetime. In contrast, many other biomedical applications require a therapeutic aid for a limited period of time, namely the healing period. The therapeutic aid is then used to help body self-repair only. Accordingly it is desirable that the temporary therapeutic aid disappear from the body after healing in order to avoid the storage of any foreign materials. Whereas permanent aids require biostable biomaterials, temporary aids should be preferably made of degradable or biodegradable compounds that can be eliminated from the body or bioassimilated after use.

Historically, biopolymers, i.e. polymers of natural origin such as polysaccharides and proteins, were primarily used as sources of wound dressings and suture threads either under their natural forms or after some chemical treatments. Because macromolecular compounds Mother Nature synthesises are usually biodegradable, i.e. degraded via biological processes, biopolymers are often regarded as suitable compounds to make bioresorbable therapeutic devices. However, biopolymers are rather difficult to process, exhibit poorly reproducible properties, can be immunogenic and are often difficult to store because of their sensitivity to humidity and to micro-organisms. Due to the advances in organic chemistry during the last century, artificial polymers were invented, and they invaded rapidly all the sectors of human activity. Over the last sixty years, this novel class of materials outclassed biopolymers in most of their industrial and domestic applications. The same evolution was observed in therapy since surgeons and pharmacologists tried to take advantages of synthetic polymers as soon as they became available. Presently, there is almost no location in the human body where polymers are not used as biomaterials. Compounds like polyethylene, poly(ethylene terephtalate), polyurethanes, silicones are equally familiar in the house, in industries and in the hospital. However, one of the main criteria that made domestic and industrial synthetic polymers better than biopolymers was biostability. This property precluded the use of industrial polymers to make degradable or biodegradable therapeutic devices. It is the reason why during the last two decades scientists have been searching for novel artificial polymers that can degrade or biodegrade when allowed to work in contact with living tissues either as surgical devices, as drug delivery systems or as matrices for tissue cultures after cell seeding. Many artificial polymers have been found biocompatible and degradable or biodegradable in an animal body. However, upon degradation a polymeric device goes from the state of material to small organic molecules and these low molecular weight by-products have to be biocompatible too. In order to minimise the problems related to biocompatibility criteria and to the elimination of degradation by-products through natural pathways, we have introduced a strategy based on the synthesis of bioresorbable polymers derived from metabolites, i.e. from chemicals normally present in biochemical processes and often similar to the building blocks forming biopolymers[1,2]. Such polymers can be considered as artificial biopolymers. Some members are currently investigated with the aim of fulfilling the list of specifications of therapeutic applications requiring bioresorbable devices or systems such as sutures, osteosynthesis devices, injectable implants and drug carriers systems that are aimed at providing controlled drug delivery such as hydrogels, microparticles or nanoparticles, self assembled polymeric micelles or aggregates, macromolecular prodrugs and even bioactive macromolecules that can be considered as a macromolecular drug[2].

In this contribution, we will successively recall the interest and the limits of biopolymers and of artificial biopolymers with regard to temporary therapeutic applications, degradable polymeric compounds being unsuitable to uses as biostable devices. However, we will first recall the problems raised by the concept of biodegradation and the use of temporary therapeutic applications, with the aim of showing that, in therapeutic applications, it is the fate of the device that is important and not its degradability nor its biodegradability.

2.   CRITERIA CONDITIONING THE FATE OF A POLYMER SENSITIVE TO A LIVING SYSTEM

A close look at the phenomena related to the degradation of polymeric compounds is necessary to better understand the nuances between the different levels of degradation and the difference between degradation, biodegradation, bioresorption and bioassimilation.

Fig 1 shows the different levels of degradation that can affect the initial integrity of a solid polymeric device.

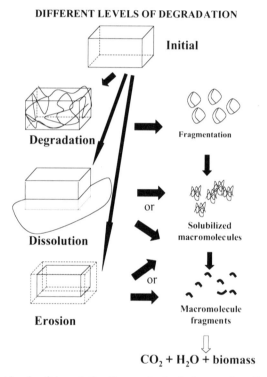

DIFFERENT LEVELS OF DEGRADATION

$$CO_2 + H_2O + biomass$$

Figure 1. Different levels of degradation from a plastic device to mineralization and biomass formation

As one can see, the parallel-sided device taken as an example can disappear from vision to the naked eye either because it is broken into tiny pieces or fragments or because it goes into solution. None of these phenomena implies that macromolecules were degraded. The case of erosion where the disappearance of the polymeric matter occurs from the surface is a special mechanism that is often related to enzymatic degradation because enzymes are known for not being able to penetrate entangled macromolecules forming solid plastics. If the disappearance of a solid macromolecular compounds involves the breakdown of macromolecules, one faces another situation with different issues, namely the formation of small molecules as end products followed by the mineralisation and/or the metabolisation of the degradation intermediates to yield carbon dioxide and water and biomass. Insofar as going from a polymeric device to its degradation up to mineralisation, metabolisation and biomass is concerned, there are two main routes (Fig 2).

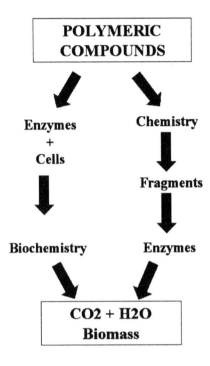

Figure 2. The two routes leading from a piece of degradable plastic to mineralisation and biomass formation (left: biotic degradation or biodegradation; right: abiotic degradation or degradation + bioassimilation)

The first route proceeds through biotic degradation and thus requires active enzymes and the living cells that produce these enzymes. Therefore, if the surrounding conditions do not allow cell life, biotic degradation cannot proceed. The second route depends on abiotic degradation, i.e. on non enzymatic chemical reactions triggered by specific phenomena such as water in the case of hydrolysis, oxygen in the case of oxidation, light in the case of photodegradation etc. Under these conditions, a diversified terminology is necessary to distinguish the different levels and routes of degradation.

Many years ago, we recommended a simple terminology that is very useful, even if it is not adopted world-wide yet (Table 2).

Table 2. Proposed terminology to distinguish the different levels of degradation and biodegradation

BIO(FRAGMENTATION)	Breakdown of a device into particles under the effect of external chemical or physical stresses without cleavage of the macromolecules
DEGRADATION	Breakdown of macromolecules through chemical or unidentified processes, regardless of their physical state
BIODEGRADATION	Breakdown of macromolecules by the action of a living system or via enzymes and thus cells
ULTIMATE BIODEGRADATION	Maximal biodegradation of macromolecules fragments
(BIO)EROSION	Elimination of matter at the surface resulting from degradation (or biodegradation) or dissolution
DISSOLUTION	Dispersion of macromolecules constituting a solid device in a solvent, to be used when the fate of the macromolecules is unknown
(BIO)ABSORPTION	Disappearance of macromolecules from the initial site of implantation. Does not imply degradation nor biodegradation nor bioassimilation. Dissolution is enough to use this term
BIORESORPTION	Degradation or Biodegradation with elimination of the by-products from an animal organism via natural pathways
BIOASSIMILATION	Degradation with metabolisation of fragments

According to this terminology, the breakdown of a device into pieces or fragments without cleavage of the macromolecules is named fragmentation or biofragmentation if part of the matter is biodegraded. A typical example is a polyethylene-starch blend. If the macromolecules go only into solution, one has to speak of dissolution. If the breakdown proceeds at the surface only, erosion or bioerosioncan be used depending on the route by which the breakdown occurs. If the breakdown involved macromolecule cleavages up to mineralisation and metabolisation, bioresorption is to be used exclusively when the elimination or the assimilation by an animal organism is demonstrated, otherwise the material is only degradable or biodegradable. In contrast, bioassimilation can be used for both animals and outdoor microorganisms or living organisms. Under these conditions, degradable reflects degradation by normal chemistry whereas biodegradable has to be reserved to those polymeric materials that can be attacked and degraded by enzymes. Last but not least, the terms degradable and biodegradable reflect mechanisms of degradation and do not say anything about the fate of the degradation by-products. In contrast bioresorption and bioassimilation do say that the degradation products were eliminated.

To summarise this first part, one can say that whether a polymeric material is degradable or biodegradable is not importantfrom the viewpoints of the respect of life and of living systems on earth. What is important is that the polymer and its degradation products be biocompatible and bioassimilated. One must keep in mind that a degradable and bioassimilable polymer has to be a material too, i.e. it has to exhibit characteristics comparable to those of the non-degradable compound to be replaced.

To achieve biodegradation and biorecycling of biopolymeric systems, Mother Nature has set up very sophisticated processes based on enzymes and thus on cells. The most attractive biodegradable polymers are the biopolymers issued from living systems. However enzymatic phenomena are very sophisticated and selective. Consequently, they rapidly fail degrading biopolymers when the corresponding macromolecules are chemically modified as it is often the case when one wants to fulfil requirements related to biofunctionality.

Presently enzymes can hardly be used to degrade artificial synthetic polymers unless it is under special conditions. It is worth noting that compounds like poly(vinyl alcohol), PVA, bacterial polymers and poly(ϵ-caprolatone), PCL, that are biodegradable under outdoor conditions are degraded abiotically and thus very slowly in an animal body where they are not biodegradable. Despite this difficulty the number of artificial polymers proposed as biodegradable biomaterial candidates to replace biopolymers or biostable polymers is increasing.

3. BIOPOLYMERS

Among the various macromolecular compounds of natural origin that can be obtained from animals, plants and micro-organisms, namely polysaccharides, proteins and bacterial polyesters, only a small number has received applications as biomaterials so far.

3.1 Polysaccharides

The family of polysaccharides is composed of a great number of polymers resulting from the combination of sugar-type building blocks. The most common polysaccharides in the biomedical field are: cellulose, alginates, dextran and chitosan.

3.1.1 Cellulose and derivatives

Cellulose is a polysaccharide based on glucose. It is not adapted to applications in deep surgery. Cellulosic compounds are primarily used as wound dressings. Hydroxyalkyl celluloses and carboxymethyl celluloses have received applications as matrices for drug delivery via the gastro-intestinal (GI) track. None of these compounds are biodegraded in the human body. Regenerated cellulose is largely used as hemodialysis membranes and hollow fibers[3] that are in contact with blood for rather short periods of time in the absence of any degradation.

3.1.2   Alginates

The family of alginates is composed of a great number of polysaccharide-type polymer chains that are composed of glucuronic and mannuronic acid, these sugar building blocks being distributed in a range of modes[4]. Alginates are primarily used as absorbing wound dressings[5]. Because they easily form hydrogels by complexation with calcium ions, alginates have been proposed as suitable for drug delivery and for cell entrapping. Alginates are involved in some pancreas prosthesis in which Langherans islets are entrapped in a hydrogel based on Ca-complexed alginates. They are not biodegradable in a human body.

3.1.3   Dextrans

The family of dextrans is composed of polysaccharides based also on glucose but the sugar building blocks are linked together via α-1-6 bonds with some other secondary linkages leading to branching[6]. Dextrans are slowly degraded by dextranases in an animal body. They are used as isotonic plasma substitutes in order to regenerate the volume of body fluids after great loss of blood. Some derivatives like dextran sulfates or oxidized dextrans are also used but the chemical modifications decrease the ability of the dextran main chain to biodegrade.

3.1.4   Chitosan

The family of chitosan-type polymers is composed of macromolecules of the glycosaminoglycan-type derived from chitin, an alternating copolymer of acetylated glucosamine and glucose. Chitosan polymers are obtained by deacetylation of chitin and their properties depend very much on the degree of deacetylation. Some members exhibit a weak polycation-type behaviour and are soluble under acidic conditions. Chitosan polymers have been proposed for many biomedical and pharmacological applications[7]. However, some members have been found hemotoxic, probably because of the presence of protonated amino groups along the chains. Whether chitosan polymers are totally biodegraded in an animal body is still questioned.

3.1.5 Hyaluronic acid

Hyaluronic acid is a glycosaminoglycan found in conjonctive tissues of any vertebrate. The macromolecule is composed of N-acetyl glucosamine and glucuronic acid. At concentrations higher than 0.2 g/L, a gel is formed. Hyaluronic acid can also be cross-linked[8]. The resulting compounds are used as temporary prostheses for sinovial liquid to treat arthrosis[9]. Hyaluronic acid is rapidly biodegraded in a human body where it is normally regenerated.

3.2 Bacterial polyesters

Some micro-organisms, especially bacteria, can synthesise aliphatic polyesters of the poly(β-hydroxy acid)-type, namely poly(β-hydroxyalkanoates) (PHA). The natural compound is poly(β-hydroxy butyrate), PHB[10].

$$\left[O-\underset{\underset{CH_3}{|}}{CH}-CH_2-CO \right]_n \qquad \left[O-\underset{\underset{R}{|}}{CH}-CH_2-CO \right]_n$$

PHB   PHA

This compound is used by the micro-organisms to store carbon and face a starving situation. Of course, this biopolymer is biodegradable and bioassimilable by microorganisms to recover the use of the stored carbon[11]. Biochemists have taken advantages of this ability to synthesise non-natural PHA. Conditions to make different homopolymers with longer alkyl chains, and copolymers of different kinds as well, have been identified but the yields are usually poor. From the viewpoint of biomedical applications, PHAs are biocompatible. However they are of limited interest when people present them as worthwhile compounds because the enzymes that are necessary to biodegrade them are not present in animal organisms. For instance, poly(β-hydroxy octanoate), PHO, was shown very resistant to biological milieu with very long post implantation lifetimes[12].

3.3 Proteins

Proteins are composed of amino acids linked together via peptide-type amide bonds. Proteins play different roles in an animal body. Some of them contribute to osseous and soft tissues scaffolds. Others serve as enzymes or as hormones. Several proteins are currently used in the biomedical field. However, they are presently regarded suspiciously because of the risk of disease transmission, regardless of their animal or human origin. Like polysaccharides, they are very sensitive to chemical modifications and can loose their biodegradability upon reaction with glutaraldehyde or other cross-linkers.

3.3.1   Collagen

Collagen is probably the most common protein in the field of biomaterials. Actually, 19 different kinds of collagen have been identified. Collagen is used in soft tissue and plastic surgery to fill up tissue defects. Collagen macromolecules are composed of rather large amounts of hydroxy proline and are surprisingly rich in glycine and proline[13]. Nowadays, collagen is regarded suspiciously because of the risk of protein-associated disease transmission. Collagen can be cross-linked by the glutaraldehyde method and by using other cross-linkers[14].

3.3.2   Fibrin

Fibrin is primarily used to make the so-called biological glue that is used in different surgical applications to control bleeding and provide air and fluid tightness. Fibrin glue is based on two components derived from plasma. The thrombin and the fibrinogen components are isolated from plasmapools of healthy donors. Regulations require that the inactivation or removal of viruses has to be demonstrated. Combination of solvent/detergent treatment and heat treatment are recommended[15]. Perfect polymerization and cross-linking by factor XIII is necessary to provide optimun mechanical strength[16,17]. Fibrin also exists under solid forms such as powder, foam, film and plain matrix[18].

4.   ARTIFICIAL BIOPOLYMERS

Modern polymer science offers the possibility to design polymeric backbone including labile bonds, to allow abiotic chemical degradation, i.e. degradation without any contribution from enzymes, a characteristic that limits the risk of specific recognition and, thus, of dramatic immune response. However, the number of such backbones is limited and one must play with secondary structural factors to enlarge the range of possible properties, a critical requirement with regard to the number of different applications which have already been identified in surgery and pharmacology. Property modulation can be achieved by taking advantage of chirality, and of side chain structures and functionality.

Our general strategy is based on the synthesis of artificial polymeric compounds which are made of naturally occurring building blocks and generate metabolites upon abiotic degradation (artificial biopolymers).

Fig 3 shows the main characteristics of macromolecules made according to this strategy, together with the various factors that can be used to diversify the properties and to adjust them to the list of required specifications of a given biomedical or pharmacological application.

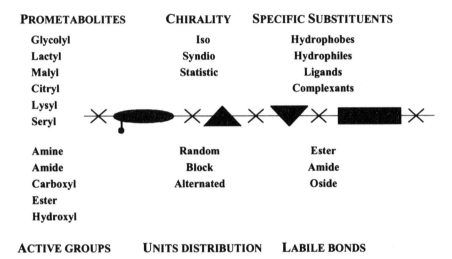

PROMETABOLITES	CHIRALITY	SPECIFIC SUBSTITUENTS
Glycolyl	Iso	Hydrophobes
Lactyl	Syndio	Hydrophiles
Malyl	Statistic	Ligands
Citryl		Complexants
Lysyl		
Seryl		
Amine	Random	Ester
Amide	Block	Amide
Carboxyl	Alternated	Oside
Ester		
Hydroxyl		
ACTIVE GROUPS	UNITS DISTRIBUTION	LABILE BONDS

Figure 3. Schematic representation of an artificial biopolymer composed of different repeating units linked through labile bonds and diversified by using various structural factors such as chirality, repeating unit distribution, substitution and functionalization.

Let us consider some examples of bioresorbable polymeric compounds synthesized according to this strategy.

4.1 Aliphatic polyesters derived from lactic and glycolic acids

Nowadays, the most attractive family of bioresorbable polymers is composed of poly(α-hydroxy acids) derived from lactic and glycolic acids (PLAGA)[19] (Table 2)

PGA is commercially available under the form of sutures and meshes, and GA-rich PLAGA as well. PLAX and PLAXGA(100-X) stereocopolymers with high X values are commercially available as osteosynthesis devices, the most successful one being the interference screw used to fix cruxiate ligaments autografts[20]. Amorphous PLAXGAY copolymers are industrially used to make microparticles for the delivery of anticancer agents.

Table 2. The family of glycolic and lactic acid-derived aliphatic polyesters (X and Y stands for the percentage of L-acid and glycolic acid units in the polymer chain, respectively, the content in D-units being given by the complement to 100%)

Lactide/DIPAGYL copolymers

However, these aliphatic polyesters are not functionalized and thus their interest is limited to implants and particulate-type vehicles. In attempts to have PLAGA polymers bearing some functional groups, polymeric compounds of the same type have been synthesized using a new substituted 1,4-dioxane 2,5-dione derived from gluconic acid, namely DIPAGYL. After suitable protection and deprotection of side chain alcohol groups, polyDIPAGYL and copolymers with lactides can be made which bear alcohol groups as side chain after acidic hydrolysis[21].

Such polymers are likely to be more hydrophilic than regular PLAGA polymers and enable crosslinking. However, they are not water soluble.

4.2 Functional aliphatic polyesters derived from malic acids

In order to make water soluble polymer and to enlarge the bioresorbable polymer family on the basis of the same strategy, carboxylic acid-bearing polymers derived from malic acid, namely (poly(β-malic acid) or PMLA, were synthesized for the first time many years ago.

Poly(R-) and poly(S-β-malic acid)s were synthesized from the two optical isomers of malic acid and of aspartic acid. The family of malic acid based polymers includes also the alpha isomers, i.e. polymers where malic acid repeating units are linked through the alpha hydroxy acid part of the molecule[22,23].

malic acid   Poly(β-malic acid)

Poly(β-malic acid) copolymers

Poly(R-malic acid) was recently found in living systems and is thus a synthetic polymer that is also a true biopolymer[24-25].

4.3   Aliphatic polyamides derived from citric acid and lysine

Bioresorbable water soluble polyamide-type polymers have been synthesised from citric acid and L-lysine.

Poly(L-lysine citramide imide)

Poly(L-lysine citramide)

Various derivatives that are based on poly(L-lysine citramide), (PLCA), and poly(L-lysine citramide imide), PLCAI) backbones have been synthesized[26,27]. These carriers can be conjugated to drugs. Doxorubicin covalently bound to PLCA carriers was able to pass the membrane of doxorubicin resistant K562 cells and reach their nucleus[28]. On the other hand, when partially hydrophobized, these polymers can form micelles or aggregates where lipophilic drugs can be incorporated temporarily through physical interactions (molecular micro-encapsulation)[29].

4.3.1   Primary amine-containing polyesters derived from serine

Recently we succeeded in preparing rather large quantities of a primary amine-bearing aliphatic polyester derived from L- and DL-serine, namely poly(amino serinate) or PAS. This polyamine is a potential carrier for oligonucleotides and DNA fragments. It enabled us to make bioresorbable polyelectrolyte complexes where drugs could be easily entrapped for sometime[30].

Serine

Poly(aminoserinate)

Poly(aminoserinate) copolymers

5. CONCLUSION

The biomaterials field has come a long way from its empirical beginnings. Now the body's response to foreign materials is better understood than ever before. From this rapid survey of biopolymers and of some artificial biopolymers in the framework of biomedical and pharmacological applications, one can conclude that designing novel artificial biopolymers derived from metabolites seems to be a strategy of interest to cover the number of relevant temporary therapeutic applications one can think about, a goal that can hardly be achieved by genuine or modified biopolymers. In parallel biopolymers, like many other biochemicals issued from natural renewable resources, can be the source of a great number of novel synthetic polymeric compounds than can potentially bioresorbed in an animal body. Based on the huge Lego® game that polymer chemistry is, one can predict that scientists will take advantages of this alternative to petrochemistry to overcome the problems raised by the risks related to the use of biopolymers extracted from human and animal bodies and those related to the biostability of more versatile but biostable synthetic polymers.

ACKNOWLEDGEMENTS

The author is indebted to his co-workers who contributed to bring the work at its present level and whose names appear in the reference list.

REFERENCES

1. Vert, M., 1986, Polyvalent polymeric drug carriers, In *"CRC Critical Reviews -Therapeutic Drug Carrier Systems"*, (S.D. Bruck Ed.), CRC Press, Boca Raton, **2**: 291-327

2. Vert, M., 1987, Design and synthesis of bioresorbable polymers for controlled release of drugs, in *"Controlled release of drugs from polymeric particles and Macromolecules"*, (S.S. Davis &L.ILLUM Eds), Wright IOP Publ. Ltd., Bristol, p.117-125

3. Hoenich, N., A., and Stamp, S., 2000, Clinical investigation of the role of membrane structure on blood contact and solute transport characteristics of a cellulose membrane, *Biomaterials* **21**: 317-324

4. Draget, K., I., Skjak-Braek, G., and Smidsrod, O., 1994, Alginic acid gels: Effects of alginate chemical composition and molecular weights, *Carbohydr. Polym.* **25**: 31-38

5. Qin, Y., and Gilding, R., K., 1996, Alginate fibers and wound dressings, *Medic. Device Technol.* **7**: 32-41

6. Mishler, J., M., 1984, Synthetic plasma volume expanders – their pharmacology, safety and clinical efficacy, *Clinics in Haematol.* **13**: 75-92

7. Muzzarelli, R., A., A., 1993, Biochemical significance of exogenous chitin and chitosan in animals and patients, *Carbohydr. Polym.* **20**: 7-16

8. Davidson, J., M., Nanney, L., B., Broadley, K., N., Whitsett, J., S., Aquino, A., M., Beccaro, M., and Rastrelli, A., 1990, Hyaluronate derivatives and their application to wound healing : Preliminary observations, in *Polymer in Medicine – 4*, (C. Migliaresi, E. Chiellini, P. Giusti and L. Nicolais Eds.), Elsevier Applied Science, London, p.171-177

9. Bourzeix, K., 2000, Biomatériaux articulaires injectables, in *"Actualités en Biomatériaux"*, (D. Mainard at al., eds.), Editions Romillat, Paris, p.141-146

10. Lee, S., Y., 1996, Bacterial polyhydroxyalkanoates, *Biotech. Bioeng.* **49**: 1-14

11. Lenz, R., W., 1993, Biodegradable Polymers, *Adv. Polym. Sci.* **107**: 1-40

12. Marois, Y., Zhang, Z., Vert, M., Deng, X., Lenz, R. and Guidoin, R., 1999, Mechanism and rate of degradation of polyhydroxyoctanoate films in aqueous media : a long term study, *J. Biomed. Mater. Res.* **49**: 216-224

13. Parkany, M., 1984, Polymers of natural origin as biomaterials. 2. Collagen and gelatin, in *Macromolecular Biomaterials*, (G.W. Hastings and P. Ducheyne Eds.), CRC Press, Boca Raton, p. 111-118

14. Casagranda, F., Ellender, G., Werkmeister, J., A., and Ramshaw, J., A., M., 1994, Evaluation of alternative glutaraldehyde stabilization strategies for collagenous biomaterials, *J. Mater. Sci. : Mater. In Med.* **5**: 332-337

15. Martinowitz, U., Spotnitz, W., D., and De-Gaetano, G.,1997, Fibrin tissues adhesives. 1997 State of the art, *Thromb. Haemost.* **78**: 661-666

16. Burnouf-Radosewich, M., Burnouf, T., and Huart, J.J., 1990, Biochemical and physical properties of a solvent-detergent-treated fibrin glue, *Vox Sang.* **58**: 77-84

17. Park, M., S., and Cha, C., A., 1993, Biochemical aspects of autologous fibrin glue derived from ammonium sulfate precipitation, *Laryngoscope* **103**: 193-196

18. G. Kerényi, 1984, Polymers of natural origin as biomaterials. 1. Fibrin in *Macromolecular Biomaterials*, (G.W. Hastings and P. Ducheyne Eds.), CRC Press, Boca Raton, p. 91-110

19. Li, S.M., and Vert, M., 1995, Biodegradation of aliphatic polyesters, in *"Degradable Polymers: Principles and Applications"*, (G. Scott and D. Gilead Eds.), Chapman & Hall, London, p. 43-87

20. Athanasiou, K., A., Agrawal, C., M., Barber, F., A., and Burkhart, S., S., 1998, Orthopaedic applications for PLA-PGA biodegradablme polymers, *Arthroscopy* **4**; 726-737

21. Marcincinova-Benabdillah, K., Coudane, J., Boustta, M., Engel, R., and Vert, M., 1999, Synthesis and characterization of novel degradable polyesters derived from D-gluconic and glycolic acids, *Macromolecules* **32**: 8774-8780

22. Braud, C., and Vert, M., 1993, Poly(β-malic acid) based biodegradable polyesters aimed at pharmacological uses, *Trends in Polym. Sci.* **3**: 57-65

23. Vert, M., 1998, Chemical routes to poly(β-malic acid) and potential applications of this water soluble bioresorbable poly(β-hydroxy alkanoate), *Polym. Degrad. And Stab.* **59**: 169-175

24. Cammas, S., Guérin, Ph., Girault, J., P., Holler, E., Gache, Y., Vert, M., 1993, Natural poly(L-malic acid): NMR shows a poly(3-hydroxy acid)-type structure, *Macromolecules* **28**: 4681-4684.

25. Fischer, H., Erdmann, S., Holler, E., 1989, An unusual polyanion from *Physarum polycephalum* that inhibits homologous DNA polymerase α in vitro, *Biochemistry* **28**: 5219-5226.

26. Boustta, M., Huguet, J., and Vert, M., 1991, New functional polyamides derived from citric acid and L-lysine: Synthesis and characterization, *Makromol. Chem., Macromol. Symp.*, **47**: 345-355

27. Henin, O., Boustta, M., Coudane, J., Domurado, M., Domurado, D., and Vert, M., 1998, Covalent binding of mannosyl ligand via 6-O position and glycolic arm to target a PLCA-type degradable drug carrier toward macrophages, *J. Bioact. Comp. Polym.* **13** 19-32

28. Abdellaoui, K., Boustta, M., Morjani, H., Manfait, M. and Vert, M., 1998, Metabolite-derived artificial polymers designed for drug targeting, cell penetration and bioresorption, *Europ. J. Pharm. Sci.* **6**: 61-73

29. Gautier, S., Boustta, M., and Vert, M., 1997, Poly(L-lysine citramide), a water soluble bioresorbable carrier for drug delivery : Aqueous solution properties of hydrophobized derivatives, *J. Bioact. Comp. Polym.* **12**: 77-98

30. Rossignol, H., Boustta, M., Vert, 1999, Synthetic poly(β-hydroxyalkanoates) with carboxylic acid or primary amine pendent groups and their complexes, *Intern. J. Biol. Macromol.* **25**: 255-264

Novel Synthesis of Biopolymers and their Medical Applications

ZBIGNIEW JEDLINSKI and MARIA JUZWA
Polish Academy of Sciences, Centre of Polymer Chemistry, 41-819 Zabrze, Poland

Abstract: Novel synthesis of biomimetic PHB polymers using supramolecular catalyst systems and preparation of artificial ion channels in cell membranes are presented.

1.   INTRODUCTION

Chemistry for health will be a major theme of chemist activities in this century. Therefore a great attention is paid to biopolymers and their potential importance in medical applications. An important goal is effective transport of drugs to tumor cells. To increase the activity of antitumor agents many attempts have been made aimed at effective increase of drug concentration in target cells. Many kinds of drug carriers including synthetic and natural polymers, liposomes, micropheres and nanospheres have been employed. Covalent attachment of the cytotoxin to drug carries seems to be very promising. Lipophylic or amphifilic polymers offer the great potential, because they can be provided of structural features allowing for an easy approach and access to a great number of cells. A variety of natural polymers, such as human serum albumin (HSA), dextran, lecitines as well as synthetic polymers: polyethylene glycol [PEG], poly(styrene-co-maleicanhydride) [SMA], poly(N-hydroxy-propylmethacrylic amid) [HPMA], poly(divinyl-ether-co-maleic anhydride) [DIVEMA] show good properties when used as carries for cytotoxic drugs.[1]

Recently hyaluronic acid [HA], a polysaccharide having alternating D-glucuronic acid and N-acetyl-D-glucosoamine-substituted units present in

Biorelated Polymers: Sustainable Polymer Science and Technology
Edited by Chiellini *et al.*, Kluwer Academic/Plenum Publishers, 2001

the human body as synovial fluid has been employed as polymer carrier for preparation of drug-polymer conjugates (e.g. coupled with drugs as mitomycin C). Same disadvantages of all these systems indentified in: 1) low biocompatibility of synthetic polymers; 2) difficulties in proper chemical characterization of high molecular weight natural polymers as e.g. hyaluronic acid, forced us to look for other polymers biocompatible, immuno neutral and exhibiting a very precisely defined structure. In this paper we will present results of studies on poly(3-hydroxybutanoic acid) used for preparation of well defined drug-polymer conjugates.

Two types of natural aliphatic polyesters having the structure of poly-(R)-3-hydroxybutanoate [PHB] are present in living systems:

- High-molecular-weight (M_w up to hundred thousands) produced in prokaryotic cells as microbial storage material
- Low-molecular-weight polymers (DP 20-120) present in prokaryotic and eukaryotic cells, forming complexes with poly(-Ca-phosphate) as building blocks of channels in cell membranes, responsible for ion transport across a membrane. The low-molecular-weight polyesters are present also in human blood plasma.

The presence of low-molecular-weight PHB polymers in living cells and their obvious importance in vital processes have attracted attention of chemists and biologists, and a lot of attempts have been made to synthesize analogues to natural PHB using various synthetic methods.

Seebach and his associates have developed an elegant method of PHB preparation by using a step-by-step polycondensation of (R)-3-hydroxybutanoic acid. However, this procedure is very laborious and time-consuming because protection and deprotection of end groups of the monomer and intermediate oligomers are necessary at each polycondenstion step.[2,3]

Another synthetic procedure was based on ring-opening polymerization of β-butyrolactone using organometallic coordinative initiators. However, the resulting polymers exhibit very broad molecular weight distribution and end groups, which are different from those found in natural polymers present in living systems.[4,5]

In this paper synthesis of biomimetic analogues of natural PHB by using (S)-β-butyrolactone as a monomer and supramolecular complexes of alkali metals as catalysts are discussed.

2. REACTIONS OF β-BUTYROLACTONE WITH ALKALI METAL SUPRAMOLECULAR COMPLEXES

Three types of initiators have been used for initiation of β-butyrolactone polymerization.

The preparation of the first one was based on discoveries of Dye[6] and Edwards[7] concerning the dissolution of alkali metals: potassium or sodium in an aprotic solvent, such as THF, containing a macrocyclic organic ligand e.g. 18-crown-6 or cryptand [2.2.2]. The specific procedure enables the preparation of an unique alkali metal supramolecular complex forming in THF solution alkali metal ion pairs, e.g. $K^+/L/K^-$ (where L = 18-crown-6 or 15-crown-5).

Such alkali metal ion pairs are capable of two electron transfer from the potassium anion towards a suitable substrate, e.g. β-butyrolactone with formation of a respective carbanion. The strong tendency to two electrons transfer is due to the unusual oxidation state of potassium anion bearing on its outer *s* orbital a labile electron doublet shielded from the positive potassium nucleus by inner orbitals. Using *S*-enantiometr of β-butyrolactone as a monomer and potassium supramolecular complex as catalyst, enolate carbanion is formed as the first reactive intermediate which induces polymerization, yielding poly-(R)-3-hydroxybutanoate[8]. The resulting biomimetic polyester has the structure similar to native PHB produced in nature, except for acetoxy-end-groups, which are formed instead of the hydroxyl ones typical for natural PHB.

By considering the fact that even subtle structural defects can change the bioactivity of a biopolymer we have been looking for another regioselective initiator, which would be able to produce poly[-(R)-3-hydroxybutyrate] bearing only –OH and –COOH end groups typical for natural PHB. It turned out that sodium salt of (R)-3-hydroxybutanoic acid activated by added crown ether can act as a very effective initiator inducing polymerization of (S)- β-butyrolactone. The hydroxybutanoate anion of the initiator attacks the chiral carbon atom of the monomer, as it is usual in ring-opening reactions of β-lactones induced by carboxylate anions. The structure of this synthetic biomimetic PHB is identical to that of the natural polymer[9] (Scheme 1).

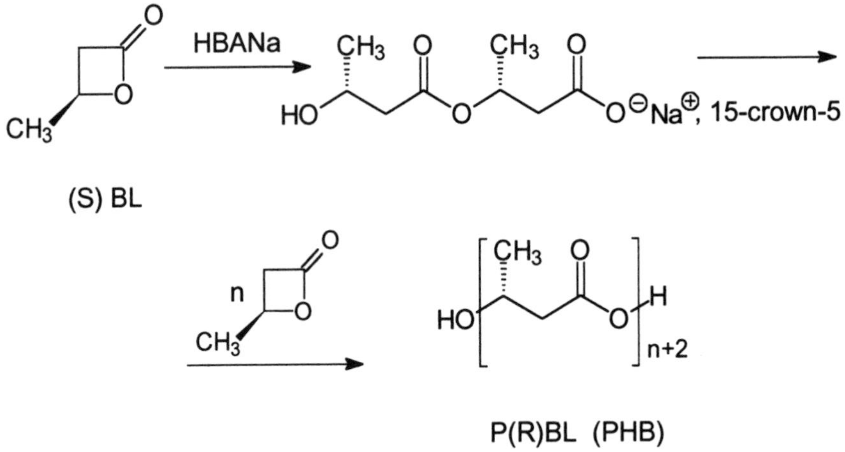

(S) BL

P(R)BL (PHB)

HBANa = ((R) 3-hydroxybutanoic acid sodium salt/15-crown-5) complex

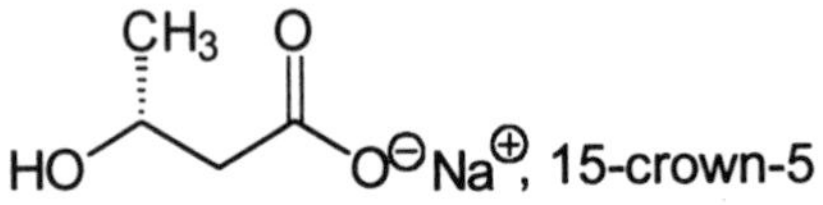

3.   PREPARATION OF BIOMIMETIC ARTIFICIAL ION CHANNELS

The artificial model of a cell membrane was prepared using biomimetic PHB and the calcium polyphosphate complex (poly-P) incorporated into lipid bilayers of 1,2-dierucoylphosphatidylcholine. It was found that PHB/poly-P channels show high conductance for Ca^{2+} and Na^+ cations[10].

Thus the low molecular weight PHB-polymer (19-20 monomer units) can be used effectively for preparation of artificial ion channels in cell membranes mimicking natural ones. The model of channels proposed contains two helices: the outer one containing poly-(R)-3-hydroxybutanoate, complexed by hydrogen bonding with the inner helix of poly(-Ca phosphonate) (Fig 1).

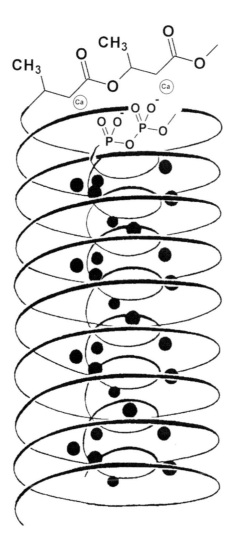

Figure 1. Structure of a P(3-HB)/Ca·PP$_i$ complex suggested by Reusch[11]

The synthetic PHB of low molecular weight is biomimetic and immuno-neutral. It is degradated by enzymatic systems in living organisms and can be used for medical applications.

REFERENCES

1. Luo, Y., Ziebell, M.R. and Prestwich G.D., 2000, A Hyaluronic Acid-Taxol Antitumor Bioconjugate Targeted to Cancer Cells. *Biomacromolecules* **1(2)**: 208-218
2. Seebach, D., Bürger, H.M., Müller, H.M., Lengweiler, U.D., Beck, A.K., Sykes, K.E., Barker, P.A. and Barcham, P.J., 1994, Synthesis of linear oligomers of (R)-3-hydroxybutyrate and solid-state structural investigations by electron microscopy and X-ray scattering. *Helv. Chim. Acta* **77**: 1099-1123
3. Lengweiler, U.D., Fritz, M.G. and Seebach, D., 1996, Synthese Monodisperser Linear und Cyclischer Oligomere der (R)-3-Hydroxybuttersäure mit bis zu 128 Einheiten. *Helv. Chim. Acta* **79**: 670-701
4. Gross, R.A., Zhang, Y., Konrad, G., and Lenz, R.W., 1988, Polymerization of β-monosubstituted-β-propiolactones using trialkylaluminum-water catalytic systems and polymer characterization. *Macromolecules* **21**: 2657-2668
5. Pajerski, A.D., and Lenz R.W., 1993, Stereoregular polymerization of β-butyrolactone by aluminoxane catalysts. *Makromol. Chem. Macromol. Symp.* **73**: 7-26
6. Dye, J.L., 1979, Compounds with alkali metal anions. *Angew. Chem.* **91**: 613-625
7. Edwards, P.P., 1981, The electronic properties of metal solutions. *Phys. Chem. Liq.* **10**:189-227
8. Jedlinski, Z., Novel Electron-Transfer Reactions Mediated by Alkali Complexed by Macrocyclic Ligand. *Accounts of Chem. Res.* **31**: 55-61
9. Jedli_ski, Z., Kurcok, P., and Lenz, R.W., 1998, First facile synthesis of biomimetic poly-(R)-3-hydroxybutyric acid *via* regioselective anionic polymerization of (S) β-butyrolactone. *Macromolecules* **31**: 6718-6720
10. Das, S., Kurcok, P., Jedlinski, Z., and Reusch R.N., 1999, Ion channels formed by biomimetic oligo-(R)-3-hydroxybutyrates and inorganic polyphosphates in planar lipid bilayers. *Macromolecules* **32**: 8781-8785
11. Reusch R.N., 1989, Poly-β-hydroxybutyrate/Calcium Polyphosphate Complexes in Eukaryotic Membranes. *Proc. Soc. Exp. Biol. Med.* **191**: 377-381

Composite Films Based on Poly(vinylalcohol) and Lignocellulosic Fibers
Preparation and Characterizations

[1]EMO CHIELLINI, [1]PATRIZIA CINELLI, [2]SYED H. IMAM and [2]LIJUN MAO
[1]*Department of Chemistry and Industrial Chemistry, University of Pisa, Via Risorgimento 35, 56126 Pisa, Italy;* [2] *Plant Polymer Research Unit, National Center for Agricultural Utilization Research, Agricultural Research Service, U.S.D.A., 1815 North University Street, Peoria, Illinois 61604, USA*

Abstract: The present contribution reports on the incorporation of agricultural waste materials as organic fillers in a film matrix based on PVA. Starch and fibers, derived from sugarcane, apple and orange waste, were cast from PVA aqueous solutions. Glycerol and urea were added as plasticizing agents and resulted effective in obtaining flexible films. Addition of cornstarch resulted in only modest loss of mechanical properties of the films, but reduced the cost of final productrs. To improve water resistance and film cohesiveness, hexamethoxymethylmelamine was added as a crosslinking agent and its effect on the mechanical properties, water resistance and biodegradation (mineralization) rate was assessed. Crosslinked films displayed improved resistance to moisture uptake and degraded slowly in soil.

1.   INTRODUCTION

Plastics produced from petroleum-based raw materials in the form of single-use consumer products are of environmental concern, as most of these materials do not degrade when disposed in the environment after their useful life is over[1]. Currently, systematic collection of plastic waste for recycling and/or disposal is expensive and is limited only to certain communities. Particularly, when plastics are contaminated with soil, foods or other chemicals their recycling become rather difficult[2]. As a result, in the last decade, significant efforts have been made to develop environmentally

Biorelated Polymers: Sustainable Polymer Science and Technology
Edited by Chiellini *et al.*, Kluwer Academic/Plenum Publishers, 2001

87

EMO CHIELLINI et al.

compatible products as an alternative to petroleum based materials. A search
of the Chemical Abstract's subject index showed an exponential increase in
both publications and patents produced on the subject of biodegradable
polymers from 1990 to 1999 (Fig 1).

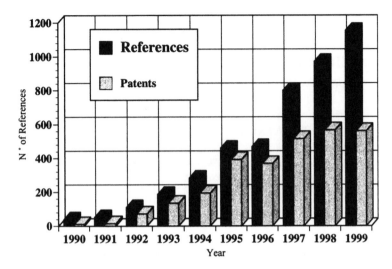

Figure 1. Number of Refences and Patents at the subject "biodegradable polymers" on
Chemical Abstract general subject index in the period 1990-1999.

Based upon their mechanical and processing properties, many natural and
renewable polymers are quite suitable to replace petroleum- based plastics in
specific applications where an extended life span of plastic product is not
desirable. Furthermore, consumers' acceptance of such bio-based plastics is
expected even if natural polymer price represents a limitation on the cost of
the final products[3]. Materials such as commodity crops, agricultural waste
and/or by-products are a good source of natural and renewable polymers and
are comparatively less expensive[4]. Agro-fibers, which represents a
considerable portion of such natural materials is available on a worldwide
basis and are being used in a variety of applications[5]. In some applications,
fibrous materials have been blended with thermoplastic matrix to develop
composites containing various percentages of the fibers[6].

In this regard, polyvinyl alcohol (PVA), a hydrolysis product of
polyvinyl acetate, is well suited for blending with natural polymers since it is
highly polar and water-soluble synthetic polymer which is also
biodegradable[7]. PVA and starch films have been prepared for use as
agricultural mulch films and as water-soluble laundry bags[8-14]. Cast films
made from PVA and cellulose, prepared in N,N-dimethylacetamide-lithium
chloride, exhibited good miscibility due to their mutual ability to form intra-

intermolecular hydrogen bonds between hydroxyl groups[15]. Cast films of PVA-Pectin from aqueous solution have also been reported[16]. Previous studies have shown that a formulation containing PVA and sugarcane bagasse was suitable for both preparing cast film and sprayable films by simply spraying aqueous suspensions of the polymeric components[17,18].

The present contribution reports on the incorporation of agricultural waste materials as organic fillers in a film matrix made-up of PVA. The organic fillers in the matrix were ligno-cellulosic fibers derived from sugarcane bagasse and from citrus fruits (apple and orange) waste recovered after juices have been extracted. Additionally, corn starch was added to the formulation to reduce the cost and to further increase the amount of natural polymers' in the formulation. Composite films were prepared by casting and characterized for their mechanical properties, water sensitivity and biodegradability. Additionally, glycerol and urea were added as plasticizing agents and their effect on mechanical properties was investigated. Hexamethoxymethylmelamine was added as a crosslinking agent, to improve water resistance and film cohesiveness. Crosslinking effect on the mechanical properties, water resistance and biodegradation (mineralization) rate was assessed.

2. EXPERIMENTAL

2.1 Materials

Poly(vinylacohol) (PVA Airvol 425) was purchased from Air Products & Chemicals Inc., Allentown, PA,USA, PVA Airvol 425 was 95.5-96.5% hydrolysed with an average molecular weight of 100,000-146,000. Hexamethoxymethylmelamine (Cymel 303) was purchased from Cytec Industries. Inc., Wallingford, CT, USA. Citric acid was obtained from Aldrich Chemical Company, Milwaukee, WI, USA. Glycerol and urea were purchased from Fisher Chemicals, St. Luis, MO, USA. Unmodified commercial-grade corn/starch (Buffalo 3401) was obtained from CPC International Inc., Argo, IL, USA., with approximately 30% amylose and 70% amylopectin. Cellulosic materials were from three different sources; sugarcane bagasse (SC) was supplied by US Sugar Corporation, Florida, USA, orange (OR) and apple (AP) pomace were the remains of fruit residue after juice extraction supplied by the Sunflo Cit-Russ Limited, Lahore, Pakistan and Tanard's Orchard, Illinois, USA, respectively. All cellulosic materials were milled, sieved to obtain 0.188 mm size particles, and analysed to determine their composition and moisture content (Table 1).

Table 1. Composition of lignocellulosic fibers utilized in the formulation of composite mixtures [a]

Fibers	Protein (%)	fat (%)	fiber (%)	Ash (%)	Cellulose (%)	Lignin (%)	Moisture (%)
SC	4.2	1.9	32.7	11.4	35.6	10.6	7.7
OR[b]	12.9	6.6	14.7	11.3	17.7	4.8	9.9
AP[c]	4.7	11.5	27.3	1.3	21.6	21.1	7.4

a) on dry weigth, b) hemicellulose and/or pectins 32%, c) hemicellulose and/or pectins 12.5%

2.2   Formulations and Sample Preparation

About 62 g of a 10% by weight PVA water solution was introduced in a 250 ml beaker, and the desired amount of glycerol, urea, starch and water was added (by weight) to bring a final concentration equivalent to 10% solids. The resulting mixture was first heated at 80 °C for 30 min under stirring and 6.2 g of fibers were added. Mixture was stirred for additional 10 min. For crosslinked samples, a desired amount of hexamethoxymethylmelamine (HMMM) and a catalytic amount of citric acid were also added in formulations and the resulting mixture was stirred at 70 °C for 45 min. After cooling at room temperature, 3 drops of BYK-019 aqueous defoamer was added and the mixture was further stirred for 5 min.

To prepare films, about 17 g of aqueous suspension (as described above) was poured in to a polypropylene plate (8 x 8 cm) and left to dry overnight at an ambient temperature (23-24°C) and finally for 3 h in a oven at 50 °C. Films were cooled to room temperature before being peeled off the plates. The prepared blends and their compositions are provided in Table 2.

Table 2. Composition of the composite films prepared by casting from water suspensions

Sample	PVA (%)	SC [a] (%)	Gly (%)	Urea (%)	Starch (%)	X [b] (%)	Cit.Ac (%)
PSCG	33.3	33.3	16.6	16.6	-	-	-
PSCSt1	31.0	31.0	15.0	15.0	8.0	-	-
PSCSt2	28.6	28.6	14.3	14.3	14.3	-	-
PSCSt3	25.0	25.0	12.5	12.5	25.0	-	-
PSCSt4	33.3	33.3	8.3	8.3	16.6	-	-
PSCSt5	22.2	22.2	22.2	22.2	11.1	-	-
PSCX1	27.3	27.3	13.7	13.7	13.7	4.0	0.4
PSCX2	26.3	26.3	13.1	13.1	13.1	7.7	0.8

[a] Natural Fibers can be respectively sugar cane (SC), orange (OR) or apple fibers (AP).
[b] X= Hexamethoxymethylmelamine (HMMM)

2.3   Water Sensitivity

Moisture content of samples was evaluated with a Sartorius MA 30 Moisture Analyser by heating about 1 g of substance at 130 ˚C up to constant weight (about 15 min.). Percent of weight loss was recorded.

Specimens were conditioned at different relative humidity at 23 °C for 28 days in a saturated LiCl-water solution atmosphere (15% RH), in a conditioned room (50% RH) and in a saturated Na_2SO_4-water solution atmosphere (95% RH) before testing.

2.4   Scanning Electron Microscopy (SEM)

For scanning electron microscopy (SEM), samples were mounted on aluminium stubs with graphite filled tape, vacuum coated with gold-palladium and examined under the scanning electron microscope (JEOL JSM 6400V, JEOL Inc., Peabody, MA).

2.5   Mechanical testing

Dog bone-shaped specimens were stamp-cut and stored under 50%-controlled relative humidity (RH) for 10 days at 23 °C before testing. The specimens, type IV ASTM D638, had a 4 mm width in the test neck, a length of 80 mm, and a neck length of 30 mm. The film thickness was determined by averaging of three measurements along the test length using a Mini Test 3000 (Elektro-Physik, Cologne, Germany).

Ultimate Tensile strength (TS), percent elongation (El) and Young Modulus (YM) for each film were measured using an Instron Universal Testing System (Instron Corp., Canton, MA). The gauge length was 25 mm, the grip distance was 45 mm and the cross-head speed of the instron was 10 mm/min. Data were collected at a rate of 20 pt/s. At least five specimens for each samples were tested and data were averaged.

2.6   Degradation Test

About 500 mg of PVA/SC/additives blends (Samples PSCSt2, PSCX1, PSCX2,) and the corresponding amounts of PVA and additives were cut in small pieces (2x2 mm) and mixed with 25 g of compost soil in the sample chamber of a closed circuit Micro-Oximax respirometer system (Columbus Instruments. Columbus OH) equipped with expansion interface, condenser and a water bath. The sample chamber was placed in the water bath thermostated at 25 °C and connected to the Micro-Oxymax respirometer.

EMO CHIELLINI et al.

The cumulative CO_2 evolution was recorded every 6 h. Experiments were carried out over a period exceeding 30 days.

3.   RESULTS AND DISCUSSION

3.1   Water Sensitivity

Fibers (SC, OR, AP) and starch showed increased moisture uptake with increased relative humidity, especially above 50% (Table 3). SC fibers had the lowest moisture uptake in comparison to OR and AP fibers. Moisture uptake in PVA, PVA/Fiber and PVA/Fiber/Starch cast films also increased with increased relative humidity. For PVA/SC blends, lower moisture uptake was recorded compared with the corresponding OR and AP containing blends. Addition of starch in the formulation reduced moisture uptake in PVA/SC and PVA/OR blends.

Table 3. Moisture Uptake in PVA/Fibers/Starch blends and Raw Materials

Sample	MU[a] at 15% RH (%)	MU at 50% RH (%)	MU at 95% RH (%)
PVA	4.5	9.3	21.7
SC	5.8	8.6	16.2
OR	7.3	9.3	21.1
AP	6.2	10.6	21.8
Starch	8.6	12.6	18.9
PCSG	6.2	18.8	31.2
PSCSt2	3.7	14.6	21.5
PSCX2	4.4	12.6	17.0
PORG	7.9	17.4	42.2
PORSt2	3.5	16.1	34.3
PORX2	6.5	17.6	41.2
PAPG	7.7	17.7	42.1
PAPSt2	5.6	15.6	42.0
PAPX2	6.1	13.3	28.4

[a] MU = Moisture Uptake

All samples, stored in 95% RH atmosphere for 28 days became wet and started to loose cohesiveness. Especially in samples containing starch and orange or starch and apple the growth of fungi on the film surface was observed. A lower moisture uptake and resistance to fungal growth was observed in crosslinked samples most likely due to steric hindrance exerted by tighter molecular entanglement.

3.2   Mechanical Tests

Shape and composition of fibers influenced film properties. Scanning electron micrographs of SC, OR and AP are provided in Fig 2, a, b, c.

SC was composed of long hollow fibers while OR fibers were more spherical shape AP fibers were thick and round.

a) SC Fibers   b) OR Fibers

c) AP Fibers

Figure 2. SEM Micrograph of SC, OR and AP fibers

While maintaining a 100/100 PVA/Fiber ratio, starch, glycerol, urea, and HMMM were introduced in the formulation. Starch was added in formulations to replace as much of PVA as possible without compromising the film properties. Addition of glycerol and urea softened the films as glycerol acted as a plasticizer for PVA in PVA/Starch or PVA/Pectin films[19,20]. Urea also had a similar plasticizing effect on PVA/SC blends[21] and blends containing urea if used as agricultural mulch films could also provide fertilizer nitrogen to soil upon disintegration.

In PVA/SC blends, elongation at break was mostly uneffected by starch amount as reported in Fig 3.

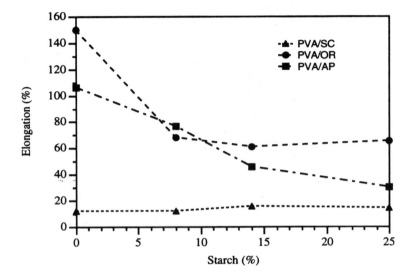

Figure 3. Variation in Elongation at Break with Increasing Amounts of Starch

Thus introduction of starch increased cohesiveness of PVA/SC films as evidenced by TS which increased with starch addition (Fig. 4).

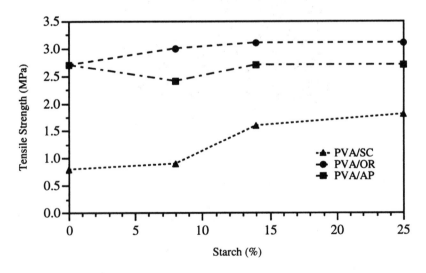

Figure 4. Variation in Tensile Strength with Increasing Amounts of Starch

In PVA/OR blends starch addition caused significant reduction in elongation at break with somewhat moderate variation in TS.

Excellent film forming properties and flexibility was observed in films even when starch concentration in formulations exceeded 25%.

Whereas, in PVA/AP blends addition of starch increased the presence of defects and small holes were detected in the films, thus indicating a loss in mechanical properties due to increased starch content.

Glycerol and urea are low molecular weight plasticizers known to disrupt inter- and intra-molecular interaction of PVA, cellulose, lignin and starch. Blends were prepared with varied contents of glycerol and urea.

Increased plasticizers content had little effect on the El of PVA/SC/Starch blends, and this effect was maintained even at elevated concentration of plasticizers. In PVA/OR/Starch and PVA/AP/Starch, El increased with increased glycerol and urea, especially at 45% (Fig. 5).

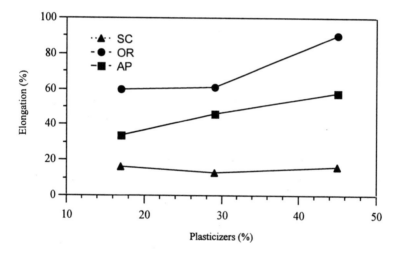

Figure 5. Variation in Elongation at Break with Increasing Amounts of Plasticizers

A reduction in TS was observed when the concentration of plasticizers reached at 45% (Fig. 6).

However, this concentration appeared to be excessive as migration of glycerol to the film surface and subsequent leaching and crystallization of urea (or whitening) was observed upon ageing.

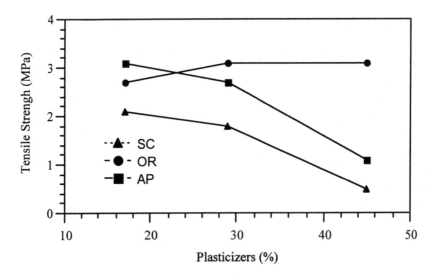

Figure 6. Variation in Tensile Strength with Increasing Amounts of Plasticizers

3.3   Degradation

The amount of CO_2 released from the additive of PSCSt2 and PSCStX1 exceeded the value corresponding to 100% mineralization (Fig.7).

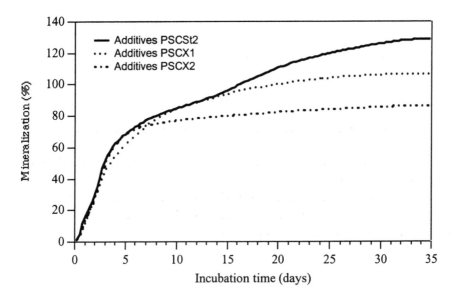

Figure 7. Mineralization of SC/Starch/Plasticizers Blends with Variable Amounts of Crosslinker

This behaviour was due to the presence of starch and urea, that acted as carbon and nitrogen source enhancing soil activity and stimulating the mineralization of the soil organic matter. This referred to "priming effect", well documented in the literature [22,23].

A very high activity was recorded for these specimens in the first days since urea, glycerol and starch are easily degraded, while cellulose and lignin degraded at a lower rate. Anyway the presence of HMMM lowered the mineralization rate in these blends prepared with Starch/Fibers and plasticizers without PVA.

In PVA/SC blends there was little difference in the mineralization rate between uncrosslinked control (PSCSt2) and the sample containing about 4% crosslinker (PSCX1). However, samples (PSCX2) degraded at much slower rate when crosslinker concentration was increased to with 8% (Fig 8).

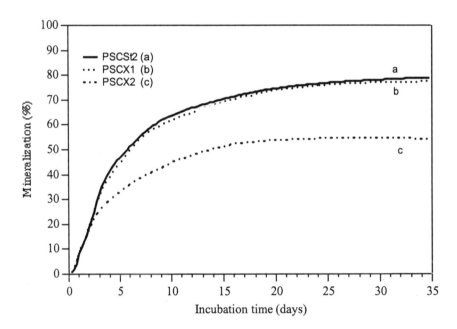

Figure 8. Time variation of the Extent of Mineralization of PVA/SC/Starch Blends with Different Amounts of Crosslinker

This mineralization rate of the blends was in accordance to that expected from PVA and additives content and degradation rate. About 50-80% mineralization was observed within 30 days. these results were in agreement with previous studies conducted on the degradation of PVA/Starch and PVA/SC blends in soil[17,24].

4.   CONCLUSION

Bio-based composite films were casted from suspensions agro-fibers in PVA aqueous solution containing starch and plasticizers. In some films, PVA was crosslinked with hexamethoxymethylmalamine. Composition and fibers size affected mechanical properties of the films.

Variations in glycerol, urea and starch contents effected mechanical properties of PVA/fibers films. Glycerol and urea were useful in obtaining flexible films. Addition of cornstarch resulted in only modest loss of mechanical properties of the films, but reduced the cost of films.

Crosslinking produced films that improved resistance to moisture uptake and degraded slowly in soil.

All components used in preparation of cast films were non-toxic and either natural renewable or biodegradable synthetic polymers. Furthermore, all ingredients are environmentally compatible and can be utilized for application as agriculture mulch. The presence of natural fillers will help maintain organic matter in the soil and urea would act as an additional source of fertilizer nitrogen.

ACKNOWLEDGMENTS

Research was done as collaboration between the University of Pisa, Italy and the Plant Polymer Research Unit of the USDA-ARS-National Center for Agricultural Utilization Research, Peoria, Illinois. The authors thank Ms. Paulette Smith, Ms. Jan. Lawton and Mr. Gary Groose for technical assistance. The financial support for Ms. Patrizia Cinelli's Ph.D. thesis research was provided by the Ministry of University and Technology of Italy and in part by the NCAUR/ARS/USDA. We are grateful to US Sugar Corporation, Florida; Sunflo Citrus Limited, Pakistan and Tanard's Orchard, Illinois for providing gratis the respective by-products materials used in this study. Assistance of the USDA International Program Office is greatly appreciated.

REFERENCES

1. Ress, B.B., Calvert, P.P., Pettigrew, C.A., Barlaza, M.A., 1998, Testing anaerobic biodegradability of polymers in a laboratory-scale simulated landfill, *Environ. Sci.Technol.*; **32**: 821-827
2. Llop, C., Perez, A., 1992, Technology available for recycling agricultural mulch film *Makromol.Chem.,Macromol.Symp.* **57**: 115-121

3. Steinbüchel, A., 1995, Use of biosynthetic, biodegradable thermoplastics and elastomers from renewable resources: the pros and cons. *J. Macromol. Sci., Pure Appl. Chem.,* **A32(4)**: 653-660

4. Young, R.A., 1994, In Vegetable Fibers, In *Kirk Othmer Encyclopedia of Chemical Technology*, 4 ed., **Vol.10**, John Wiley & Sons, NY, pp 727-744

5. Narayan, R., 1994, Polymeric materials from agricultural feedstocks, In *Polymers from Agricultural Coproducts,* (Fishman M.L., Friedman R.B., Huang, S.J., eds.), American Chemical Society, Washington DC, pp 2-28

6. Young, R.A., 1996, Utilization of natural fibers: characterization, modificationand applications, *First International Lignocellulosics-Plastics Composites*, (Leao, A.L.; Carvalho, F.X.; Frollini, E. eds.), March 13-15 , Sao Paolo Brazil, pp. 1-22

7. Matsumura, S., Tomizawa, N., Toki, A., Nishikawa, K., Toshima, K., 1999, Novel poly(vinyl alcohol)-degrading enzyme and the degradation mechanism, *Macromolecules,* **32**: 7753-7761

8. Otey, F. H., Mark, A. M., Mehltretter, C. L., and Russell, C. R., 1974, Starch-based film for degradable agricultural mulch, *Ind.Eng.Chem.Res.* **13**: 90-92.

9. Otey, F. H, and Mark, A. M., 1976, Degradable starch-based agricultural mulch film, *U.S. Patent* **3,949,145**

10. Westhoff, R .P., Kwolek, W. F., and Otey, F. H., 1979, Starch-Polyvinyl alcohol films-effect of various plasticizers, *Starch/Stärke.* **5**: 163-165

11. Otey, F. H., Westhoff, R. P., Doane, and W. M., 1987, Starch-based blown films. 2, *Ind.Eng.Chem.Res.* **26**: 1659-1663

12. Akashah, S. A., Lahalih, SM., and Al-Hajjar, F.H., 1986, Degradable agricultural mulch film, *E.P.* **0,224,990**

13 Lahalih, S. M., Akashah, S. A., and Al-Hajjar F. H., 1987, Development of degradable slow release multinutritional agricultural mulch film, *Ind.Eng.Chem.Res.* **26**: 2366-2372

14. Otey, F. H., and Westhoff, R. P., 1989, Biodegradable starch-based blown films, *U.S.Patent* **4,337,181**

15. Nishio, Y., Hratani, T., Takahashi, T., and Manley, R. S., 1989, Cellulose/Poly(vinyl alcohol) blends: an estimation of thermodynamic polymer-polymer interaction by melting point depression analysis *Macromolecules.* **22**: 2547-2549

16. Coffin. R., Fishman. M. L., Ly, T. V. 1996, Thermomechanical properties of blends of pectin and poly(vinyl alcohol), *J.Appl.Polym.Sci.* **57**: 71-79

17. Chiellini, E., Cinelli, P., Corti, A., Kenawy, E.-R., Grillo Fernandes, E., and Solaro R., Environmentally sound blends and composites based on water-soluble polymer matrices, 2000, *Mac.Symp.* **152**: 83-94,

18. Cinelli, P., Palla, C., Miele, S, Magni, S., Solaro R., and Chiellini, E., 1999, Compositi biodegradabili a matrice polivinilica e cariche da fonti rinnovabili, In XIV Abstracts of *XIV Italian Conference of Macromolecolar Science and Technology (AIM)*, Salerno (Italy), 13-16 Sept.1999, pp.865-689

19. Lawton W., Fanta G.F., 1994, Glycerol-plasticized films prepared from starch-poly(vinyl alcohol) mixtures: effect of poly(ethylene-co-acrylic acid), *Carbohydrate Polymers.* **23**: 275-280

20. Fishmann ML, Coffin D R. 1996, Pectin/starch/glycerol films: blends or composites? *J. Macromol. Sci. Pure Appl. Chem.,* **A33** (5): 639-654

21. Cinelli P., 1999, PhD Thesis *Formulation and Characterization of Envrionmentally Compatible Polymeric materials for Agriculture Applications.* University of Pisa

22. Sharabi N.E., Bartha R., 1993, Testing of some assumptions about biodegradability in soil as measured by carbon dioxide evolution, *Appl.Environm.Microbial,* **59**: 1201

23. Shen J., Bartha R., 1996, Priming effect of substrate addition in soil -based biodegradation tests, *Appl.Environm.Microbial*, **62**: 1428-1430
24. Chen L., Imam SH, Gordon SH, Greene RV. 1997, Starch-polyvinyl alcohol crosslinked film- performance and biodegradation, *J.Environ. Polym.Degrad.* **5**: 111-117

Composite Materials Based on Gelatin and Fillers from Renewable Resources
Thermal and Mechanical Properties

EMO CHIELLINI[1], PATRIZIA CINELLI[1], ELIZABETH GRILLO
FERNANDES[1], EL-REFAIE KENAWY[2] and ANDREA LAZZERI[3]
*[1]Department of Chemistry and Industrial Chemistry, University of Pisa, Via Risorgimento 35,
56126 Pisa, Italy; [2]Department of Chemistry, Faculty of Science, University of Tanta, Tanta,
Egypt; [3]Department of Chemical Engineering, Industrial Chemistry and Material Science,
University of Pisa, via Diotisalvi 2, 56126 Pisa, Italy*

Abstract: Scraps of animal gelatin (WG), generated in pharmaceutical industry, were
             filmed by casting of water suspensions. To improve consistence of the films,
             gelatin scraps were blended with a synthetic biodegradable polymer such as
             poly(vinyl alcohol) (PVA) and fillers from renewable resources such as sugar
             cane bagasse (SCB) that is the lignocellulosic residue derived from sugar cane
             juice extraction. Glutaraldehyde was used as crosslinking agent to improve
             water resistance. In order to investigate the interactions among the components
             and the effect of composition on properties of the final items, thermal, thermo-
             mechanical and mechanical properties of some composite films based on WG,
             PVA and SCB were investigated.

1.   INTRODUCTION

As a partial solution to the global issue of plastic waste, in recent years much interest has been devoted to the formulation of environmentally degradable plastic materials[1]. In particular the use of natural polymers presents several advantages such as biodegradability, utilizing of renewable resources, recyclability. At the same time water sensitivity and degradability of natural polymers limit their possible applications. Consequently bio-plastics cannot replace synthetic plastics in every application but they can result appropriate in specific products especially for those applications in which recovery of plastics is not economically feasible, viable and controllable such as plastic items for one time use[2]. Biodegradability of bio-polymers is the base for their applications, but their processing, performances and in particular price are by far more important issues since bio-polymers should compete with synthetic low cost polymers[3].

Among naturally occurring polymers, gelatin offers good processability properties both in aqueous media and in the melt, good film forming properties and adhesion to various substrates[4]. Gelatins are high molecular weight polypeptides produced by denaturation and physical and chemical degradation of collagen, which is the primary protein component of animal connective tissues, such as bone, skin, tendons[5,6].

Gelatin is characterized by having a high content of the amino acid glycine (33 mol%) and the presence of the amino acid hydroxyproline (10 mol%) and hydroxylysine (0.5 mol%)[7]. The glycine residues spaced throughout the chain create a profusion of spaced "hinges" around which relatively unrestricted rotation can occur. The presence of chain-chain bonding induces restrictions to rotation around the skeletal bonds. The addition of components such as water or glycerol which are able to separate effectively the chains from each other, eliminates these constraints[8.] The diluent is supposed to occupy the space between collagen molecules forming hydrogen bonding regularly along the molecular length. Thereby linking one molecule to another while also promoting the arrangement of the molecules in parallel[7].

Gelatin is currently used in various applications comprising manufacturing of pharmaceutical products, x-ray and photographic films development and food processing[6]. The higher price of gelatin, over 4 USD/Kg for animal derived proteins[4], as compared to some other biopolymers, especially starch, represent a limit for a research on technical applications. Gelatin scraps generated in the different manufacturing processes constitute a no-value material suitable for productive applications. As a part of a research program aimed at the preparation and evaluation of environmentally degradable polymers for various applications in agricultural practices, with specific reference to the in situ formulation of self-fertilizing mulching films[9], we started to consider the possibility of utilizing scraps of animal gelatin, generated in pharmaceutical industry[10]. To improve consistence of the films and regulate time of degradation in the soil gelatin scraps were blended with a synthetic biodegradable polymer such as poly(vinyl alcohol) (PVA) and fillers from renewable resources such as sugar cane bagasse (SCB) that is the lignocellulosic residue derived from sugar cane juice extraction. PVA water solubility and soil structuring properties are important characteristics for agricultural applications[11,12]. The use of waste gelatin (WG) and SCB allows for low cost and in situ formulation. When WG based cast films were applied on the soil the introduction of SCB in the formulations increased time of permanence conferring higher cohesiveness to the samples and lowering mineralization rate of the composites in soil burial tests[13,14]. Glutaraldehyde was used as crosslinking agent to improve water resistance and regulate degradation rate in soil[14].

In order to investigate interactions among the components and the effect of composition on properties of the final items, thermal, thermo-mechanical and mechanical properties of some composite films based on WG, PVA and SCB were investigated.

2.   EXPERIMENTAL

2.1   Materials

Poly(vinyl alcohol) (PVA) was Hoechst product, Mowiol 08/88 with an average molecular weight of 67 kD and 88% hydrolysis degree, scraps of animal gelatin (WG), not anymore suitable for in house recycling, deriving from production of pharmaceutical capsules was kindly supplied by Rp Scherer Egypt. WG elemental analysis gave a content of C 49.5%, H 6.3% and N 11.8%. WG contain several additives from original formulation such as pigments and glycerol. Sugar cane bagasse (SCB) fibers were dried in an oven at 50 °C for 24 h and then ground with a blade grinder. The ground SCB was sieved and the fraction passing through a mesh sieve (0.212 mm) was collected. Elemental analysis on SCB and WG is reported in Table 1. Glutaraldehyde was Aldrich product, 50% by weight aqueous solution.

Table 1. Elemental Analysis of SCB

Sample	C (%)	N (%)	Protein (%)	Fat (%)	Fibers (%)	Ash (%)
SCB	48.7	0.3	9.1	2.6	42.6	6.0

2.2   Films Preparation

WG has been suspended in water at 50 °C after stirring for 30 min. For blends preparation a 10% by weight PVA water solution was introduced in a 10% by weight WG water suspension and the resulting mixture was stirred at 70 °C for 20 min. Composites with SCB were prepared by adding the required amount of SCB in a WG water suspension. The mixture was stirred for 30 min. at 50 °C. Crosslinked films were obtained by addition of the desired amount of glutaraldehyde water solution in a WG water suspension. The mixture was stirred for 5 min. at room temperature.

Cast films were prepared by slow evaporation of a water suspension of the ingredients in teflonated aluminum trays. Composition of the tested blends is reported in Table 2.

Table 2. Composition of Blends Based on WG, PVA and SCB

Sample	WG (%)	PVA (%)	SCB (%)	Glutaraldehyde (%)
PVA	-	100	-	-
WG	100	-	-	-
WGP10	90	10	-	-
WGP20	80	20	-	-
WGP30	70	30	-	-
WGP50	50	50	-	-
WGP80	20	80	-	-
WGP90	10	90	-	-
WGP800X	20	79.75	-	0.25
WGX1	99.75	-	-	0.25
WGX2	99	-	-	1.00
WGX3	97.5	-	-	2.50
WGSCB20	80	-	20	-
WGSCB20X	79.75	-	20	0.25

2.3   Methods

2.3.1   Thermal Analysis

A Mettler TA4000 System consisting of a TG50 furnace, a M3 microbalance, a DSC-30 cell and a TA72 Graph Ware was used for thermal measurements and evaluations. For thermal gravimetric analysis (TGA) samples (about 10 mg) were heated from 25 to 600 °C at a scanning rate of 10 °C/min. under nitrogen atmosphere, with a flow rate of about 200 ml/min. Onset temperature T_{on} was evaluated as the temperature corresponding to the crossover of tangents drawn on both sides of the decomposition trace. The temperature of the decomposition peak (TPeak) was evaluated such as the minimum of the first derivative curve.

Differential Scanning Calorimetry (DSC) analysis was performed under nitrogen flow (about 100 ml/min); heating samples from -100 °C to 220 °C at a scanning rate of 10 °C/min. All samples were run in duplicate.

2.3.2   Dynamic Mechanical Thermal Analysis (DMTA)

Gelatin sheets were analyzed by means of a dynamic-mechanical analyzer from Perkin-Elmer DMA-7. Three point bending scan were performed using a static-to-dynamic stress ratio of 110% at 1 Hz frequency and a heating rate of 5 °C/min. Samples were maintained in a conditioned room, 50% relative humidity and 23 °C, for 10 days before testing.

2.3.3   Tensile Test

The tensile properties of the prepared films were determined by using an Instron Universal Tensile Testing Machine (Mode 4300). Dog-bone shaped specimens were stamp cut and then maintained in a room, conditioned at 50% relative humidity (Rh) and 23 °C, for 10 days before testing. Specimens, type IV ASTM D638, had a rectangular cross-section of 6 mm and a gauge length of 40 mm with a thickness of 0.3 mm. Experiments were run out at a crosshead speed of 20 mm/min. Values were averaged on at least five specimens.

3.   RESULTS AND DISCUSSION

3.1   Thermal Properties WG/PVA Cast Films

Thermal characteristics and weight loss temperature of WG, PVA, SCB and composites of the same are reported in Table 3.

Thermogravimetric analysis of cast films based on WG and blends with PVA under nitrogen atmosphere showed a first loss of weight, usually placed in the 25-150 °C range attributed to loss of water and/or volatile components.

The thermal stability of PVA (T_{on}= 276 °C) was appreciably larger than that of WG (T_{on}= 237 °C). The thermal stability of WG/PVA blends occurred between that of each individual component decreasing as increased the less stable WG component. Anyway all blends decomposed at higher temperature than that recorded for WG itself.

This behavior suggests that the more stable PVA acted such as a protective component.

TGA also showed that all the blends resulted suitable for processing or application up to 200 °C which are largely sufficient for agriculture applications or processing from the melt in presence of the appropriate content of plasticizers.

Table 3. Thermal Characteristics of PVA and WG Cast Films Evaluated by TGA.

Sample	Volatile (%-wt)	T_{on} (°C)	TPeak1 (°C)	WL DP1 (%-wt)	TPeak2 (°C)	WL DP2 (%-wt)	Residue (%-wt)
PVA	6.8	276.3	305.0	68.6	412.3	21.5	3.1
WGP90	6.5	274.2	305.0	68.7	412.3	17.1	7.7
WGP80	5.9	272.4	305.5	64.5	412.3	20.1	9.5
WGP70	6.4	266.1	307.3	64.7	403.1	18.4	10.5
WGP50	8.0	252.2	288.7	60.0	400.7	20.6	11.4
WGP20	9.0	250.0	295.7	58.0	397.0	16.6	16.4
WG	10.5	237.5	287.5	65.6	-	-	23.9
WGX1	11.4	236.9	286.3	67.6	-	-	21.0
WGX2	10.2	235.1	271.7	68.8	-	-	21.0
WGX3	9.7	235.7	281.7	67.0	-	-	23.3
WGSCB20	9.3	223.4	312.0	66.3	-	-	24.4
SCB	5.5	238.8	342.3	74.3	-	-	20.2

Ton = Decomposition temperature corresponding to temperature taken in correspondence of the crossover of tangents drawn on both sides of the decomposition trace,

TPeak = Temperature of the Decomposition Peak evaluated as the minimum of the first derivative curve, WL DP1, WLP2 = % of Weight loss in respectively the first and the second decomposition peak, Residue = % of Weight Residue at 600°C

In Fig 1 WG change of storage modulus and heat flow as a function of temperature are compared.

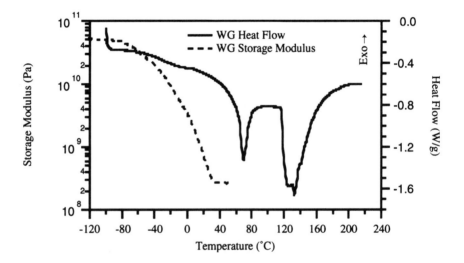

Figure 1. DSC and DMTA Traces of WG cast films

DSC trace shows at about -30°C a change in the base line that corresponded to the drop of the storage modulus. In gelatin cast films the first glass transition has been associated with gelatin soft blocks composed of sequences in which are present mainly α-amino acids, including glycine at every third position.

The second glass transition was associated with gelatin rigid blocks composed of sequences mainly made up of the imino acids proline and hydroxyproline including glycine at every third position which are predominant in mammalian gelatin[15]. So the transition at -30°C can be attributed at the glass transition of plastified soft blocks in WG.

Above 60 °C WG cast film started to soft and it was not possible to continue DMTA measurement. In DSC the second baseline shift with energy absorption can be interpreted as the Tg of the plastified rigid blocks. A third transition is observed indicating that WG has more than the two common phases.

So it is possible that this third transition corresponds to several thermal events such as evaporation, structural reorganization and the non-equilibrated Tg of the rigid blocks.

Thus TGA results reported the evaporation of volatile at this range of temperature. In DSC analysis of WG/PVA blends a plasticizing effect of WG on PVA was observed (Fig.2)

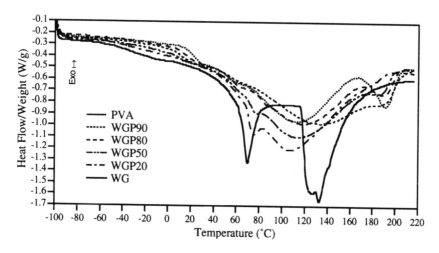

Figure 2. DSC Traces of WG and PVA Based Blends

The step associated with PVA glass transition, placed at about 22 °C for pure PVA cast films conditioned at 50% Rh, slightly moved towards lower temperatures with increasing WG content in the blends and was placed at about 7 °C for WGP50. In WGP20 the PVA content was not sufficient to detect its glass transition any more. PVA melting temperature moderately decreased passing from pure PVA cast films (191 °C) to WGP50 (187 °C).

The melting peak became broader with increasing of WG content and was not detectable in WGP20.

With increasing of PVA content in WG/PVA blends the Tg relative to WG, at 71 °C, moved to higher temperature, while the transition at 130 °C was no more individuated due to the broader endhotherm that, as discussed above, corresponds to the overlaying of more than one thermal process. As a consequence in WGP80 and WGP90 was observed a single broad peak placed at about 120° C overlaid with water evaporation.

3.2   Mechanical Properties in WG/PVA Blends

WG cast films appeared red coloured, translucent, flexible, with small amounts of insoluble residues dispersed in the gelatin matrix. WG values of mechanical properties corresponded to WG plasticized by water and glycerol present in the scraps. WG cast films presented 116% Elongation at Break (El), 11 MPa for Ultimate Tensile Strength (UTS) and 78 MPa for Young's Modulus (YM) as reported in Table 4.

Table 4 Mechanical Properties and Relevant Standard Deviations (StDv) of WG based Blends.

Sample	El (%)	StDv	UTS (MPa)	StDv	YM (MPa)	StDv
WG	116	22	11	1.2	78	12
WGP50	133	13	8.3	0.9	79	17
WGP70	132	14	14	0.5	86	14
WGP80	257	28	22	2.3	133	13
WGP90	235	24	26	0.8	200	45
PVA	211	26	35	3.1	387	63
WGP80X	241	23	22	1.9	128	15
WGX1	146	16	10	1.6	33	1
WGX2	140	29	10	2.5	33	7
WGX3	300	17	13	2.0	10	3
WGSCB20	14	2.1	10	0.9	228	27
WGSCB20X	14	2.6	11	1.5	283	21

El= Elongation at Break, UTS= Ultimate Tensile Strength, YM= Young Modulus,

Addition of up to 20% of WG to PVA (WGP90, WGP80) increased the El from 211% for pure PVA up to 257% for WGP80 and reduced UTS and YM, as reported in Table 4. This behavior can be attributed to the plasticizing effect of glycerol present in WG. At higher amount of WG in the blends, mechanical properties approached the values recorded for pure WG.

3.3   Composites Based on WG and SCB

The addition of SCB fibers to WG hardened the resulting films and darkened film color. Accordingly films based on WG containing 10-30% SCB were rather flexible whereas films containing 40-50% SCB resulted very hard and brittle. A WG/SCB 80/20 ratio (WGSCB20) appeared to be the most interesting composition as far as filler content and mechanical properties are concerned. This WG/SCB weight ratio appeared to be adequate to completely cover and aggregate SCB fibers in the continue matrix.

When submitted to TGA, SCB decomposed in one step with a minimum in the first derivative curve at 342°C, while WG decomposed mainly in two overlaid steps with minimum peaks at 276 °C and 305 °C respectively. WGSCB20 thermal stability (Ton = 223 °C) was sligthly lower than WG (Ton = 237 °C) and WGSCB20 minimum peak (T= 312°C) occurred between WG and SCB minimum peaks (Fig. 3).

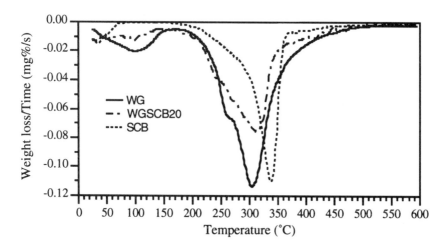

Figure 3. First Derivative of the Decomposition Traces of WG, SCB and WGSCB20

In DSC traces of WGSCB20 the thermal transition associated with the Tg of plasticized gelatin soft blocks moved from -30 °C of WG to -35 °C (Fig. 4). The other two thermal events present in WG resulted in a single broad one with minimum at about 90°C. This thermal behavior results from the overlaying of the evaporation of volatile present in the composite with the transition related to the Tg of gelatin rigid blocks.

In DMTA analysis WGSCB20 storage modulus decreased as compared with parent WG. The thermal transition associated with WG soft blocks moved from 33 °C of WG to 57 °C for WGSCB20. In addition WG in the composites with SCB began to flow at higher temperature (ca. 100 °C).

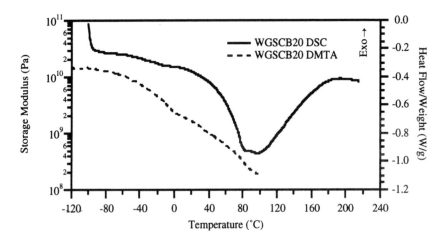

Figure 4. DSC and DMTA Traces of a Blend 80/20 WG and SCB

Due to the casting process, SCB fibers presented a random distribution and formed fibers aggregates thus reducing El as showed by tensile tests. Indeed WG cast film presented 116% El while in WGSCB20 El was 14%. UTS were almost the same and YM increased from 78 MPa for WG to 228 MPa for WGSCB20.

3.4   Waste Gelatin (WG) Crosslinked with Glutaraldehyde

Hardening of gelatin with low molecular weight aldehydes is well documented in the literature[16-19]. Crosslinking is predominantly due to Schiff's base formation by condensation of the formil group and the ε-amino groups present in lysine and hydroxylysine residues.

In TGA no significant differences were recorded for crosslinked sample in comparison with uncrosslinked ones.

In DSC analysis (Fig. 5) the Tg of the plasticized WG rigid blocks, at 71 °C, shifted to higher temperature with increasing of glutaraldehyde content and became less evident. Also the third transition became broader with the minimum moving to lower temperature. In WGX3 it was almost overlaid with the Tg. In DSC analysis of hot cast films crosslinked with formaldehyde Fraga[15] observed a broader temperature range of gelatin glass transition attributed at the crosslinking of gelatin chains through the α-amino acid present in the soft blocks with a consequent increase of rigidity in these blocks.

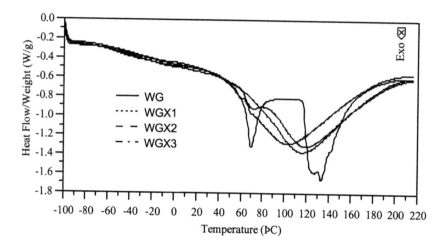

Figure 5. DSC Traces of WG Based Films Crosslinked with Glutaraldehyde

DMTA analysis showed a reduction in storage modulus with increasing of glutaraldehyde content and an increase in softening point. For WGX2 and especially for WGX3 it was possible to perform measurements up to 100° C while in WG and WGX1 the softening of the samples did not permit to go over 60 °C (Fig.6).

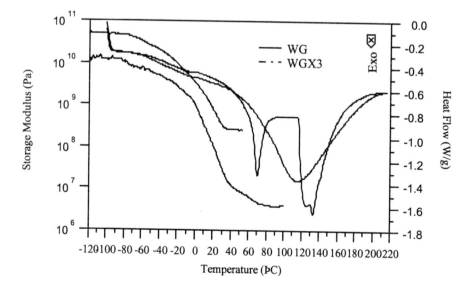

Figure 6. DSC and DMTA Traces of WG and WGX3 Cast Films

El in WG cast films increased significantly when the glutaraldehyde was introduced in the formulations as reported in Table 4. Thus depending upon crosslinking density, temperature and diluent content a gelatin specimen can behave as a rubber capable of extending by 700% or as a viscous liquid[6]. For a crosslinker content of 0.25%, as in WGX1, El was 140 % increasing to 300 % for 2.5% of glutaraldehyde as in WGX3. At the same time YM decreased from 78 MPa in WG up to 10 MPa for WGX3, whereas UTS did not significantly change. This behavior of gelatin can be attributed to the disruption of the helical structure allowing the chains to assume a random coil conformation with glutaraldehyde acting as a crosslinker among the chains[19]. On the other hand, WGP80X and WGSCB20X did not show any significant difference in comparison with WGP80 and WGSCB20, thus indicating that 0.25% of glutaraldehyde has a negligible effect on blend properties, also in accordance with results from previous studies of water sensitivity and mineralization rates [10,13,14].

4.   CONCLUSION

Gelatin scraps from pharmaceutical farm appear suitable to be used for the production of cast films or water suspensions for self-fertilizing mulching films or liquid mulches in blends with PVA and/or SCB.

The presence of a substantial amount of water and glycerol in WG allowed for a partial compatibility between gelatin and PVA. Particularly in WG/PVA blends containing an amount of WG up to 20% by weight PVA-glycerol interactions caused a marked effect on PVA mechanical properties.

Introduction of SCB in WG conferred a dark color to the films and hardened the blends, thus increasing YM. A WG/SCB 80/20 ratio showed an interesting balance between fillers amount and mechanical properties.

Crosslinking with glutaraldehyde increased elongation to break and conferred an elastomeric behavior to WG cast films for amount of glutaraldehyde of 2.5%.

Thermal and mechanical properties can be modulated by suitably tuning composition. These results indicate that by-products such as animal gelatin scraps and SCB can be used for the production of interesting specific items for applications in which the intrinsic characteristics of the waste material do not constitute a problem and in some cases may be considered an advantage. SCB and WG represent a humic substance source and a nitrogen rich material respectively.

ACKNOWLEDGMENTS

The financial support by MURST through the PRIN'98 project entitled "Synthesis of Polymeric Materials for Unconventional Applications" is gratefully acknowledged. Grateful recognition has to be devoted also to the ICS-UNIDO for having acted as proactive mandator trough his action on Environmentally Degradable Polymers & Plastics of the research activity quoted in the present manuscript.

REFERENCES

1. 5th International Scientific Workshop on Biodegradable Plastic and Polymers, Stockholm, Sweden, June 9-13, 1998, (Albertsson, A.C., Chiellini E., Feijen J., Scott G., Vert M., eds.), *Macromol. Symp.* 1999, **144**: 479 pp.

2 Doane, W. M., 1994, Biodegradable Plastics. *J. Polym. Mater.* **11**: 229-237

3. Steinbüchel A., 1995, Use of biosynthetic, biodegradable thermoplastics and elastomers from renewable resources: the pros and cons., *J. Macromol. Sci., Pure Appl. Chem.*, **A32(4)**: 653-660

4. de Graaf L.A., Kolster P., 1998, Industrial proteins as a green alternative for "petro" polymers: potentials and limitations, *Macromol. Symp.*, **127**: 51-58

5. Sperling L.H., Carraher C.E., 1988, Polymers from renewable resources in *Encyclopedia of Polymer Science and Engineering* (H. F. Mark, N. M. Bikales, C. G. Overberger, and G. Menges, eds.), Wiley, New York, **Vol.12**, pp.658-690

6. Yannas I.V., 1972, Collagen and gelatin in the solid state, *J. Macomol. Sci. Macromol. Chem.*, **C7**: 49-104

7. Rose P.I., 1987, Gelatin in *Encyclopedia of Polymer Science and Engineering*, (Mark, H. F., Bikales N. M., Overberger C. G., and Menges G., eds.) Wiley, New York, **Vol.7**, pp.488-513

8 Yannas, I.V. Tobolsky A.V., 1968, Stress relaxation of anhydrous gelatin rubbers, *J. Appl. Polym. Sci.*, **12**: 1-8

9 Cinelli P., Palla C., Bargiacchi E., Miele S., Magni S., Solaro R., Chiellini E., 1999, Compositi biodegradabili a matrice polivinilica e cariche da fonti rinnovabili, In XIV Italian Conference of Macromolecolar Science and Technology, Salerno (Italy), 13-16.Sept.1999, pp. 865-868

10. Cinelli P., 1999, *Formulation and Characterization of Envrionmentally Compatible Polymeric materials for Agriculture Applications*. PhD Thesis, University of Pisa.

11. Stefanson R. C., 1974, Soil stabilization by poly(vinyl alcohol) and its effect on the growth of wheat, *Aust .J. Soil Res.*, **12**: 59-62

12. Oades J.M., 1976, Prevention of crust formation in soils by poly(vinyl alcohol), *Aust. J. Soil Res.*, **12**: 139-148

13. Chiellini E., Cinelli P., Corti A., Kenawy E.R., Grillo F. E., Solaro R., 2000, Environmentally sound blends and composites based on water-soluble polymer matrices *Macromol. Symp.* **152**: 83-94

14. Kenawy E. R., Cinelli P., Corti A., Miertus S., Chiellini E., 1999, Biodegradable composite films based on waste gelatin, *Macromol. Symp.*, **144**: 351-364

15. Fraga A.N., Williams R.J.J., 1985, Thermal properties of gelatin films, *Polymer*, **26**: 113-118

16. Akin H., Harisci N., 1995, Preparation and characterization of crosslinked gelatin microspheres, *J. Appl. Polym. Sci.*, **58**: 95-100
17. Chatterji P. R., 1989, Gelatin with hydrophilic/hydrophobic grafts and glutaraldehyde crosslinks, *J. Appl. Polym. Sci.*, **37**: 2203-2212
18. Digenis G. A., Gold T. B., Shah V. P., 1994, Cross-linking of gelatin capsules and its relevance to their in vitro-in vivo performance, *J. Pharm. Sci.*, **83**: 915-921
19. Olde Damink L.H.H., Dijkstra P.J., Van Luyn M.J.A., Van Wachem P.B., Nieuwenhuis P., Feijen J., 1995, Glutaraldehyde as a crosslinking agent for collagen-based biomaterials, *J. Mat. Sci. Mat. In Medicine*, **6**: 460-472

Properties of PHAs and their Correlation to Fermentation Conditions

FLORIAN SCHELLAUF[1], ELIZABETH GRILLO FERNANDES[2],
GERHART BRAUNEGG[1] and EMO CHIELLINI[2]
[1] Institut für Biotechnologie, TU-Graz, Petersgasse 12, A-8010 Graz, Austria, [2]Dipartimento di Chimica e Chimica Industriale, Università di Pisa, Via Risorgimento 35, I-56126 Pisa, Italy

Abstract: This paper presents part of the results obtained during an ICBT short term scientific exchange mission at the University of Pisa, Dipartimento di Chimica e Chimica Industriale. Several samples of 3-hydroxybutyrate-co-3-hydroxyvalerate (PHBV) with comonomer contents in the range of 2,5 to 76 % were produced in fed batch fermentations with different strains under varying cultivation conditions. These samples were characterized by DSC, TGA, GPC, ^{1}H-NMR. Under certain PHA producing conditions, PHA quality and processability are rather bad due to an inhomogeneous distribution of the 3HV units in the polymer while polymers from other fermentations have excellent properties. Analysis of these fermentation conditions resulted in very important hints for improvement of existing and development of future fermentation processes. A strong dependence of the molecular weight distribution of the polymer on the main carbon source used in the fermentation was also found.

1.   INTRODUCTION

Polyhydroxyalkanoates (PHAs) are polyesters formed by many prokaryotic micro-organisms when unbalanced nutritional conditions are chosen for the producing cells[1-3]. Up to more than 90% of the cell dry weight can be accounted for as polymer[4]. Besides the homopolyester poly-R-3-hydroxybutanoate, consisting of 3-hydroxybutanoate (3HB) only, two main types of copolyesters can be formed by different microorganisms[5]. The first type of PHAs always contains C3 units in the polymer backbone, but the

Biorelated Polymers: Sustainable Polymer Science and Technology
Edited by Chiellini et al., Kluwer Academic/Plenum Publishers, 2001

side chains can contain either H-, methyl- or ethyl-groups if prepared with micro-organisms like *Ralstonia eutropha*, or propyl- to nonyl-groups found in the side chains of copolyester prepared with *Pseudomonas oleovorans*. In the latter case branchings[6], double bonds[7], epoxides[8], and aromatic structures[9] can be introduced into the side chain. Furthermore copolyesters containing halogenalkanoates (F, Cl, Br) can be produced[10-12]. In the case of *P. oleovorans* and other strains from the group of fluorescent pseudomonads PHA formation only occurs, when the organisms are grown either with fatty acids (butanoate to hexadecanoate) or with alkanes (hexane to dodecane).

The second type of PHA is a short side chain polyester, containing hydrogen, methyl-, or ethyl groups in the side chains, and having C3, C4, and C5 units in the backbone of the polymer[13,14]. Carbohydrates, alcohols, and low fatty acids are typical substrates for growth and PHA formation for these micro-organisms. In most cases, cosubstrates have to be fed to the producing cultures as precursors for copolyester formation[14,15]. Typical precursors that have been used are propionate, valerate, or 1,4-butanediol, leading to analogues of 3HB such as 4-, and 5-hydroxyalkanoates.

Polyhydroxyalkanoates (PHAs) can be biodegradable substitutes to fossil fuel plastics and can be produced from renewable raw materials such as saccharides, alcohols and low-molecular-weight fatty acids[16]. They are completely degradable to carbon dioxide and water through natural microbiological mineralisation. Consequently, neither their production nor their use or degradation have a negative ecological impact. By keeping closed the cycle of production and re-use, PHAs can enable at least part of the polymer-producing industry to switch from ecologically harmful end-of-the-pipe production methods towards sounder technologies. It is evident that such materials must have at least the same or even better properties than those they will replace in future. Therefore a thorough characterization of the polyesters produced is necessary. Many of the physical properties like the crystalline melting point, elongation to break, glass transition temperature and molecular weight distribution can be influenced by the microbial strains used for PHA production and/or by the conditions chosen during the PHA production phase of the fermentation. However, little is known about these relations, mostly because biotechnological laboratories, which investigate the production of such biopolyesters do not always have the necessary equipment for their characterization.

2.   EXPERIMENTAL PART

2.1   Equipment and Chemicals

Characterization of all the samples was carried out with the following equipment:

DSC: Mettler DSC 30 with Low Temperature Cell.

Usually four consecutive runs were performed:

- from 30°C to -100°C at 100°C/min
- from -100°C to 220°C at 10°C/min
- from 220°C to -100°C at 100°C/min
- from -100°C to 220°C at 10 °C/min

TGA: Mettler TG 50

One run from 25°C to 600°C at 10°C/min.

DSC and TGA measurements were performed under nitrogen atmosphere

GPC: Pump: Jasco PU 1580
        For RI-detection: Jasco 830-RI
        For UV-detection: Perkin Elmer LC75

   Two PLGel MixedC 5µm columns (Polymer Laboratories)
   Flow: 1 ml/min, chloroform

NMR: Varian Gemini 200

IR: Perkin Elmer 1600 Series FTIR

Optical Microscopy: Reichert Jung Polyvar equipped with a Mettler
                FP52 Heating Stage

All chemicals were obtained from Carlo Erba Fine Chemicals Co.

3.   RESULTS

3.1 ^{1}H-NMR

The following Figure shows the spectrum of the homopolymer PHB.

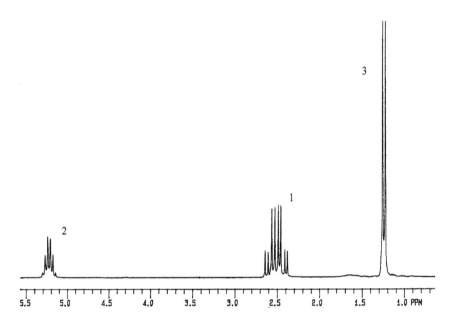

Figure 1. Typical ^{1}H-NMR of the homopolymer PHB

The doublet (3) at ca. 1,25 ppm is the side chain, a methyl group. The sextet (2) at 5,25 ppm is the chiral carbon atom in the backbone with 5 vicinal hydrogens and near to the oxygen of the ester bond. The CH_2 in the backbone gives the signal at 2,5 ppm. The vicinal coupling of the proton resonance is due to the rotation of the CH_2 - CH backbone bond.

The spectrum of a copolymer of PHB and PHV is shown in Fig 2. The triplet at 0,85 ppm (7) is the methyl group of the side chain of the comonomer. Some small peaks overlapping with the main peaks can be detected. This could be caused by some impurities. The peaks at 1,8 ppm (6) are from the CH_2 group of the side chain. The peaks of the protons of the backbone of 3HV are slightly shifted in regard to the peaks of the 3HB backbone and cause overlapping and distortion (1+4 for CH_2 and 2+5 for CH). This distortion gets worse with increasing 3HV content till a maximum 50 wt% of 3HV is reached and then decreases as the 3HB content gets lower.

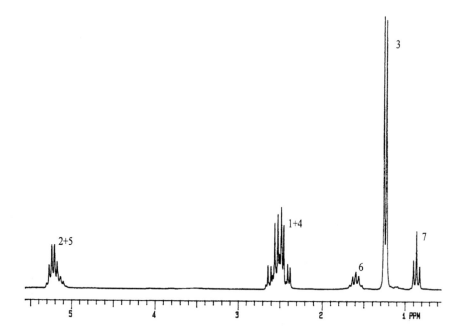

Figure 2. A ^{1}H-NMR spectrum of a copolymer with 17,6 wt% P3HV

3.2 Determination of the molecular weight distribution by GPC

For GPC analysis 2 mL of a 0,5% solution of the polymer in chloroform was prepared. Where higher molecular weight was expected a 0,2% solution was prepared. This solutions were filtrated with 0,45 μm teflon filters to remove insoluble particles and then injected into the GPC. Selected results are shown in table 1.

The use of glycerol as main carbon source for the production of polymer B led to the very low Mw of 106.000. Madden et al.[17] found, that hydroxy-compounds act as chain termination agents during polymerisation, and that the use of this carbon source leads to such low molecular weights. Since glycerol has also positive effects on the polymer accumulation a compromise has to be found. Indeed when using a mixture of glucose and glycerol the molecular weight reaches high values (E).

The polymer C produced with lactate as main carbon source in the accumulation phase has a Mw as high as one of those produced with glucose (A). Both carbon sources do not have a negative influence on the molecular weight.

The highest molecular weight has been found for polymer D. This polymer was produced by the strain *Alcaligenes paradoxus* already known to produce polymers with a molecular weight higher than 10^6. Any other connection between the molecular weight and the different strains could not be found. It seems that the influence of the carbon source on the molecular weight is by far the strongest. However all the fermentations were carried out at the same pH. A different pH in some of the fermentations should have led to strong changes of the molecular weight distribution[18].

In the samples which were not purified, some high molecular weight compounds could be found by UV detection. In most cases, no relation between the UV peaks and the polymer peaks detected by RI detection could be found. Purification of the samples seemed to remove these compounds and it is assumed that there were some proteins adsorbed to the polymer causing UV absorption by the aromatic groups of the amino acids.

Table 1. Selected results of the GPC analyses (in Da):

Sample ID	Mw	Mn	PD
A	839.000	153.000	5.46
B	106.000	24.000	2.41
C	904.000	276.000	2.48
D	1.138.000	443.000	2.57
E	852.000	236.000	3.61

The polydispersity of sample A is rather high. This is due to a fungal infection of the extracted polymer. It seems that the polymer chains did undergo some cleaving by extracellular depolymerases from the fungi, which led to increased PD and a rather low value of Mn. This is a mechanism also responsible for the biodegradation of these polymers in a natural environment.

3.3   Analysis of the polymers by DSC and TGA

3.3.1   DSC

Fig 3 shows the DSC of the homopolymer PHB. This material has a melting temperature of 177.1°C with a ΔHm of 93.8 J.g^{-1} In a second run, (after quenching) the Tg of 0.4°C and recrystallisation at 50.4°C were found.

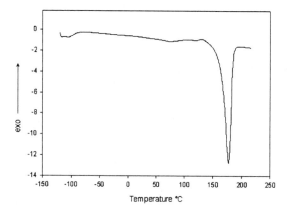

Figure 3. DSC of the homopolymer PHB (first heating)

Fig 4 shows a typical DSC of a copolymer. One can clearly see, that there are three overlapping melting peaks at very low temperatures probably indicating parts of the polymer with a very high content of 3HV and one at 174.7°C for the homopolymer. Madden and Anderson[19] found, that this type of polymer seems to be actually a mixture or intracellular blend of polymers with different 3HV content, which can be separated by fractionated precipitation. The 3HV content measured by GC and NMR is only the average value of all the fractions. Until now it was thought that there are different regions with differing distribution of the comonomers in one and the same polymer chain.

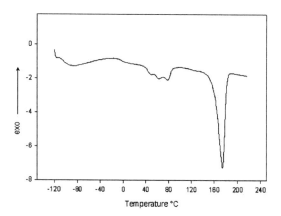

Figure 4. DSC of a copolymer (sample A)

Whatever the exact structure of the polymer chains is, it is evident that a polymer with such a melting behaviour is more or less useless for melt-processing, since one of the major positive effects of the introduction of the 3HV monomers (the lowering of the melting temperature) is lost and the polymer has to be heated almost to the degradation temperature of 185°C for a complete melting.

Not all the polymer samples were that heterogeneous. Fig 5 shows the DSC of sample B. It was produced with the strain *Alcaligenes latus* DSM 1124 usually known for growth associated polymer production. However by a special feeding strategy almost no homopolymer was produced during the growth phase as can be seen by an only very small melting peak at 171°C. Considering the much higher heat of melting of the homopolymer in comparison to that of the copolymer, the fraction of homopolymer in this sample is almost neglectable. These results are in accordance with the results from the optical microscopy. It was observed that most of the polymer melted in the range of 70°C to 100°C but some very small domains still remained in solid state until the heat stage reached about 170°C.

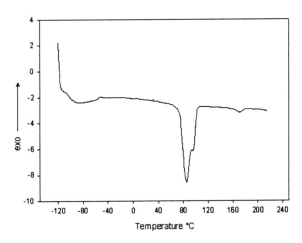

Figure 5. DSC of sample B

3.3.2   TGA

The PHAs are decomposing in one single step. All the PHA samples start to lose weight at about 230 to 250°C and maximum decomposition takes place in the range from 270 to 290°C. Interestingly the polymer with the lowest molecular weight (sample B) turned out to be the most stable (Tp = 291,3°C), while those with high molecular weight (D, E) are degraded at temperatures, which are about 20°C lower (see Table 2). Judging from these results the thermal stability of the material seems to be affected by homogeneity, too. Homogeneous samples seem to have a higher stability, while inhomogeneous samples, even if they have a much higher molecular weight, are already degraded at lower temperatures. The only exception is sample A with the same peak decomposition temperature as sample B but with a disadvantageous melting behaviour (Fig. 4).

Table 2. Selected results of the TGA analysis

Sample	T_1 °C	T_5 °C	T_p °C	T_d °C	Volatiles wt%	Residue[a] wt%
A	248.0	266.5	291.0	272.2	0.00	traces
B	252.5	269.3	291.3	274.2	0.29	0.00
C	242.4	260.0	285.7	268.0	0.40	0.00
D	239.3	252.0	274.3	258.5	2.06	0.77
E	229.0	243.6	270.0	251.6	0.00	0.00

T_1 and T_5 are temperatures where 1 and 5 wt % of weight loss, respectively, is observed, taken as reference temperature at which no weight loss related to evaporation process is occurring; T_P is the peak temperature of maximum rate decomposition; and T_d is the onset decomposition temperature at the crossover of tangents drawn on both sides of the decomposition trace.
[a] At 580°C.

4.   CONCLUSION

Many properties of PHAs can be influenced by the fermentation conditions during the production process. Influences of carbon source and pH on the molecular weight distribution are already known. Not much attention was paid to the homogeneous distribution of the comonomeric units in the polymer so far.

As shown here melt-processability and arguably thermal stability of the polymer are much affected by an inhomogeneous distribution of the 3HV units.

By comparing the properties of the polymers with the conditions of their production several hints for a better production process could be found. Much further work is needed to find the exact parameters for an optimal production of completely homogeneous polymers, which will help to improve existing and develop future polymer production processes.

REFERENCES

1. Dawes, E. A., and Senior, P. J., 1973, The role and regulation of energy reserve polymers in micro-organisms. *Adv. Microbiol. Physiol.* **10**: 135

2. Anderson, A. J., and Dawes, E. A., 1990, Occurrence, metabolism, metabolic role, and industrial uses of bacterial polyhydroxyalkanoates. *Microbiol. Rev.* **54**: 450

3. Braunegg, G., Lefebvre, G., and Genser, K., 1998, Polyhydroxyalkanoates, Biopolyesters from renewable resources. *J. Biotechnol.* **65**: 127

4. Braunegg, G., and Bogensberger, B., 1985, Zur Kinetik des Wachstums und der Speicherung von Poly-D(-)-3-hydroxybuttersäure bei *Alcaligenes latus. Acta Biotechnol.* **4**: 339

5. Braunegg, G., Lefebvre, G., Renner, G., Zeiser, A., Haage, G., and Loidl-Lanthaler, K., 1995, Kinetics as a tool for polyhydroxyalkanoate production optimisation. *Can.J. Microbiol.* **41**: 239

6. Fritsche, K., Lenz, R.W., and Fuller, R.C., 1990, Bacterial polyesters containing branched poly(β-hydroxyalkanoate) units. *Int.J.Biol.Macromol.* **12**: 92

7. Fritsche, K., Lenz, R.W., and Fuller, R.C., 1990, Production of unsaturated polyesters by *Pseudomonas oleovorans. Int.J.Biol.Macromol.***12**: 85

8. Bear, M. M., Leboucherdurand, M. A., Langlois, V., Lenz, R. W., Goodwin, S., and Guerin, P., 1997, Bacterial Poly-3-Hydroxyalkanoates with Epoxy Groups in the Side-Chains. *React. Funct. Polymers* **34**: 65

9. Kim, Y. B., Lenz, R. W., and Fuller,R. C., 1991, Preparation and characterization of poly(β-hydroxyalkanoates) obtained from *Pseudomonas oleovorans* grown with mixtures of 5-phenylvaleric acid and n-alkanoic acids. *Macromolecules* **24**: 5256

10. Abe, C., Taima, Y., Nakamura, Y., and Doi, Y., 1990, New bacterial copolyesters of 3-hydroxyalkanoates and 3-hydroxy-ω-fluoroalkanoates by *Pseudomonas oleovorans. Polym. Commun.* **31**: 404

11. Doi, Y., and Abe, C., 1990, Biosynthesis and characterization of a new bacterial copolyester of 3-hydroxyalkanoates and 3-hydroxy-ω-chloroalkanoates. *Macromolecules* **23**: 3705

12. Kim, Y. B., Lenz, R. W., and Fuller, R. C., 1992, Poly(3-hydroxyalkanoate) copolymers containing brominated repeating units produced by *Pseudomonas oleovorans. Macromolecules* **25**: 1852

13. Saito, Y., Nakamura, S., Hiramitsu, M., and Doi, Y., 1996, Microbial Synthesis and Properties of Poly(3- Hydroxybutyrate-co-4-Hydroxybutyrate). *Polym. Int.* **39**: 169

14. Doi, Y., Tamaki, A., Kunioka, M., and Soga, K., 1987, Biosynthesis of terpolyesters of 3-hydroxybutyrate, 3-hydroxyvalerate, and 5-hydroxyvalerate in *Alcaligenes eutrophus* from 5-chloropentanoic and pentanoic acids. *Makromol. Chem., Rapid Commun.* **8**: 631

15. Kunioka, M., Nakamura, Y., and Doi. Y., 1988, New bacterial copolyesters produced in *Alcaligenes eutrophus* from organic acids. *Polym. Commun.* **29**: 174

16. Doi, Y., 1990, Microbial polyesters, VCH Publishers Inc., New York

17. Madden L.A. et al., 1999, Chain termination in polyhydroxyalkanoate synthesis: involvement of exogenous hydroxy-compounds as chain transfer agents. *Int. J. Biol. Macromol.* **25**: 43

18. Hori K. et al., 1994, Effects of culture conditions on molecular weights of poly(3-hydroxyalkanoates) produced by *Pseudomonas putida* from octanoate. *Biotechnol. Lett.* **16** (7): 709

19. Madden L.A. and Anderson, A.J., 1998, Synthesis and characterization of poly(3-hydroxy-butyrate and poly(3-hydroxybutyrate-co-3-hydroxyvalerate) polymer mixtures produced in high-density fed-batch cultures of *Ralstonia eutropha* (*Alcaligenes eutrophus*). *Macromolecules* **31**: 5660

PART 3

(BIO)SYNTHESIS AND MODIFICATIONS

Synthesis of Biopolymers

GERHART BRAUNEGG, RODOLFO BONA, GUDRUN HAAGE,
FLORIAN SCHELLAUF and ELISABETH WALLNER
Institut für Biotechnologie, TU Graz, Petersgasse 12, A-8010 Graz, Austria

Abstract: A number of polymers can be produced via fermentation, using special
consòrtia of microorganisms to convert renewable raw materials (surplus or
waste products from agriculture or foresting) into useful products within
sustainable processes. The main groups are either extracellular polysaccharides
produced by many eukaryotic or prokaryotic microorganisms, or
polyhydroxyalkanoates stored intracellularly as reserve products for carbon
and energy in a variety of bacteria. In order to lower production costs for these
biopolymers, continuous processes have to be designed for the future.

1.   INTRODUCTION

Quite a number of naturally occurring biopolymers is produced in
technical scale biotechnological processes. Others show such interesting
features that they are under discussion as future sustainable industrial
products. For all of them, production starts from renewable resources and all
of them can be mineralized by microbial activity. Studying their chemical
nature it can be seen that they are either polysaccharides (e.g. alginates,
dextran, cellulose, starch), proteins (e.g. collagen, silk, elastin), polyesters
(polyhydroxyalkanoates) or others (e.g. lignin, melanins). Whilst
biopolymers belonging to the group of proteins in most cases can be isolated
from agricultural products or wastes, or are produced by insects (silk), many
polysaccharides (e.g. alginates, dextran), all polyhydroxyalkanoates, and
some melanins are produced by either prokaryotic or eukaryotic
microorganisms in bioreactors under controlled conditions. In this paper an
overview about biosynthesis and production of polysaccharides and
polyhydroxyalkanoates will be given.

2.   SYNTHESIS AND PRODUCTION OF BIOPOLYMERS

2.1   Polysaccharides

High molecular mass polysaccharides are formed by condensation of large numbers of small units of activated sugars. If they are composed with one type of sugars, they are called *homopolysaccharides* (e.g. dextran, curdlan), if they differ from each other, *heteropolysaccharides* are the polymerization reaction results (e.g. xanthan). Starch, a widely used plant reserve polymer consists of amylose (long unbranched chains of α [1→4] linked D-glucose, molecular weight 2×10^5 to 2×10^6) and amylopectin (α [1→4] linked D-glucose in the backbone and α [1→6] linked D-glucose in the branch points, molecular weight up to 4×10^8) is a raw material that can be made thermoplastic for production of compostable items, or can be complexed with ethylen-vinyl alcohol copolymers[1]. Cellulose, produced by plants or microorganisms is another important natural polysaccharide consisting of β [1→4] linked D-glucose residues. In the following pages biotechnological production of dextrans and xanthans are discussed as examples of modern industrially produced polysaccharides. Dextran has been used mainly as a blood plasma expander and for many other applications shown in table 1[2]. Xanthan gums are widely used in food industry as additives[3], as well in oil drilling due to the fact that the viscosity of xanthan solutions is nearby not effected by pH changes in the range of 1 to 13 or by temperature changes between 0 and 200° F[4].

2.1.1   Dextran[2]

Even though dextran can be prepared by polymerizing 1,6-anhydro-2,3,4-tri-*O*-benzyl-ß-D-glucopyranose using phosphorous pentachloride as a catalyst and subsequently removing benzyl groups, all commercially available dextrans are biotechnological products. 96 strains have been described as useful to form the polysaccharide, but only *Leuconostoc mesenteroides* and *Leuconostoc dextranicum* are used commercially. These microorganisms produce the enzyme complex dextransucrase, responsible for the dextran formation according to the overall equation:

$$(1,6\text{-}\alpha\text{-D-Glucosyl})_n + C_{12}H_{22}O_{11} \xrightarrow{\text{Dextran-Sucrase}} (1,6\text{-}\alpha\text{-D-Glucosyl})_{n+1} + nC_6H_{12}O_6$$

| sucrose | *Sucrase* | dextran | fructose |

The enzyme glycoprotein releases fructose from sucrose and transfers the glucose residue to the reducing end of the growing dextran chain on an acceptor molecule, which is bound to the enzyme[5,6]. During polymerization, the growing dextran chain is always bound to the enzyme. The degree of polymerization increases until an acceptor molecule releases the polymer chain from the enzyme.

Table 1. Further Uses of Dextrans[2]

Product	Dextran Function
Pharmaceutical grade	
Cryoprotective	Inhibits cell damage on freezing
X-ray opaque compositions	Suspending agent
Water-insoluble vitamin preparations	Stabilizing agent
Tablets	Binding agent
Sustained-action tablets	Protracts dissolution
Chloral-dextran complex	Supresses taste and stomach irritant action
Microcapsules of kerosene, menthol, aspirin, etc.	Methylcellulose-dextran, encapsulating agent
Cosmetic preparations	Wrinkle smoothing
Food grade	
Syrups and candies	Improves moisture retentivity and body and inhibits crystallization
Gum and jelly confections	Gelling agent
Ice cream	Prevents shrinkage and ice formation
Icing compositions	Stabilizing agent
Pudding compositions	Bodying agent
Industrial grade	
Oil drilling fluids	Dextran-aldehyd complex, inhibits water loss and coats well wall
Solution for flooding underground reservoirs	Increases viscosity of water
Drilling muds	Protective colloid
Olefinically polymerizable resins	Filler and modifier
Alumina manufacture	Sedimentation agent
Purification of caustic soda	Iron-dextran complex precipitates
Metal powder production	Gel precipitation suppresses crystal growth
Nuclear fuel production	Complexing agent

Dextran can be produced either directly by batch fermentation or indirectly by the use of the enzyme complex dextransucrase mentioned above. Direct production is, of course, the simpler process, but molecular weights of dextrans are varying. The fermentation process is performed in stirred and aerated bioreactors with typical volumes of up to 200 m^3. Bacterial growth requires amino acids and growth factors. Fermentation is started at $25 - 30$ °C and a pH of $6.5 - 7.0$, dropping to about 4.5 at the end of the production cycle (after about 48 hours) due to formation of lactic acid as a by-product. Maintenance of an adequate dissolved oxygen concentration is problematic due to the non-Newtonian fluid behavior of the nutritional broth becoming more and more viscous during fermentation. During dextran production phase, no more oxygen is needed, and bioreactors are stirred only, fermentation control proceeds by measuring fructose concentration in the broth. From 0.5 g of bacterial biomass about 80 g of dextran are formed during the process.

The enzymatic process permits better reaction control leading to a more uniform material with molecular weights depending on the nature of the glucosyl acceptor employed. Very often, a dextran with a molecular weight of about 75.000 is produced. The process consists in two stages, the aerobic production of the extracellular enzyme complex typically at low sugar concentrations (2 % sucrose) at 25 °C and a pH of 6.7, and the enzymatic dextran synthesis under reducing conditions. Ammonium ions depress the yield of enzyme produced, pH adjustment is made with caustic alkali. The enzyme is separated from the bacterial cells by centrifugation (pH = 5.0) and can be stored at 15 °C for as long as 30 days. The expected enzyme yield is about 40 dextransucrase units per ml, converting 40 mg of sucrose to dextran in 1 hour under standard conditions.

For dextran production, 10% w/v sucrose is supplied as substrate, the incubation is conducted at 15 °C or below, in the presence of about 30 units/ml of dextran sucrase. Molecular weight of dextran produced can be controlled by sucrose concentration, enzyme concentration, and by the temperature and time used for incubation.

After its production, dextrans are precipitated by addition of either methanol or acetone (1 °C), and the supernatant is decanted. Precipitated dextran is resolubilized in distilled water at 60 to 70 °C and precipitated again for cleaning purposes. The typical yield for dextran is about 60-70% of the glucose part of sucrose. About 2000 tons of dextranes are consumed worldwide per year. Depending on product qualities they are sold for a price of 35 – 2800 $US/kg.

2.1.2   Xanthan

Xanthan gum, an extracellular heteropolysaccharide synthesized by *Xanthomonas campestris* is another biopolymer produced commercially in quantities above 20.000 tons per year[3]. Properties of xanthans used in foods are shown in table 2. Xanthan gum has a high molecular weight ($2x10^6$ – $5x10^7$) and contains D-glucose (2.8 mol), D-mannose (3.0 mol), D-glucuronic acid (2.0 mol), acetic acid (approximately 4.7%) and pyruvic acid (approximately 3%). The structure of xanthan is a pentasaccharide repeating unit consisting of a β[1→4]-linked D-glycosyl backbone ("cellulosic backbone") with α [1→3]-linked trisaccharide side chains (D-mannose-β [1→2]-D-glucuronic acid-β[1→4]-mannose). Most commercial xanthans are fully acetylated on the internal D-mannose residue and carry pyruvate ketals on about 30% of the side chain terminal mannose residues.

Xanthan gums are produced aerobically in bioreactors (200 m^3 reactor volume) in a strictly monoseptic batch process at 28 °C and pH equal to 7.0 from carbon sources like flours, starch, starch hydrolyzates, glucose or sucrose. Initial carbohydrate concentrations may vary from 2 to 5% depending on the substrate type[8]. Dissolved oxygen concentration is kept above 20 % of air saturation. Often organic nitrogen sources are used (meat-peptone, soy peptone dried distillers'soluble, urea). Sometimes, ammonium nitrate is used without loss of polymer yields. Xanthan yields can be increased if the pH value of the nutritional broth is controlled during fermentation by addition of ammonium hydroxide. Rheology of the increasing viscous fermentation broth is a rather complex problem, because xanthan solutions exhibit pseudoplastic behavior and display a yield stress and visco-elasticity. Because of the high final viscosity mechanically agitated bioreactors are used. As the turbulence and shear rate in vessels equipped with rotating straight bladed impellers is higher near the impeller, special agitators may be used, and an optimal reactor design is required to provide sufficient turbulence for good air dispersion and bulk mixing. In about 36 to 84 hours, 40 – 70% of the carbon source used is transformed to xanthan. The maximum xanthan concentration in the broth should not be higher than 5% in the broth. The whole bioreactor has to be pasteurized carefully now in order to kill the xanthan producing bacteria, that are plant pathogens prior to the downstream processing.

Table 2. Different properties of Xanthan used in Foods (3)

Function	Application
Adhesive	Icings and glazes
Binding agent	Pet foods
Coating	Confectionery
Emulsifying agent	Salad dressing
Encapsulation	Powdered flavors
Film Formation	Protective coatings, sausage casings
Foam stabilizer	Beer
Stabilizer	Ice cream, salad dressings
Swelling agent	Processed meat products
Syneresis inhibitor	Cheeses, frozen foods
Thickening agent	Jams, sauces, syrups, and pie fillings

The final fermentation broth is diluted with water in order to decrease viscosity and centrifuged for partial removal of the cells. Then the product is precipitated by addition of methanol or i-propanol in the presence of 2%w/w potassium chloride. Other recovery methods proposed are drum-drying or spray-drying for a technical grade product. $10\text{-}20 \times 10^3$ tons of xanthan are produced worldwide and sold for a price of $10 - 14$ \$US/kg.

2.2   Polyhydroxyalkanoates

Polyhydroxyalkanoates (PHAs), a general formula is given in Fig 1, are polyesters formed by many prokaryotic microorganisms when unbalanced nutritional conditions are chosen for the producing cells[9]. Up to more than 90% of the cell dry weight can be accounted for as polymer[10]. Beside the homopolyester   poly-R-3-hydroxybutanoate,   consisting   of   3-hydroxybutanoate (3HB) only, two main types of copolyesters can be formed by different microorganisms[11]. The first type of PHAs always contains C_3 units in the polymer backbone, but the side chains can contain H-, methyl- or ethyl- groups if prepared with microorganisms like *Ralstonia eutropha*, or propyl- to nonyl groups are found in the side chains if the copolyester is prepared with *Pseudomonas oleovorans*. In the latter case branchings[12], double bonds[13], epoxides[14], and aromatic structures[15] can be introduced into the side chain. Furthermore copolyesters containing ω-chloroalkanoates (F, Cl, Br) can be produced[16-18]. In the case of *P. oleovorans* and other strains from the group of fluorescent pseudomonads PHA formation only occurs, when the organisms are grown either with fatty acids (butanoate to hexadecanoate) or with alkanes (hexane to dodecane). The second type of PHA is a short side chain polyester, containing hydrogen, methyl-, or ethyl groups in the side chains, and having C_3, C_4, and C_5 units in the backbone of the polymer[19, 20]. Carbohydrates, alcohols, and low fatty acids are typical substrates for growth and PHA formation for these microorganisms. In most cases, cosubstrates have to be fed to the producing cultures as precursors for copolyester formation[20, 21]. Typical precursors that have been used are propionate, valerate, or 1,4-butanediol, leading to analogues of 3HB such as 4-, and 5-hydroxyalkanoates.

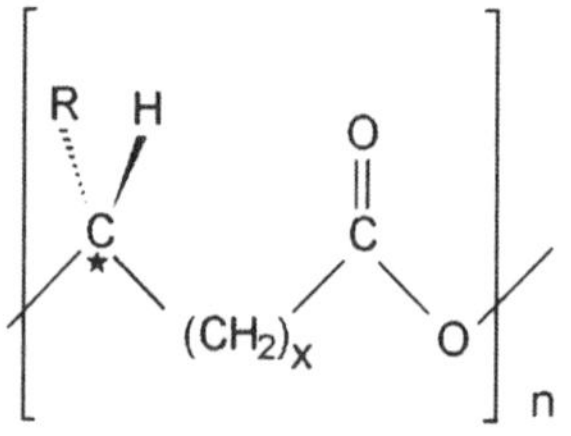

Figure 1. General formula of PHAs

Formation of random copolyesters results in many physical changes in the PHAs, including liquid-crystalline-amorphous forms, and a variety of piezoelectric, thermoplastic, elastomeric and other properties[22].

The "mixed" polyesters formed depend on the organisms used to produce them, and on the carbon sources and polyester precursors. For PHAs formed by prokaryotes the main physiological role accepted is, that the polyesters function as carbon and energy reserve materials. When growth is limited by exhaustion of nitrogen, phosphate, sulfur, oxygen, etc. in the nutritional broth, excess carbon is channeled into PHAs, leading to polyesters with molecular weights of up to about 3.4 MD[23]. Recently other physiological roles have been recognized for PHAs. Relatively small molecular weight PHAs (up to about 35 KD) are incorporated into membranes and their tertiary helical structures (back-bone inside) such that ion pores are putatively used for transport of ions into and out of the cells[24-26.]

Polyhydroxyalkanoates (PHAs) can be biodegradable substitutes to fossil fuel plastics that can be produced from renewable raw materials such as saccharides, alcohols and low-molecular-weight fatty acids. They are completely degradable to carbon dioxide and water through natural microbiological mineralization. Consequently, neither their production nor their use or degradation have a negative ecological impact. By keeping closed the cycle of production and re-use, PHAs can enable at least part of the polymer-producing industry to switch from ecologically harmful end-of-the-pipe production methods towards sounder technologies.

2.2.1   Kinetics of PHA accumulation

The new findings reported above and those reviewed from the literature point out the existence of three distinct types of growth and PHA accumulation behavior, each one typified here by one or more microorganism-carbon source combinations:

1) PHA synthesis occurs in association with growth (ex.: *A. latus* with sucrose);

2) PHA synthesis occurs in partial association with growth (ex.: *R. eutropha* G[+3] with glucose, *A. latus* with glucose);

3) PHA is hyperproduced after a carbon starvation period (ex.: *Pseudomonas* 2 F with glucose).

In principle, the three behaviours can be exploited for PHA production in batch culture, but due to its higher productivity, a continuous production process is of higher commercial interest, especially for strains with a high maximum specific growth rate. To prove this point, the overall productivity of a batch system will be compared to that of a continuous culture in the following way[27]:

$$\frac{\mathrm{Pr_{CSTR}}}{\mathrm{Pr_{DSTR}}} = \ln\frac{X_e}{X_i} + t_0\,\mu_{max}$$

[1]

where $\mathrm{Pr_{CSTR}}$ and $\mathrm{Pr_{DSTR}}$ are the productivities of a continuous stirred-tank reactor (CSTR) and a discontinuous stirred-tank reactor (DSTR), respectively, X_e is the maximum biomass concentration, X_i is the initial biomass concentration, and t_0 is the period of time between the end of a production run and the start of the next one.

The use of values of μ_{max} for *A. latus* and *R. eutropha* G^{+3} and a maximum biomass concentration of 30 gL^{-1} in Eq. 1 gives a productivity ratio of 8.2 with *A. latus* and 5.25 with *R. eutropha* if t_0 is set to a low 10 h. This means that for a fixed desired amount of product per unit of time, the bioreactor volume can be substantially reduced if a continuous culture is chosen over a batch process. From an engineering point of view, the reactor performance would also be easier to control, as lower fermentor volumes lead to less segregation through better mixing at inferior energy expenditure[28].

But if a continuous process is considered, the issue of kinetics must be addressed. The data from our experiments suggest the following possibilities for a PHA production process:

1)   an autocatalytic process of biomass growth and polymer production. Example: *A. latus* with sucrose;

2)   an autocatalytic process of biomass growth and polymer production followed by hyperproduction after carbon starvation. Example: *Pseudomonas* F 2 with glucose;

3)   an autocatalytic process of biomass (and PHB) formation followed by a phase of linear PHA accumulation. Example: *A. latus* with glucose, *R. eutropha* G^{+3} with glucose.

Consideration of the basic differences between these scenarii lead to the present proposition that for a multi-stage continuous PHA-production process, the use of a plug-flow tubular reactor, or PFTR (in which Reynolds numbers are large), brings substantial increases in productivity when compared to a system consisting of CSTRs only. Support of this assertion comes from the works of Levenspiel (29), where mean residence times for the two types of reactors are compared under the restriction of a desired goal:

For a CSTR in steady-state, where the concentrations in the reactor are the same as those in the outflow, the mass balance for biomass is given by

[2] $F(X_i - X_e) + V(r_X) = 0$

where F is the flow rate, X_i is the incoming biomass concentration, X_e is the effluent biomass concentration, V is the reactor volume and r_X is the absolute rate of biomass increase through growth, or dX/dt (amount per unit volume per unit time).

The mean residence time τ_{CSTR}, during which X_i is converted to X_e, is therefore simply

$$[3] \qquad \tau_{CSTR} = \frac{V}{F} = \frac{X_e - X_i}{r_x}$$

The mean residence time necessary for a CSTR to produce an effluent containing the concentration X_e can thus be determined graphically when the relationship between X and $1/r_X$ is known.

In the case of an PFTR, which can be looked upon as a series of small CSTRs, the mean residence time is given by the integration

$$[4] \qquad \tau_{PFTR} = \int_{X_i}^{X_e} \frac{1}{r_X}\, dX$$

Here again, τ can be determined graphically for a known process.

Data from fermentations with *R. eutropha* G^{+3} were used to plot $1/r_X$ as a function of the biomass concentration[5]. Graphical determinations of τ_{PFTR} and τ_{CSTR} revealed that for the autocatalytic process of simple biomass growth, a CSTR is the optimal system. For strains storing PHA growth-associatedly, a CSTR is therefore the best system for growth and PHA storage. If hyperproduction of PHA after a phase of carbon starvation is expected, a two-step continous system, consisting of two CSTRs in series, should be employed. In most cases however, growth is followed by a distinct PHA accumulation phase. This is the case for *R. eutropha* G^{+3}, and a PFTR is clearly the superior solution for the second stage. In other words, a combination of the two systems, i.e., a CSTR followed by a PFTR, allows a minimal total bioreactor volume to yield the same productivity as a two-CSTR arrangement of higher total volume. In the case of *R. eutropha* just described, the ratio τ_{CSTR}/τ_{PFTR} is 8,9 for the accumulation. A second-stage PFTR needs thus only have 11,4 % the volume of a CSTR to achieve the same results. In the case of *A. latus*, the ratio is 5,1, or a PFTR 19,8 % the size of a CSTR. The CSTR-PFTR arrangement not only guarantee maximum productivity, but also minimizes cosubstrate loss, and might be a tool for enhancing product quality, since very narrow residence time distributions (and therefore uniform cell population) are characteristic of plug-flow tubular reactors[30].

REFERENCES

1. Degli-Innocenti, F., and Bastioli, C, 1998, EDPs based on starch-Mater-Bi. In: *ICS UNIDO Environmentally Degradable Polymers*, V.98-54243-August 1998-500, pp. 35-42
2. Murphy, P.T., and Whistler, R.L., 1978, Dextrans. In *Industrial Gums*, Academic Press Inc., pp. 513 – 542
3. Sutherland, I.W., 1996, Extracellular Polysaccharides. In: *Biotechnology* Vol. 6, VCH Verlagsgesellschaft Weinheim, FRG, pp. 613-657
4. Kang, K.S. and Cottrell, I.W, 1979, Polysaccharides. In *Microbial Technology* Vol.1, Microbial Process, Academic Press New York, p.481
5. Robyt, J.F., and Walseth, T.F., 1978, The mechanism of acceptor reactions of Leuconostoc mesenteroides B512F dextransucrase. *Carbohydr. Res.* **61**: 433-445
6. Robyt, J.F., and Walseth, T.F., 1979, Production, purification and properties of dextransucrase from Leuconostoc mesenteroides. *Carbohydr. Res.* **68**: 95-111
7. Slodki, M.E., and Cadmus, M.C., 1978, Production of microbial Polysaccharides. *Adv.Appl.Microbiol.* **23**: 19-54
8. Casas, J.A.,Santos, V.E.,Garcia-Ochoa F., 2000, Xanthan gum production under several operational conditions: molecular structure and rheological properties. *Enzyme Microb. Technol.* **26**: 282-291
9. Braunegg, G., Lefebvre, G., and Genser, K., 1998, Polyhydroxyalkanoates, Biopolyesters from renewable resources. *J. Biotechnol.* **65**: 127-161
10. Braunegg, G., and Bogensberger, B., 1985, Zur Kinetik des Wachstums und der Speicherung von Poly-D(–)-3-hydroxybuttersäure bei *Alcaligenes latus. Acta Biotechnol.* **4**: 339
11. Braunegg, G., Lefebvre, G., Renner, G., Zeiser, A., Haage, G., and Loidl-Lanthaler, K., 1995, Kinetics as a tool for polyhydroxyalkanoate production optimization. *Can.J. Microbiol.* **41**: 239
12. Fritsche, K.,.Lenz, R.W., and Fuller, R.C., 1990, Bacterial polyesters containing branched poly(β-hydroxyalkanoate) units. *Int.J.Biol.Macromol.* **12**: 92
13. Fritsche, K.,.Lenz, R.W., and Fuller, R.C., 1990, Production of unsaturated polyesters by *Pseudomonas oleovorans. Int.J.Biol.Macromol.* **12**: 85
14. Bear, M. M., Leboucherdurand, M. A., Langlois, V.,. Lenz, R. W., Goodwin, S., and Guerin.. P., 1997, Bacterial Poly-3-Hydroxyalkenoates with Epoxy Groups in the Side-Chains. React. *Funct. Polymers* **34**: 65
15. Kim, Y. B., Lenz, R. W., and Fuller, R. C., 1991, Preparation and characterization of poly(β-hydroxyalkanoates) obtained from *Pseudomonas oleovorans* grown with mixtures of 5-phenylvaleric acid and n-alkanoic acids. *Macromolecules* **24**: 5256
16. Abe, C., Taima, Y., Nakamura, Y., and Doi, Y., 1990, New bacterial copolyesters of 3-hydroxyalkanoates and 3-hydroxy-ω-fluoroalkanoates by *Pseudomonas oleovorans. Polym. Commun.* **31**: 404
17. Doi, Y., and Abe, C., 1990, Biosynthesis and characterization of a new bacterial copolyester of 3-hydroxyalkanoates and 3-hydroxy-ω-chloroalkanoates. *Macromolecules* **23**: 3705

18. Kim, Y. B., Lenz, R. W., and Fuller, R. C., 1992, Poly(3-hydroxyalkanoate) copolymers containing brominated repeating units produced by *Pseudomonas oleovorans*. *Macromolecules* **25**: 1852

19. Saito, Y., Nakamura, S., Hiramitsu, M., and Doi, Y., 1996, Microbial Synthesis and Properties of Poly(3- Hydroxybutyrate-*co*-4-Hydroxybutyrate). *Polym. Int.* **39**: 169

20. Doi, Y., Tamaki, A., Kunioka, M., and Soga, K., 1987, Biosynthesis of terpolyesters of 3-hydroxybutyrate, 3-hydroxyvalerate, and 5-hydroxyvalerate in *Alcaligenes eutrophus* from 5-chloropentanoic and pentanoic acids. *Makromol. Chem., Rapid Commun.* **8**: 631

21. Kunioka, M., Nakamura, Y., and Doi, Y., 1988, New bacterial copolyesters produced in *Alcaligenes eutrophus* from organic acids. *Polym. Commun.* **29**: 174

22. Doi. Y., 1990. Microbial polyesters, VCH Publishers Inc., New York

23. Akita, S., Einaga, Y., and Fujita,. H., 1976, Solution properties of Poly(D-β-hydroxybutyrate).1. Biosynthesis and characterization. *Macromolecules* **9**: 774

24. Reusch, R. N., 1992, Biological complexes of poly-β-hydroxybutyrate. *FEMS Microbiol. Rev.* **103**: 119

25. Müller, H.M., and Seebach, D., 1993, Poly(hydroxyfettsäureester), eine fünfte Klasse von physiologisch bedeutsamen organischen Biopolymeren? *Angew.Chem.* **105**: 483

26. Seebach, D., Brunner, A., Bürger, H. M., Schneider, J., and Reusch,. R. N., 1994, Isolation and ¹H-NMR spectroscopic identification of poly(3-hydroxybutanoate) from prokaryotic and eukaryotic organisms. Determination of the absolute configuration (R) of the monomeric unit 3-hydroxybutanoic acid from *Escherichia coli* and spinach. *Eur. J. Biochem.* **224**: 317

27. Aiba, S., Humphrey, A. E., and Millis, N. F., 1973, *Biochemical engineering*. Second edition. Academic Press, Inc., New York

28. Zlokarnik, M., 1967, Eignung von Rührern zum Homogenisieren von Flüssigkeitsgemischen. *Chemie-Ing.-Techn.* **39**: 539-548

29. Levenspiel, O., 1972, *Chemical reaction engineering*. Second edition. John Wiley & Sons, New York

30. Steiner, W., 1980, Zum Mischverhalten in chemischen Reaktoren und Bioreaktoren. PhD thesis, University of Technology Graz, Graz, Austria

The Production of Poly-3-hydroxybutyrate-co-3-hydroxyvalerate with *Pseudomonas cepacia* ATCC 17759 on Various Carbon Sources

ELISABETH WALLNER, GUDRUN HAAGE , RODOLFO BONA, FLORIAN SCHELLAUF, and GERHART BRAUNEGG
Institute for Biotechnology, Graz University of Technology, Petersgasse 12, Graz, Austria

Abstract: Polyhydroxyalkanoates (PHAs) are biodegradable polyesters which can substitute conventional thermoplastics from non-renewable fossil resources. They are accumulated by a vast number of microorganisms as an intracellular storage product. The material properties of PHB (Polyhydroxybutyrate), the most commonly found PHA, can be improved by adding precursors as for instance 3-Hydroxyvalerate or 4-Hydroxybutyrate during the polymer accumulation phase to the medium generating Poly-3-hydroxybutyrate-co-3-hydroxyvalerate   and   Poly-3-hydroxybutyrate-co-4-hydroxyvalerate respectively. The main obstacle for a wide use of Polyhydroxyalkanoates as thermoplastics is the high production costs, which can be lowered significantly by using cheap carbon sources, since the cost of the carbon source contributes almost 50 % to the production costs. Cheap carbon sources are waste materials from agricultural and food industry. Whey accumulates in high amounts in cheese-production and therefore causes high waste management costs. Since strain *Pseudomonas cepacia* ATCC 17759 is capable of both producing PHA and utilizing whey composites lactose and its monomers glucose and galactose as carbon sources the whey could be used for this purpose. Growth kinetics and storage behavior of *P. cepacia* on various carbon sources and valerate as precursor have been investigated.

1.   INTRODUCTION

Whey is a major by-product from the manufacture of cheese and casein. It represents about 85-96 % of the milk volume and retains 55 % of milk nutrients, one of the most abundant of these being lactose with 4,5-5 % w/v[6]. Disposal costs in Europe 1997 were about 390×10^6 ECU[1]. On the other hand, cheap carbon sources for cost efficient polyhydroxybutyrate production are sought to make this biodegradable polymer economically producible. Most efforts to utilise the whey component lactose for PHA production have been made with genetically engineered *Escherichia coli*[7,3]. Only few micro-organisms are known to utilise lactose as carbon source and accumulate polyhydroxyalkanoates as storage material[8,9], one of them being *Pseudomonas cepacia*.

2.   MATERIALS AND METHODS

Strain *Pseudomonas cepacia* ATCC 17749 was stored under liquid nitrogen for longer storing periods at the institutes strain collection. For culture maintenance the strain was grown on solid media for two to three days at 30 °C on LB medium and on mineral medium with various carbon sources (glucose, galactose and lactose) respectively. Then the plates were kept at 4 °C. Fresh plates were inoculated every two to three weeks.

The basic synthetic mineral medium for the fermentations and plates was described earlier[5]. Ammonium sulfate was the sole nitrogen source. Maximum concentrations of ammonium sulfate were 2,4 g/L. Carbon sources for the plates and fermentations were glucose, galactose and lactose. The initial concentration of carbon sources was about 10 g/L. Sodium valerate was added in concentrations not exceeding 2 g/L during polymer accumulation phase.

For pre-cultures, small shaking flasks with 100 mL medium with 2 g/L ammonium sulfate and 10 g/L glucose were inoculated and incubated for 12 h at 30 °C and 120 rpm. 20 mL of the cultures were transferred to eight shaking flasks each containing 250 mL of the same medium. After incubation for another 12 h under the same conditions, the culture was transferred into the reactor.

The fermentations were carried out in stirred tank reactors with 10 L working volume in discontinuous fed-batch mode which means that carbon and nitrogen sources were fed to the media in several portions but not continuously. The temperature was maintained at 30 °C. The pH was regulated automatically to maintain 7 ± 0.1 by addition of 20 wt% NaOH and 20 wt% H_3PO_4 respectively. Agitation and aeration were controlled manually.

Fermentations were performed in a two-stage mode, which is characterized by a growth phase with all necessary nutrition composites available in sufficient amounts and a nitrogen-source limited polymer accumulation phase. In the second phase, valerate was added to the medium as precursor for the formation of copolymer.

The concentrations of the carbon sources and the precursors were determined by HPLC (Animex HPX87-H, RI-Detector). Ammonium sulfate was detected with an ammonia-sensitive electrode (Orion Research Inc., Boston, USA). Polymer concentrations were determined after Braunegg´s method described earlier[2]. To determine the cell dry weight (CDW) 5 mL of the culture broth were centrifuged at 4000 rpm for ten minutes in pre-weighted glass tubes. After lyophilisation of the pellets, the CDW was calculated from the weigh difference.

Kinetic data (specific growth rate, yield coefficient) correlating to the biomass were taken from the residual biomass Xr, which is the biomass without polymer content.

3. EXPERIMENTS AND RESULTS

3.1 Fermentation with glucose and valerate

This fermentation was carried out with glucose as sole carbon source. In order to obtain 3-hydroxybutyrate-co-3-hydroxyvalerate copolymer, valerate was added when polymer accumulation phase began. Results are presented in Fig. 1 and 2 and Table 1.

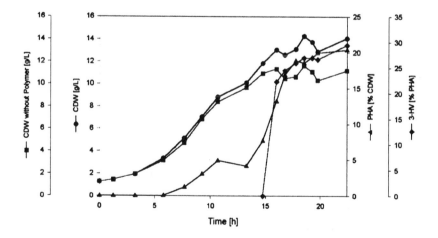

Figure 1. Fermentation on glucose as carbon source and production of P(3HB-co-3HV) with sodium valerate as precursor and *Pseudomonas cepacia* ATCC 17759

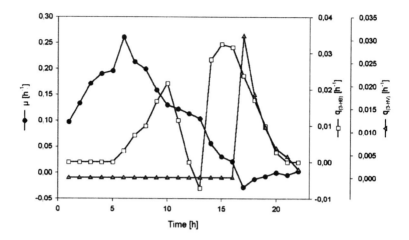

Figure 2. Fermentation on glucose as carbon source and production of P(3HB-co-3HV) with sodium valerate as precursor and *Pseudomonas cepacia* ATCC 17759, development of kinetic data

Table 1. Kinetic data calculated from the fermentation on glucose as carbon source

$\mu = 0{,}26\ h^{-1}$	$q_{P(3HV)} = 0{,}031$
$X_r = 11{,}2\ g/L$	PHA = 2,87 g/L (20,4 % of CDW)
$Y_{Xr/S} = 0{,}33$	P(3HV) = 0,84 g/L (29,4 % of PHA)
$q_{P(3HB)} = 0{,}032$	

Only small amounts of polyhydroxyalkanoates are formed during growth phase, which are due to partial nitrogen limitation being removed after supplementation with ammonium sulfate. Maximum specific growth rate was reached at 6 hours after beginning the fermentation with 0,26 h^{-1}. After about 13 hours accumulation phase began, which is recognizable at rapidly increasing values of specific production rate of polyhydroxybutyrate. Maximum value is reached after 15 hours with 0,032 h^{-1}. With the beginning of accumulation phase, sodium valerate was added, but rapid production of hydroxyvalerate comonomer started with a retardation at about 14 h after beginning of the fermentation. But nevertheless maximum specific polymer accumulation rate of hydroxyvalerate is almost as high as for hydroxybutyrate with 0,031 h^{-1} which lead to a hydroxyvalerate portion of nearly 30 % at the end of the fermentation. Both production rates showed rather fast decreasing values after reaching their maximum. After 22,4 h fermentation was finished because no substantially higher amount of accumulation could be expected with further performance. Total amount of polyhydroxyalkanoates is rather low with 20,4 % of the cell dry weight.

3.2   Fermentation with glucose and galactose

The carbon source for this fermentation consisted of a 50 to 50 wt% mixture of glucose and galactose that would correspond to an enzymatically hydrolysed lactose solution with the same concentration. Results are shown in Fig. 3 and 4 and Table 2.

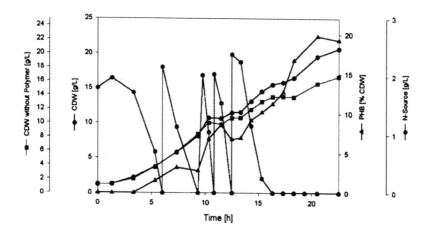

Figure 3. Fermentation on glucose and galactose as carbon source with *Pseudomonas cepacia* ATCC 17759

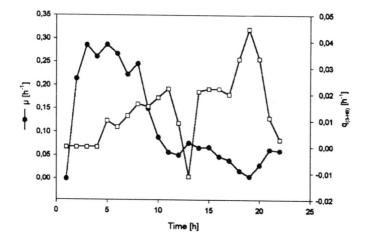

Figure 4. Fermentation on glucose and galactose as carbon source with *Pseudomonas cepacia* ATCC 17759, development of kinetic data

Table 2. Kinetic data calculated from the fermentation on glucose and galactose as carbon source

$\mu = 0,29$ h^{-1}	$q_{P(3HB)} = 0,045$
Xr = 16,7 g/L	PHA = 4,0 g/L (19,3% of CDW)
$Y_{Xr/S} = 0,32$	

Like in the first fermentation small amounts of polyhydroxybutyrate accumulation during growth phase are due to temporary nitrogen limitations. Ammonium sulfate was added in several portions. Specific growth rate for the residual biomass reaches even slightly higher values for this medium composition with 0,29 h^{-1} then growth on glucose as sole carbon source. During accumulation phase maximum value for specific polyhydroxybutyrate production reaches 0,045 h^{-1}, which is also higher than in the other fermentation. Increase of residual biomass after depletion of ammonium sulfate could be due to excretion of exopolysaccharides.

4.   DISCUSSION

P. cepacia shows satisfying specific growth rates on both carbon sources. Kinetic data of both fermentations are quite similar. Significant amounts of polyhydroxyalkanoate production start after nitrogen limitation. Specific polyhydroxybutyrate production rates show average values in contrast to the relative high specific hydroxyvalerate production rate which lead to a comonomer part of approximately 30 %. This is a sufficient amount to change polymer properties to eliminate negative physical properties of pure polyhydroxybutyrate such as high brittleness and high glass transition point. Production of poly-(3-hydroxybutyrate-co-3-hydroxyvalerate) was reported before with propionic acid as precursor and fructose as carbon source until 30,4 % of total polymer[4] which corresponds to our results. In contrast to propionic acid, sodium valerate has got a lower toxicity and can be present in higher amounts in the medium. The disadvantage of using sodium valerate as precursor for production of copolyester of course is its high price, which would not contribute to the reduction of production costs.

Total amounts of accumulated polyhydroxyalkanoates are rather low but could probably be improved by adjusting medium composition. In shaking flask experiments this strain is known to accumulate approximately 50 to 60 % PHA of cell dry weight with glucose and lactose as carbon sources respectively[9].

P. cepacia ATCC 17759 is able to grow on lactose as carbon source. Production of β-galactosidase when grown on lactose was shown by using a laser-based fluorescent probe for β-galactosidase[9]. Nevertheless specific growth rates on lactose are found to be significantly lower than on glucose as carbon source $(0,10 \text{ h}^{-1})$[9]. Enzymatic hydrolysis of whey lactose can be performed with the enzyme lactase, which is commercially available from species of yeast and microfungi[6]. Our experiments show that it is possible to use the hydrolysation product for PHA production with *P. cepacia* having kinetic behaviour similar to pure glucose. By using lactose hydrolysate costs for separation of the sugar mixture would be dropped.

REFERENCES

1. Braunegg, G., 1997, Project Application within the Frame of the EU Biotechnology Research Program, Dairy Industry Waste as a source of Biodegradable Polymeric Materials, Wheypol.
2. Braunegg, G., Sonnleitner B. and Lafferty R. M., 1978, A rapid gaschromatographic method for the determination of poly-β-hydroxybutyric acid in microbial biomass. *European Journal of Applied Microbiology* **6**: 29-37
3. Lazar, Z., 2000, Versuche zur Produktion von Polyhydroxyalkanoaten mit Lactose als Kohlenstoffquelle. Diplomarbeit, Graz University of Technology
4. Ramsay, B.A., Ramsay, J.A and Cooper, D. A., 1989, Production of Poly-β-Hydroxyalkanoic Acid by *Pseudomonas cepacia, Applied Environmental Microbiology* **55**: 584-589
5. Renner, G., Schellauf, F. and Braunegg, G., 1998, Selective enrichment of bacteria accumulating polyhydroxyalkanoates in multistage continuous culture. *Food technol. Biotechnol.* **36** (3): 203-207
6. Siso, G.M.I., 1996, The Biotechnological Utilization of Cheese Whey: A Review. *Bioresource Technology* **57**: 1-11
7. Wong, H.H. and Lee, S.Y., 1998, Poly-(2-hydroxybutyrate) production from whey by high-density cultivation of recombinant *Escherichia coli. Applied Microbiology and Biotechnology* **50**: 30-33
8. Yellore, V. and Desai, A., 1998, Production of poly-3-hydroxybutyrate from lactose and whey by Methylobacterium sp. ZP24. *Letters in Applied Microbiology* **26**: 391-394
9. Young, F.K., Kastner, J.R. and Sheldon, W.M., 1994, Microbial Production of Poly-β-Hydroxybutyric Acid from D-Xylose and Lactose by *Pseudomonas cepacia. Applied and Environmental Microbiology*, **Nov.**: 4195-4198

Production of Poly-3-hydroxybutyrate-co-3-hydroxy-valerate with *Alcaligenes latus* DSM 1124 on Various Carbon Sources

GUDRUN HAAGE, ELISABETH WALLNER, RODOLFO BONA, FLORIAN SCHELLAUF and GERHART BRAUNEGG
Institut für Biotechnologie, TU Graz, Petersgasse 12, A-8010 Graz, Austria

Abstract: Polyhydroxyalkanoates are thermoplastic polyesters which are produced as storage compounds by a vast number of bacterial strains under growth limiting conditions. These polymers can be produced in different qualities with a wide range of physical properties, either as homo-polymer (Poly-3-hydroxybutyrate) or, with different co-substrates, as copolymers (e.g. Poly-3-hydroxybutyrate-co-3-hydroxyvalerate, Poly-3-hydroxybutyrate-co-4-hydroxybutyrate). They are completely biodegradable and therefore an interesting alternative to the mineral-oil derived plastics as packaging materials. Nevertheless up to now PHAs have not been able to compete with the conventional plastics because their price is still much too high. One possible way to reduce the production costs is to use cheaper carbon sources like agricultural waste materials. The strain *Alcaligenes latus* DSM 1124 is able to accumulate high amounts of PHAs and therefore very interesting for a future production. The growth and the accumulation of PHAs on various carbon sources like glucose, glycerol and Na-valerate have been investigated in a labscale fermenter. PHA contents and kinetic data like the specific growth rate and yield coefficients were determined.

1.   INTRODUCTION

Many bacteria can use glycerol as a carbon source. As it is a by-product of the production of bio-diesel it can be achieved cheaply. Therefore it seems to be a good alternative to glucose for the production of Polyhydroxyalkanoates. Sodiumvalerate is often used as a precursor in order to obtain the copolymer Poly-3-hydroxybutyrate-co-3-hydroxyvalerate. As

Biorelated Polymers: Sustainable Polymer Science and Technology
Edited by Chiellini *et al.*, Kluwer Academic/Plenum Publishers, 2001   147

the costs of the precursor contribute a lot to the overall production-costs growth through degradation via the fatty acid pathway is undesirable.

The strain *Alcaligenes latus DSM 1124* is known to be able to accumulate high ammounts of PHAs (more than 80% of CDW) during growth on glucose. Therefore the investigation of this strain in respect of the two substrates mentioned above was of great interest.

2.   MATERIAL AND METHODS

2.1   The strain *Alcaligenes latus*

The strain *Alcaligenes latus* DSM 1124 (ATCC 29713) was isolated from an Australian soil sample by Palleroni et Palleroni[9]. Nearly at the same time two related strains were isolated from a Californian soil sample (*A. latus* DSM 1122 (type strain) and *A. latus* DSM 1123). On a solid mineral medium colonies are round, greyish pink and opaque. In fresh isolates they are wrinkled but can become smooth upon subcultivation. The cells are short, gram negative rods and motile by means of 5 to 10 flagella arranged in peritrichous fashion. The metabolism is respiratory, and molecular oxygen is the final electron acceptor. *A.latus* ist capable of facultatively autotrophic growth. It is known to be able to accumulate high amounts of PHAs already during exponential growth.

2.2   Maintenance, Media, Pre-cultures

The strain was originally obtained from the DSM, Braunschweig, Germany. For culture maintenance the strains were grown on solid media for two days and then kept at 4°C. Single colonies were transferred to fresh plates every five weeks. As a backup and for longer periods the cells were stored in liquid nitrogen.

For all fermentations a synthetic mineral salt medium was used[11]. The sole nitrogen source was ammonium sulfate. The maximum concentrations for glucose, glycerol and Na-valerate have been 10 g/L, 5,5 g/L and 2 g/L respectively.

Pre-cultures were inoculated in 100 mL shaking flasks containing 70 mL medium with 2 g/L ammonium sulfate, 1 g/L Yeast-extract and 10 g/L glucose and 5 g/L glycerol and 2 g/L Na-valerate respectively and incubated over night at 30°C and 130 rpm. 10 mL of the culture were transferred to eight shaking flasks each containing 250 mL of the same medium. After incubation for another 12-24 h under the same conditions the culture was transferred into the reactor.

2.3 Fermentation conditions

All fermentations were carried out in stirred tank reactors with 10 L working volume. The fermentation with glucose was performed in batch mode the ones with glycerol and Na-valerate in discontinuous fed-batch mode. For that the respective nitrogen and carbon sources were fed to the media in several portions in order not to get too high concentrations in the fermentation broth. The Temperature was maintained at 30°C. The pH was kept at 7 ± 0.1 by automatic addition of 20 wt% NaOH and 20 wt% H_3PO_4. Agitation and airation were controlled manually to keep the dissolved oxygen concentration above 45% of maximum saturation.

The fermentations were divided in two stages. First the growth phase, in which balanced nutrition conditions exist and second, after depletion of the nitrogen source the PHA accumulation phase. In this phase precursors were added in order to obtain copolymers.

Kinetic data correlating to the biomass were calculated from the residual biomass Xr, which is the biomass without polymer content.

2.4 Analytics

The concentration of the biomass was determined by measureing the cell dry weight (CDW). 5 mL of the culture broth were centrifuged at 4000 rpm for 10 minutes in pre-weighed glass tubes. The pellets were lyophilised and the CDW was calculated from the weight difference. Polymer concentrations were determined from the dried pellets after Braunegg's method described earlier[2]. From the supernatant the following analyses were performed: The concentrations of the carbon sources and the Na-valerate were determined by HPLC (Aminex HPX87-H, RI-Detector). The amount of γ-butyrolactone was detected by GC. Ammonium sulfate concentrations were determined with an ammonia-sensitive electrode (Orion Research Inc., Boston USA).

3. EXPERIMENTS AND RESULTS

3.1 Fermentation with glucose and γ-butyrolactone

This fermentation was carried out in batch mode with glucose as the sole carbon source. In order to obtain the copolyester P(3HB-co-4HB) γ-

butyrolactone was added after the depletion of ammonium sulfate. Results
are shown in Fig 1 and table 1.

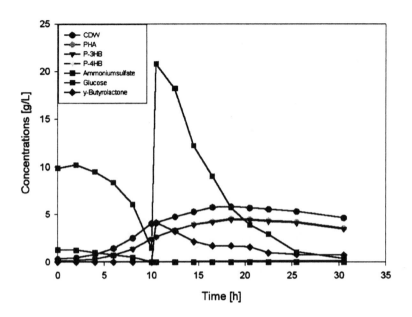

Figure 1: Fermentation on glucose as carbon source and production of P(3HB-co-4HB) with
γ-butyrolactone as precursor and *Alcaligenes latus* DSM 1124

Table 1. Kinetic data calculated from the fermentation on glucose as main carbon source.

μ_{max} = 0,31 h-1	$Y_{P3HB/Glu}$ = 0,12 g/g
X = 5,8 g/L	$Y_{P4HB/Lac}$ = 0,033 g/g
$Y_{Xr/S}$ = 0,45 g/g	PHA = 4,53 g/L (78,1% of CDW)
q_{P3HB} = 0,25 h-1	P-4HB = 0,11 g/L (2,97% of PHA)
q_{P4HB} = 0,013 h-1	

As one can expect growth on glucose is rather fast (μ_{max}= 0,31 h^{-1}) and
exponential after a lag phase of 4 hours. The yield coefficient for the
biomass of 0,45 g/g is rather high and in the same range as in other
experiments in our working group[7]. During the growth nearly 60% PHB of
CDW was accumulated. However after the depletion of the nitrogen source
γ-butyrolactone was added and specific polymer accumulation rates for
P(3HB) and P(4HB) of 0,25 h^{-1} and 0,013 h^{-1} respectively were reached.
After 22 hours of total fermentation time polymer accumulation starts to
decrease although more than 2 g/L glucose and appr. 1 g/L γ-butyrolactone
are still present in the reactor. The PHA content at the end of the

fermentation is 78% of CDW and therefor lower than average (normally around 90%) also the P(4HB) fraction of the polymer of 3% is rather low[6].

Considering the impact of the costs of the precursors on the overall production costs of copolyesters the yield coefficient for P(4HB) is much too low.

3.2 Fermentation with glycerol

The main source in this experiment was glycerol. The growth phase lasted for 27 hours and only small amounts of P(3HB) were accumulated. This is very surprising since *A. latus* is well known accumulating large amounts of polymer during growth. As known from other strains, we have a completely changed behavior of a strain with a carbon source other than glucose. The results are presented in Fig 2 and table 2.

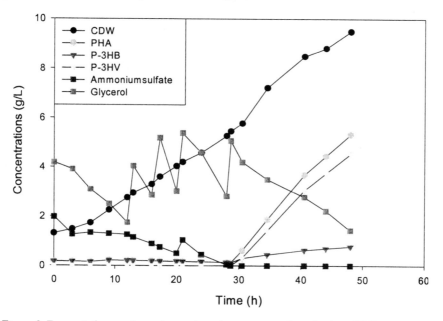

Figure 2. Fermentation on glycerol as main carbon source and production of P(3HB-co-3HV) with *Alcaligenes latus* DSM 1124.

Table 2. Kinetic data calculated from the fermentation on glycerol as main carbon source.

μ_{max} = 0,075 h-1	$Y_{P3HB/Glyc}$ = 0,13 g/g
X = 5,26 g/L	$Y_{P3HV/Val}$ = 0,55 g/g
Yx_r/s = 0,43 g/g	PHA = 5,31 g/L (56,0% of CDW
q_{P3HB} = 0,01 h-1	P-3HV = 4,53 g/L (85% of PHA)
q_{P3HV} = 0,05 h-1	

The specific growth rate of 0,075 h-1 is significantly lower than in the other experiments. The yield coefficient for the biomas is again quite high. After the depletion of ammonium sulfate, Na-valerate was added to obtain the copolyester P(3HB-co-3HV). Due to problems with the analytical equipment no data for valerate could be obtained. The specific production rate of P(3HV) was five times higher than that of P(3HB) an resulted in a P(3HV) fraction of 85 %. This unusually high fraction is only surpassed by the work of Doi et al.[5] who reached a P(3HV) fraction of 90 % by using pentanoic acid as sole carbon source. The inhibition of P(3HB) accumulation during growth phase and the low production rate during the accumulation phase suggest a strong negative effect of glycerol on the P(3HB) accumulation while the production of P(3HV) seems to be stimulated considerably by this substrate.

3.3   Fermentation on sodium valerate

Sodium valerate normally is only used as a precursor. Bergey's manual[1] indicates that *Alcaligenes latus* DSM 1122 shows no growth on valerate.
    A.latus DSM 1124 however shows growth on valerate The results are given in Fig 3 and table 3.

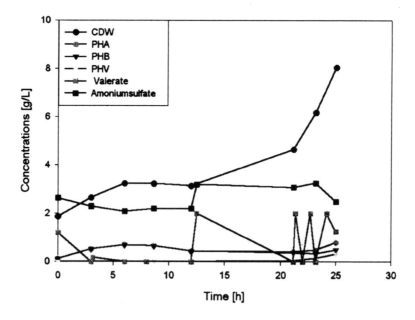

Figure 3. Fermentation on Na-valerate as sole carbon source and precursor for the production of P(3HB-co-3HV) with *Alcaligenes latus* DSM 1124.

Table 3. Kinetic data calculated from the fermentation on Na-valerate as carbon source.

μ_{max} = 0,13 h-1	q_{P3HV} = 0,013 h-1
X = 8,0 g/L	PHA = 0,79 g/L (9,9% of CDW)
$Y_{Xr/S}$ = 0,65 g/g	P-3HV = 0,3 g/L (38,7% of PHA)
q_{P3HB} = 0,07 h	

During the first part of the fermentation the Na-valerate concentration in the fermenter was nearly zero and therefore growth was limited. The PHA content also stayed low. Due to the valerate-limitation, ammonium sulfate was not consumed either. Only in the second part of the experiment, when the feed of valerate was more regularly, growth became exponential and the ammonium sulfate concentration decreased.

The specific growth rate of 0,13 h^{-1} is significantly lower than in the experiment with glucose (0,31 h^{-1}), but still higher than in the fermentation with glycerol (0,075 h^{-1}). The yield coefficient for the biomass is in comparison very high. As the nitrogen source never depleted one cannot distinguish between a growth- and an accumulation phase. The obtained amount of PHA also is rather low, but the trends in the last measure points indicate that higher values are possible. It is notable that from the valerate not only P(3HV) is accumulated, but also P(3HB).

4. DISCUSSION

The strain *Alcaligenes latus* DSM 1124 is able to grow and to accumulate PHAs on many different carbon sources. In order to lower the production costs it is important to find alternative, cheap carbon sources. Bogensberger showed that the strain is able to use green syrup and molasses[4]. Further possible raw materials are starch hydrolysates, glycerol, and whey as carbon sources, and corn steep liquor as nitrogen source.

The big disadvantage of this strain is that it accumulates a substantial amount of PHA already during the growth phase. So first a block of homopolymer (P3HB) is formed followed by a block of copolymer leading to a bad product quality. Addition of the precursor already in the growth phase led to strong inhibition of growth (Haage G. unpublished data). An other possible strategy to diminish or avoid the homopolymer block is the introduction of an starving phase between growth and accumulation phase. Studies on other strains brought satisfying results[10].

Without any doubt the best substrate for PHA production with *Alcaligenes latus* DSM 1124 is glucose. The specific growth rate is the best obtained in these experiments. The PHA content reached is also very high and no negative effects on the polymer quality have to be expected from the use of this substrate. The Copolymer fraction is relatively low here but with some optimisation better levels can be reached.

The growth rate on glycerol is rather low. But the change from growth associated accumulation on glucose to non growth associated accumulation on glycerol is very interesting. It might be possible to boost growth on glycerol and then switch to a glucose medium for the accumulation phase thus obtaining a polymer with good quality. Further investigation in the very fast production of P(3HV) in the accumulation phase should lead to substantially lower costs for valerate as precursor with glycerol as main carbon source.

A further draw back for the use of glycerol in the accumulation phase is the fact that glycerol acts as a chain transfer agent in the chain termination step of the polymerisation. Thus the molecular weight of the polymer is lowered considerable[8]. The optimal glycerol concentration to avoid the negative effect on the molecular weight still has to be found. For certain applications low molecular weight PHAs nevertheless are of interest.

Unlike as in the literature reported, *Alcaligenes latus* DSM 1124 is also able to grow on Na-valerate as sole carbon source which is an economical draw back in respect of the costs for valerate as a precursor. Again the specific growth rate is much lower than on the growth with glucose. Unfortunately the experiment was broken off too early so it is not possible to make more detailed conclusions. More experimental work has to be done here.

As a conclusion, one could state that although the growth associated accumulation of PHAs in *Alcaligenens latus* DSM 1124 comes out to be a disadvantage, the strain is still of some interest for future PHA production.

REFERENCES

1. Bergey's Manual of systematic Bacteriology 1984. Vol. 1-4 (Ed. In Chief Holt J.G.) Williams& Wilkins Baltimore/London
2. Braunegg, G., Lefebvre G., and Genser K., 1998, Polyhydroxyalkanoates, biopolyesters from renewable resources: physiological and engineering aspects, *J. Biotech.* **65**: 127-161
3. Braunegg, G., Sonnleitner, B. and Lafferty R.M., 1978, A rapid gaschromatographic method for the determination of poly-β-hydroxybutyric acid in microbial biomass. *European Journal of Applied Microbiology* **6**: 29-37
4. Bogensberger B., 1985, Doctoral thesis. *Poly-D(-)-3-Hydroxybuttersäure Studien an Alcaligenes latus unter Verwendung technischer Substrate*, Graz University of Technology
5. Doi, Y., Tamaki, A., Kunioka, M. and Soga, K., 1987, Biosynthesis of an unusual copolyester (10 mol% 3-hydroxybutyrate and 90 mol% 3-hydroxyvalerate units) in *Alcaligenes eutrophus* from pentanoic acid. *J. Chem. Soc. Chem. Commun.*: 1635-1636
6. Haage G., 1995, Diploma thesis. *Poly-(3-HB-co-4-HB)-Produktion bei Vertretern der genetischen r-RNA-Supergruppe 3 (β-Gruppe)*, Graz University of Technology

7. Latal B., 1992, Diploma thesis *P(3HB-co-3HV)-Produktion mit Alcaligenes latus DSM 1124*, Graz University of Technology

8. Madden, L.A., Anderson, A.J., Shah, D.T., and Asrar, J., 1999, Chain termination in polyhydroxyalkanoate synthesis: involvement of exogenous hydroxy-compounds as chain transfer agents. *Int. J. Biol. Macromol.*, **25**: 43-53

9. Palleroni, N.J., and Palleroni, A.V., 1978, *Alcaligenes latus*, a New Species of Hydrogen-Utilizing Bacteria. *Int. J. Syst. Bacteriol.* **28**: 416-424

10. Renner, G. et al. 1994, Effect ofa carbon starvation period on the kinetics of copolymer production by Alcaligenes latus DSM 1122. International Symposium on bacterial PHA, Mc.Gill University, Montreal, Canada, p.68

11. Schlegel, H.G., Gottschalk, G., and Von Bartha R., 1961, Formation and Utilization of Poly-β-hydroxybutyric acid by Knallgasbacteria (Hydrogenomonas), *Nature* **191**: 463-465

Biosynthesis of Polyhydroxyalkanoates and their Regulation in Rhizobia

SILVANA POVOLO and SERGIO CASELLA
Dipartimento di Biotecnologie Agrarie, Università di Padova, Agripolis, Padova, Italy

Abstract: During free living reproductive growth *Rhizobium* spp. is capable to accumulate poly-β-hydroxybutyrate (PHB) and to synthesize intracellular glycogen. These bacteria can also produce and excrete exopolysaccharides and β-1,2-glucan. Rhizobia provide an excellent model to investigate the connection between cellular metabolism and polyester accumulation. We have shown that under oxygen-limiting conditions, free-living cells of *Sinorhizobium meliloti* 41 can use intracellular glycogen to generate ATP, while maintaining their PHB content. PHB synthesis serves as an alternative pathway for storage/regeneration of reducing equivalents. We have described genes involved in PHB biosynthesis in *S. meliloti* encoding for β-ketothiolase (*phaA*), acetoacetyl-CoA reductase (*phaB*) and PHA-synthase (*phaC*) together with an open reading frame, referred to as *aniA*. Under oxygen-limiting conditions (such as conditions in the bacteroid state) *aniA* is actively expressed, and a mutation in this gene generates an overproduction of extracellular polymeric substances (EPS). This finding suggests that the production of EPS could be directly or indirectly regulated by *aniA*. Therefore, in *S. meliloti*, *aniA* is likely to be involved in carbon/energy flux regulation that, in turn, is dependent upon oxygen availability. By hybridization studies we revealed, in various soil bacteria the presence of genes with sequence similarity to *aniA* of *S. meliloti* 41. These results will be important to gain a deeper insight into *aniA* function in the control of PHB (more generally PHA) and EPS biosynthesis.

1.   INTRODUCTION

Polyhydroxyalkanoates (PHAs), a complex class of naturally occurring polyesters, are produced as intracellular granules by a large variety of bacteria[1]. These polymers have recently attracted considerable attention by scientists from academia and industry mainly because they are biodegradable thermoplastics and elastomers that can be manufactured to materials useful for various technical applications[2,3,4]. The diversity of different monomers that can be incorporated into PHAs provide an enormous range of new polymers[5], and the advent of genetic engineering combined with modern molecular microbiology will ensure that environmentally friendly polyesters are available for future generations. The physiology, enzymology and molecular genetics of PHAs have been studied for a decade[2,5]. It is known that PHA accumulation occurs in most bacteria in the absence of a nutrient, such as nitrogen, sulphur, phosphate or oxygen, and if a carbon source is provided in excess. However, not much is known about the regulatory mechanism. The extent of PHA accumulation is dependent on the relative rates of synthesis and degradation, which are controlled by growth conditions[5]. Intracellular PHA degradation is poorly understood. PHA formation is regulated at physiological and genetic level. The first is obtained through the availability of metabolites and cofactor inhibition of the enzymes. The second through alternative σ-factors, two component regulatory systems, and autoinducing molecules. Another level of regulation concerns granule size and molecular weight control by levels of PHA polymerase and other granule-associated proteins[2,5]. The control of PHAs accumulation is important to better understand the functioning of the biosynthetic machinery and for industrial applications.

2.   *RHIZOBIA* AND PHA

Rhizobia are symbiotic bacteria that elicit the formation on leguminous plants of specialized organs, root nodules, in which they fix nitrogen. *Rhizobium*, *Sinorhizobium*, *Bradyrhizobium* and *Azorhizobium* accumulate poly-β-hydroxybutyrate (PHB) in free life[6,7,8], and in symbiosis[9,10, 11,12]. *Rhizobium* spp. are capable to synthesize intracellular glycogen and can also produce and excrete exopolysaccharides and β-1,2-glucan[13]. Production of these polymers requires a fine regulation during free life and in symbiosis. Oxygen seems to play an important role in this process. In *Rhizobium* ORS571 during free life the lack of carbon source does not result in breakdown of PHB, but other factors (oxygen and nitrogen concentrations) are very important[6]. In this bacteria, PHB degradation occurs at high oxygen concentrations[6]. No breakdown is observed at oxygen limitation, no matter if an exogenous carbon source is present or not. In *Rhizobium etli* CE3 (auxotrophic for biotin and thiamin, cofactors for pyruvate dehydrogenase and α-ketoglutarato dehydrogenase) in the absence of vitamins PHB accumulates to high levels in aerobic free-living state. TCA cycle is not completely active as a result of these auxotrophies. PHB formation appears to take over the role of TCA cycle as an overflow for carbon and reducing equivalents[14]. During bacteroid state PHB is accumulated in *R. etli*. A PHB-negative mutant of *R. etli* shows higher and prolonged nitrogen fixation[15]. This mutant has an impaired growth on glucose and pyruvate, but not on succinate as carbon source. During growth on succinate it excretes increased levels of organic acids and has a lower level of the ratio NAD to NADH, when compared to the parental strain[15]. Mutations in the nitrogenase-encoding *nifD*, *nifK* and *nifH* genes of *Bradyrhizobium japonicum* producean increased PHB accumulation[16], resulting in an alternative pathway for regeneration of reducing equivalents.

In *R. leguminosarum* there is a link between amino acid metabolism and PHB formation[17]. Glutamate synthesis and secretion appears to be important to balance carbon and reducing equivalents, especially in the absence of a functional TCA cycle or PHB formation[17].

PHA formation in *Rhizobium* spp is studied because it provides an excellent model to follow the interplay between cellular metabolism and polyester formation. PHB biosynthesis in *Rhizobium* spp proceeds through the action of at least three enzymes as first described for *Ralstonia eutropha*[18] and *Zoogloea ramigera*[19]: a β-ketothiolase, an NADPH-dependent acetoacetyl CoA reductase and a PHB synthase[20]. The genes involved in PHB biosynthesis in *S. meliloti* encoding for β-ketothiolase (*phaA*), acetoacetyl-CoA reductase (*phaB*) and PHA-synthase (*phaC*) together with an open reading frame, referred to as *aniA* were characterized[20,21]. In a previous paper, mutants of *Sinorhizobium meliloti* 41 unable to synthesize PHB were described. The symbiotic traits of the mutant are similar to those of the parent strain[22]. The degradative portion of the PHB cycle was recently characterized in *Sinorhizobium meliloti* 1021[23,24].

3. PHB SYNTHESIS AND REGULATION IN *SINORHIZOBIUM MELILOTI* 41

Sinorhizobium meliloti could accumulate PHB during free life, but not in symbiosis (Figure 1). It was hypothesised that this could be due to a low activity of the NADH-dependent malic enzyme[25]. *S. meliloti* has two malic enzymes, one NADH dependent (Dme) and the other NADPH dependent (Tme). While Dme and Tme are both expressed in the free-living state, Tme is repressed in bacteroids and Dme is inhibited by excess of acetyl-CoA. As a consequence PHB is not accumulated in the bacteroid because the levels of NAD(P)H are too low.

Free living *S. meliloti* 41 accumulates up to 50% of its weight as poly- β-hydroxybutyrate (PHB) in yeast salts medium[7] with mannitol as carbon source. When other carbon sources are used, *S. meliloti* 41 still produces high amounts of PHB under the same growth conditions (Table 1). This strain can also accumulate a copolymer when propionate or n-valerate were supplied to the culture (data not shown).

Table 1. Accumulation of poly-β-hydroxybutyrate in Sinorhizobium meliloti 41 grown under standard conditions in different carbon sources

Carbon Source	% PHB (w/w)
Glucose	71,1
Lactic acid	48,5
Lactose	23,4
Galactose	58,4
Sucrose	28,4

As shown in previous studies oxygen concentration can modulate PHB production in rhizobia[6]. It was demonstrated that oxygen limitation can enhance PHB synthesis in *Rhizobium* spp cells[6]. The incubation of exponential phase cells of *S. meliloti* 41 under low oxygen generate a clear decrease of the cell glycogen content while maintaining PHB level almost steady (Fig 2).

In addition, cell ATP content increased during the first few hours of anoxic incubation (Fig 2), although no cell growth or death was observed (data not shown). Protein content did not decline during the first hours of incubation (data not shown) indicating that ATP increase could not be ascribed to amino acid catabolism. This suggests that anoxic conditions do not inhibit the synthesis of PHB. Other reserve compounds (i.e. glycogen) are metabolized inside the cell in order to meet the ATP requirement necessary to survive under such an adverse situation. Under these culture conditions a gene, designed *aniA*, was found to be expressed[21]. The predicted AniA protein is cytoplasmatic, with a putative DNA-binding domain, frequently found in transcription factors. A mutation in *aniA* was shown to cause an overproduction of extrapolymeric substances (EPS)[21]. As *aniA* transcription is activated under low oxygen conditions, these data suggest that this gene plays a role in the physiology of the cell when oxygen supply is limited. Incubation under anoxic conditions for several hours triggers repression of EPS production in *S. meliloti* 41, while this did not happen to the *aniA*-mutant (Fig 3). This finding indicates that the synthesis of EPS could be directly or indirectly regulated by *aniA*. Hybridization studies were performed to check the presence of the *aniA* gene in other rhizobia. Southern blot analysis of total DNA from different strains showed DNA bands with sequence similarity to *aniA* of *S. meliloti* 41 (Fig 4).

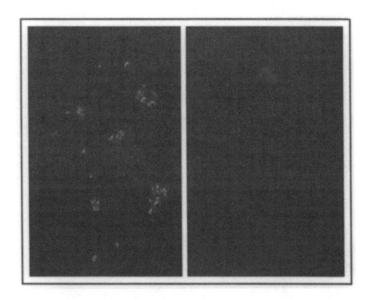

Figure 1. Fluorescence microscopy of *Sinorhizobium meliloti* stained with Nile red. (left) free-living cells containing multiple yellow-fluorescent intracellular inclusions of poly-β-hydroxybutyrate; (right) bacteroid cells extracted from nodules of *Medicago sativa* with no granules.

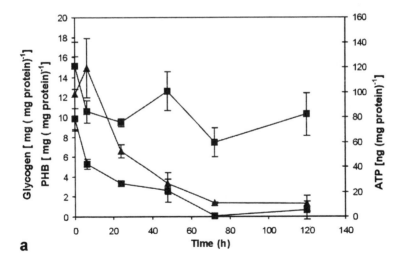

Figure 2. Glycogen (●), ATP (▲) and PHB (■) content of *Sinorhizobium meliloti* 41 incubated under low-oxygen conditions. The values drawn are means of three independent experiments ± standard deviation.

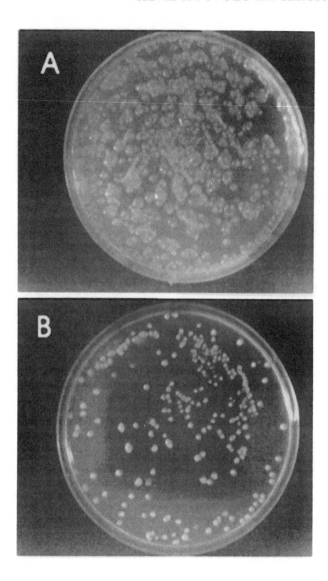

Figure 3. Colonies formed on Yeast Mannitol Agar plates by *Sinorhizobium meliloti* 41 (B) and *aniA*-mutant strain (A). Cells were plated after 70 h anoxic treatment.

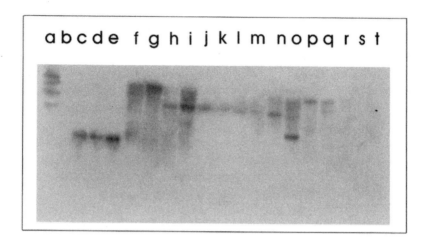

Figure 4. DNA-DNA hybridization with a digoxygenin-labeled DNA probe containing *aniA* gene. Total DNA was *Hind*III-digested. Digoxygenin-labeled molecular weight markers, (b) *Sinorhizobium meliloti* 41100, (c) *Sinorhizobium meliloti* 41, (d) *Sinorhizobium meliloti* CM2, (e) *Sinorhizobium meliloti* CM12, (f) *Rhizobium galega*, (g) *Rhizobium trifolli*, (h) *Rhizobium trifolii* 8ST, (i) *Rhizobium hedysarii* HCNT1, (j) *Rhizobium hedysarii* IMAP801, (k) *Rhizobium hedysarii* IMAP837, (l) *Rhizobium hedysarii* IMAP810, (m) *Rhizobium hedysarii* IMAP885, (n) *Rhizobium etli* CFN 42, (o) *Rhizobium tropici* IIA CFN 299, (p) *Rhizobium tropici* IIB CIAT 899, (q) *Rhizobium loti* USDA 3471, (r) *Bradyrhizobium japonicum* USDA 6, (s) *Bradyrhizobium japonicum* USDA 110, (t) *Rhizobium hedysarii* IS 123.

An open reading frame (designed *phaQ*) was also located on the opposite strand of *phaC* gene of *S. meliloti* 41, but no transcription of this putative gene has been detected yet. Although this gene was shown to be present in the same loci in other bacteria[5] we did not find any protein in GeneBank database with sequence similarity to *phaQ*. The role in PHA regulation of a presumed protein or RNA derived from this gene needs to be investigated.

4.   CONCLUSIONS AND PERSPECTIVES

We have shown that maintenance of *Sinorhizobium meliloti* 41 under oxygen limiting conditions enables PHB accumulation. Under very low oxygen concentrations TCA cycle does not function at normal rate and exogenous carbon sources can not be metabolised, so cellular compounds or other reserve materials (like glycogen) are probably used for energy conservation while carbon storage is ensured by PHB synthesis. *S. meliloti* 41, indeed, seems to use glycogen to generate ATP while forming PHB to safeguard TCA cycle[21]. The role of PHB in free life for these bacteria should therefore be to regenerate reductive power and to allow surviving under oxygen shortage. In *S. meliloti* 41 the gene *aniA* is likely to be involved in carbon/energy regulation, especially important under anoxic conditions. The ubiquitous presence of *aniA* sequence similarities in other soil bacteria indicates its importance during their physiology. In fact, many soil bacteria are able to shift from aerobic to anaerobic life and a regulation gene acting to control carbon flux may likely represent a key surviving strategy. Moreover, it is well known that a number of soil bacteria are PHAs producers, representing a wide reserve of biological material to be potentially used for industrial applications.

Since PHA synthesis could be part of a regulon, probably controlled by environmental conditions, other oxygen inducible genes may be present and need to be investigated in *S. meliloti* through powerful examination, e.g. by proteome analysis or microarray application. Detailed knowledge of biosynthetic pathways and their regulation will contribute to develop new processes for PHAs production. An efficient control of the carbon/energy flux and optimal conditions for PHA accumulation is a requisite for its economic future production.

ACKNOWLEDGEMENTS

This work was supported by University of Padova 60% projects and by Special Grant "Assegno di Ricerca 1999-2000".

REFERENCES

1. Steinbüchel A., 1991, Polyhydroxyalkanoic acids. In *Biomaterials* (Byron D., ed.) Macmillan Publishers Ltd and ICI Biological Products Bussiness, pp. 123-214.
2. Anderson A. J. and Dawes E. A., 1990, Occurrence, metabolism, metabolic role, and industrial uses of bacterial polyhydroxyalkanoates. *Microbiol. Rev.* **54**: 450-472.
3. Hocking P.J. and Marchessault R.H., 1994, In *Chemistry and technology of biodegradable polymers* (Griffin G.J.L., ed.), Blackie Academic, pp. 48-96.
4. Steinbüchel A., 1996, Biopolymers. In *Biotechnology* (Roehr M., ed.) Wiley-VCH, pp. 403-464.
5. Madison L.L. and Huisman G.W., 1999, Metabolic engineering of poly(3-hydroxyalkanoates): from DNA to plastics. *Microbiol Mol. Biol. Rev.* **63**: 21-53.
6. Stam H., van Verseveld H.W., de Vries W., Stouthamer A.H., 1986, Utilization of poly-β-hydroxybutyrate in free-living cultures of *Rhizobium* ORS571. *FEMS Microbiol. Letters* **35**: 215-220.
7. Tombolini R., Nuti M.P. 1989. Poly (β-hydroxyalkanoate) biosynthesis and accumulation by different *Rhizobium* species. *FEMS Microbiol. Letters* **60**: 299-304.

8. Chiellini E., Solaro R., Casini E., Casella S., Leporini C., Picci G., 1989, Biosynthetic degradable polymers. Study of the activity of *Rhizobium "hedysari"* strains in the production of poly(β-hydroxybutyrate). *J. Bioact. Compat. Polym.* **4**: 296-303.
9. Goodchild D.J., 1977, The ultrastructure of root nodules in relation to nitrogen fixation. In *International review of cytology* (Bourne G. H., Danielli J.F. and Jeon K.W., eds.). Academic Press, New York, pp 235-288.
10. Karr D.B., Waters J.K., Suzuki F., Emerich D.W., 1984, Enzymes of the poly-β-hydroxybutyrate and citric acid cycles of *Rhizobium japonicum* bacteroids. *Plant Physiol.* **75**: 1158-1162.
11. N'doye I., Debilly S. F., Vasse J., Dreyfus B. and Truchet G., 1994, Root nodulation of *Sesbania rostrata*. *J. Bacteriol.* **176**: 1060-1068.
12. Wong P.P. and Evans H.J., 1971, Poly-β-hydroxybutyrate utilization by soybean (*Glycine max. Merr.*) nodules and assessment of its role in mantainance of nitrogenase activity. *Plant Physiol.* **47**: 750-755.
13. Zevenhuizen L.P.T.M., 1981, Cellular glycogen, β-1,2-glucan, poly-β-hydroxybutyric acid and extracellular polysaccharides in fast-growing species of *Rhizobium*. *Antonie van Leeuwenhoeck* **47**: 481-497.
14. Encarnacion S., Dunn M., Willms K. and Mora J., 1995, Fermentative and aerobic metabolism in *Rhizobium etli*. *J. Bacteriol.* **177**: 3058-3066.
15. Cevallos M.A., Encarnacion S., Laija A., Mora Y., Mora J., 1996, Genetic and physiological characterization of *Rhizobium etli* mutant strain unable to synthesize poly-β-hydroxybutyrate. *J Bacteriol.* **178**: 1646-1654.
16. Hahn M., Meyer L., Studer D., Regensburger B. and Hennecke H., 1984, Insertion and deletion mutations within the *nif* region of *Rhizobium japonicum*. *Plant Mol. Biol.* **3**: 159-168.
17. Walshaw D.L., Wilkinson A., Mundy M., Smith M. and Poole P.S., 1997, Regulation of the TCA cycle and the general aminoacid permease by overflow metabolism in *Rhizobium leguminosarum*. *Microbiology* **143**: 2209-2221.
18. Peoples O.P. and Sinskey A.J., 1989, Poly-β-hydrxybutyrate (PHB) biosynthesis in *Alcaligenes eutrophus* H16. Identification and characterization of the PHB polymerase gene (*phbC*). *J. Biol. Chem.* **264**: 15298-15303.
19. Peoples O.P., Masamune S., Walsh C.T. and Sinskey A.J., 1987, Biosynthetic thiolase from *Zoogloea ramigera*. III. Isolation and characterization of the structural gene. *J. Biol. Chem.* **262**: 97-102.
20. Tombolini R., Povolo S., Buson A., Squartini A., Nuti M.P., 1995, Poly-β-hydroxybutyrate (PHB) biosynthetic genes in *Rhizobium meliloti* 41. *Microbiol.* **141**: 2553-2559.
21. Povolo S. and Casella S., 2000, A critical role for *aniA* in energy-carbon flux and symbiotic nitrogen fixation in *Sinorhizobium meliloti*. *Arch. Microbiol.* **174**: 42-49.
22. Povolo S., Tombolini R., Morea A., Anderson A.J., Casella S., Nuti M.P., 1994, Isolation and characterization of mutants of *Rhizobium meliloti* unable to synthesize poly-β-hydroxybutyrate. *Can. J. Microbiol.* **40**: 823-829.
23. Cai G., Driscoll B.T. and Charles T.C., 2000, Requirement for the enzymes acetoacetyl Coenzyme A synthase and Poly-3-hydroxybutyrate (PHB) synthase for growth of *Sinorhizobium meliloti* and PHB cycle intermediates. *J. Bacteriol.* **182**: 2113-2118.
24. Charles T.C. and Aneja P.A., 1997, Megaplasmid and chromosomal loci for the PHB degradation pathway in *Rhizobium* (*Sinorhizobium*) *meliloti*. *Genetics* **146**: 1211-1220.
25. Driscoll B. and Finan T.M., 1997, Properties of NAD⁺- and NADP⁺-dependent malic enzymes of *Rhizobium*(*Sinorhizobium*) *meliloti* and differential expression of their genes in nitrogen-fixing bacteroids. *Microbiology* **143**: 489-498.

Polyhydroxyalkanoates Production by Activated Sludge

LUÍSA S. SERAFIM, PAULO C. LEMOS, JOÃO G. CRESPO, ANA M. RAMOS and MARIA A. M. REIS
Departamento de Quimica - CQFB, Faculdade de Ciências e Tecnologia, Universidade Nova de Lisboa, 2825-114 Caparica, Portugal

Abstract: Polyhydroxyalkanoates (PHA) are intracellular polymers stored by many bacterial species. Presently PHA are industrially produced by pure cultures fermentation where high quality substrates are used. Mixed cultures using raw substrates are able to produce PHA when submitted to transient conditions like oscillations on substrate feeding or on oxygen supply. The yield on PHA produced by activated sludge submitted to these dynamic conditions reach values comparable to those obtained by the pure cultures, being the first process less cost intensive than the last one. The chain length of the polymer produced in both processes is similar.

1. INTRODUCTION

Polyhydroxyalkanoates (PHA) are polymers synthesised by bacteria as intracellular carbon and energy sources. PHA are industrially produced by pure cultures using as main substrates glucose and propionic acid. The major expenses in the PHA production are determined by the cost of substrate and extraction of polymer from inside the cells[1].

In biological wastewater treatment processes microorganisms are able to store large amounts of polymers, which play an important role in their metabolism. In some processes the amount of PHA accumulated inside the biomass can account for almost 60 % of the dry weight. This yield is lower than the maximum value reported for industrial PHA production by pure cultures (85 % of the cell dry weight)[2]. However, PHA production by

Biorelated Polymers: Sustainable Polymer Science and Technology
Edited by Chiellini *et al.*, Kluwer Academic/Plenum Publishers, 2001

activated sludge from wastewater treatment plants requires less expensive substrates (raw materials present in the influent) and simpler processes (no reactor sterilisation and less process control are needed) allowing for a significant reduction of the operating costs. Therefore, PHA production by activated sludge is an attractive alternative to the use of pure cultures.

Activated sludge processes experiment periods of external substrate availability alternated with periods with substrate limitation. Under these dynamic conditions activated sludge is able to accumulate internal storage products namely PHA, lipids or glycogen. PHA is the most dominant storage polymer in activated sludge. It was recently shown that PHA production by activated sludge, in terms of quantity and quality, depends on the nature of the substrate fed[3] and on the type of process used[4]. Based on these facts an intensive research was carried out to select processes and conditions that favour selection of cultures enriched in microorganisms with high capacity for PHA production.

This work discusses the use of different processes for PHA production by activated sludge: enhanced biological phosphorus removal (EBPR), microaerophilic-aerobic process and aerobic periodic feeding (feast/famine conditions).

2.   PHA PRODUCTION BY BIOLOGICAL PHOSPHORUS REMOVAL PROCESS

One of the most interesting processes for phosphorus removal from wastewaters is the enhanced biological phosphorus removal (EBPR). This process is based on the capacity of polyphosphate accumulating microorganisms (Poly-P bacteria) to store inside the cells large amounts of phosphorus, when submitted to alternation of anaerobiosis and aerobiosis periods. In these bacteria, phosphorus and carbon metabolisms are closely associated. The metabolism of Poly-P bacteria was recently elucidated[5-7] and is shown in Fig 1.

Under anaerobic conditions, these bacteria are able to convert short chain fatty acids into PHA with simultaneous phosphorus release to the external medium[7]. The energy needed for carbon activation comes from the breakdown of intracellular polyphosphate and, at a smaller extent, from glycogen degradation. The reducing equivalents needed for PHA production are originated from glycogen consumption[8] and substrate degradation in the tricarboxylic acid cycle[5]. Under aerobic conditions, the PHA accumulated in anaerobiosis is metabolised for microbial growth and to restore the glycogen reserves. Part of the energy produced is used for synthesis of intracellular polyphosphate.

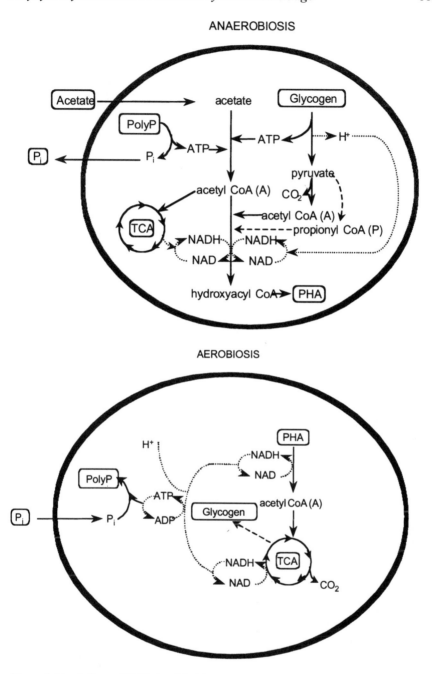

Figure 1. Metabolism of EBPR (modified from5)

Fig 2 and 3 show granules of PHA inside Poly-P bacteria stained with Nile blue by bright field microscopy and by epyfluorescence microscopy, respectively. As described by[9], the polymer inclusions stain bright orange in a dark background.

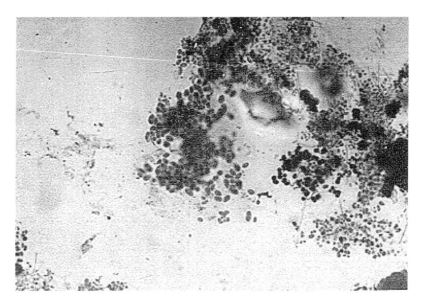

Figure 2. Activated sludge (dark spots) stained with Nile blue (bright field microscopy)

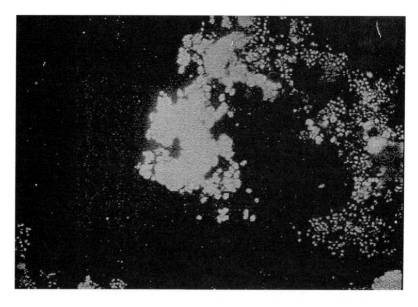

Figure 3. Granules of PHA (orange bright) produced by Poly-P bacteria stained with Nile blue (epyfluorescence microscopy).

PHA production by Poly-P bacteria remains at relatively low levels (around 20% of the sludge dry weight). The reported values were obtained in processes where optimisation of phosphorus removal was the main concern, polymer production being a side stream goal[3, 10]. If the objective of the process is the PHA production, cells should be harvested at the end of anaerobiosis, when they are full of PHA.

The PHA production by Poly-P bacteria is strongly dependent on the reactor operating parameters and on the carbon source[3]. The concentration and the kind of carbon source fed determine the amount and the composition of the polymer produced. By using acetate, propionate or butyrate (the usual compounds present in fermented wastewaters) copolymers containing hydroxybutyrate (HB) and hydroxyvalerate (HV) units were always obtained (Table 1). It is important to stress that these copolymers were obtained without additional co-substrate feeding, the opposite procedure for copolymer biosynthesis by pure cultures. In this process, when mixed substrates were used, copolymers with different composition were still produced[3]. Physical properties of these polymers are highly determined by the proportion of HB and HV units that constitute the copolymer. As illustrated in Table 1, the ratio of these monomers can be tailored by adjusting the substrate composition fed to the microorganisms. In fact, HB units are mainly produced from acetate while propionate promotes the formation of HV units.

Table 1. Composition on HB and HV, average molecular weight (M_w), polydispersity (M_w/M_n) and PHA yield produced by Poly-P bacteria using acetate, propionate and butyrate (320 mg COD.l^{-1}).

Substrate	HB (mg.g^{-1} cell dry wt)	HV (mg.g^{-1} cell dry wt)	Molar ratio (HB/HV)	$M_w \times 10^{-5}$	M_w/M_n	PHA Yield (% cell dry wt)
Acetate	129.79	62.77	3.04	6.5	1.9	19.2
Propionate	52.56	96.26	0.39	6.0	2.1	14.9
Butyrate	56.08	58.01	1.48	4.0	1.6	11.4

As was previously referred, the polymer yield obtained in this process does not exceed 20 % of cell dry weight being considerably lower than the values for commercial PHA production, which can reach 80 % [11].

Concerning the copolymer characterisation, the average molecular weight, M_w, and the polydispersity (M_w/M_n, being M_n the number average molecular weight) were also influenced by the type and concentration of carbon source (Table 1). These results accord perfectly with others previously reported in a more extensive study preformed about the influence of carbon source and feeding conditions on the PHA obtained in an enhanced biological phosphorus removal process[3]. The characteristics of

LUÌSA S. SERAFIM et al.

PHA obtained by this process can be compared with those of the polymer usually obtained by using pure cultures (Table 2).

Table 2. Values of average molecular weight, polydispersity and yield of PHA obtained from pure cultures of bacteria

Microorganism	Type of PHA	Substrate	M_wx10^{-5}	M_w/M_n	PHA yield (% cell dry wt)	Ref.
Alcaligenes eutrophus	PHB	N-alkanoates of C_2-C_{22}	2.3 - 15.5	1.4 - 2.5	55	[12]
	3PHB-co-PHV	N-alkanoates of C_2-C_{22}	0.36 - 30	1.7 - 2.5	57	[12]
	PHB	Butyric acid	12 - 33	2.3 - 7.7	80	[13]
	PHB	Radiolabelled glucose	13 - 20	3.0 - 4.1	70	[14]
	PHB	Fructose	17.3	2.7	29	[15]
	PHB	Glucose	18	2.9	65	[15]
	PHB	Sodium succinate	18.2	2.4	30	[15]
	PHB	Sodium pyruvate	16.4	3.2	70	[15]
	PHB	Sodium acetate	18.5	3.2	45	[15]
Alcaligenes latus	3PHB-co-4PHB	Sucrose and γ-butyrolactone	2.9 - 3.9	1.5 - 2.0	60	[16]
Azotobacter vinelandii	PHB	Glucose and fish peptone	17 - 28	1.7 - 3.0	85	[2]
	PHB	Beet molasses	10 - 45	1.2 - 1.8	73	[17]
Pseudomonas 135	PHB	Methanol	2.6 - 3.7	10.4-11.1	55	[18]
Pseudomonas cepacia	PHB	Glucose, xylose and lactose	2.2 - 8.7	1.6 - 3.9	56	[19]
Methylobacterium	PHB	Methanol	2.0 - 6.0	4.3 - 5.0	54	[20]
extorquens (pseudomonas sp. AM1)	PHB	Sodium succinate	9.0 - 17.0	2.9 - 5.7	60	[20]
	PHB	Methanol	1.7 - 6.2	4.0 - 8.3	21	[15]
	PHB	Sodium succinate	7.2 - 16.6	2.9 - 5.7	47	[15]
	PHB	Methanol and fructose	11.1 - 11.3	2.9 - 3.0	8	[15]
	PHB	Methanol and glycerol	2.6 - 3.9	3.3 - 4.5	5	[15]
	PHB	Methanol and ethanol	3.2	4.2 - 5.7	8	[15]

The range of molecular weight of the polymers obtained is similar to those produced by pure cultures of Pseudomonas spp. (M_w around $6 \cdot 10^5$), and lower than the values for the polymers produced by *Alcaligenes spp.* (M_w around $1\text{-}3\cdot10^6$). Nevertheless, the order of magnitude of molecular weight obtained is in the range of M_w for common applications of PHA.

3. MICROAEROPHILIC-AEROBIC PROCESS

The extraction and purification steps in the industrial PHA production are very expensive, leading to the development of processes where PHA content inside the cells is maximised. With this purpose a new process for PHA production by activated sludge was recently described[10]. The "microaerophilic-aerobic" process is mainly a modification of the anaerobic-aerobic one. In this process an aerobic period alternates with microaerophilic conditions, where a very limited amount of oxygen is supplied. It was proposed that during the microaerophilic period the amount of oxygen supplied should be just enough for the energy production required for PHA accumulation, preventing assimilative activity such as the synthesis of proteins and glycogen or other cellular materials. PHA production requires less energy than needed for glycogen synthesis. According to[10] the oxygen supplied should be around 0.51 % of the chemical oxygen demand (COD) provided, in order to allow bacteria to store high amounts of PHA intracellulary. By using this process a PHA content of 62 % of cell dry weight was achieved, a value that is comparable with the yield, reported for pure cultures used industrially[10].

This seems to be a very promising process for PHA production, deserving further research, namely on elucidation of the mechanisms of PHA storage, on the characterisation of microorganisms and on process engineering.

4. AEROBIC PERIODIC FEEDING

In biological wastewater treatment processes, microorganisms are usually submitted to dynamic conditions as a consequence of sludge recycling between zones with different substrate concentration. Under these transient conditions microorganisms improve their storage capacity. Storage occurs when an excess of available carbon source is supplied after a previous period of growth limitation. PHA are the most common storage polymers under conditions of excess of carbon source[4].

The mechanism of storage under dynamic conditions is illustrated in Fig 4. Immediately after a substrate pulse, cells previously starved start to synthesise PHA and the oxygen uptake rate (OUR) reaches a maximum value, remaining constant until exhaustion of the carbon source. When the external carbon source is completely consumed, cells start to metabolise PHA, changing from an easily degradable substrate to a more slowly degradable substrate leading to a strong decrease of their OUR value[21].

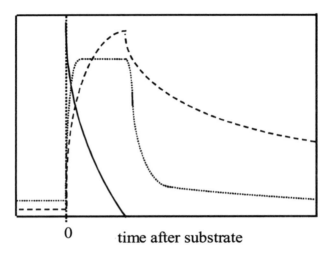

0   time after substrate

Figure 4. Evolution of PHA (-------), carbon substrate (————) and OUR (·········) for aerobic period feeding, (adapted from [4, 22])

Bacteria use their storage capacity to provide substrate for growth when the external substrate is depleted, having a strong competitive advantage over microorganisms without this ability[23]. PHA storage is the first and faster response to a transient situation and only a slight growth is detected. The reason for this preference is that during growth a more complex "machinery" is required for the synthesis of proteins, RNA and other polymers (polysaccharides and lipids) than for PHA storage[22].

According to[21] activated sludge are able to accumulate large amounts of PHB when cultivated under alternation of feast and famine of external substrate (acetate). The yield of PHB production by this culture was up to 50% of its dry weight, which is comparable with the value referred above for the cultures cultivated under microaerophilic-aerobic conditions. The influence of operating conditions, namely frequency of feeding and the ratio between the substrate and initial biomass concentration on PHA production was evaluated[24]. The quality of polymer produced under these conditions was not determined.

5.   CONCLUSIONS

Activated sludge is able to produce high amounts of PHA as internal reserves. In contrast to the current processes, where pure cultures and specific substrates are used, production of PHA by activated sludge might be more economical, since cheap organic waste streams or agricultural products can be used as substrates. As mentioned above, the physical properties of the polymer are closely dependent on substrate composition. Volatile fatty acids seem to be the best substrates for PHA production. The ratio between acetate and propionate determines the copolymer composition. Therefore, if a waste stream shall be selected as a substrate for the process, its composition in volatile fatty acids and its potential of acidogenic fermentation should be previously evaluated. In the latter, effluents rich in carbohydrates, fermented in an earlier step to a defined composition of volatile fatty acids by setting suitable reactor operating conditions, might be of interest.

In all the processes described, the PHA production always occurs as a response to a transient behaviour: either alternation of anaerobic/aerobic conditions, or microaerophilic-aerobic conditions, or intermittent substrate feeding. In the last decade, interest in activated sludge for PHA production led to intensive research and to better understanding of the PHA production mechanisms and to processes developments, where selection of microorganisms with high storage capacities can occur. The processes described in this context show a high potential for application with special interest for the microaerophilic-aerobic and intermittent substrate feeding, both revealing higher yields of PHA production, than EBPR. Poly-P bacteria have a very complex metabolism and it is not clear whether they will be able to reach the yield values reported for the other PHA accumulating bacteria.

The high PHA storage capacity by activated sludge and the good chain lengths and composition of the polymers produced by these processes deserve further research with the purpose of process optimisation and final application at full scale.

ACKNOWLEDGEMENTS

The authors would like to thank Dr. Valter Tandoi and Dr. Caterina Levantesi from CNR/IRSA, Rome, for the photos presented. Luísa S. Serafim acknowledges Fundacão para a Ciência e Tecnologia for grant PRAXIS XXI BD/18287/98 and Paulo C. Lemos acknowledges for grant BD/1217/91-IF.

REFERENCES

1. Anderson, A. J. and Dawes, E. A., 1990, Occurrence, metabolism, metabolic role, and industrial uses of bacterial polyhydroxyalkanoates. *Microbiol. Rev.*, **54**: 450-472.
2. Page, W. and Cornish, A., 1993, Growth of *Azotobacer vinelandii uwd* in fish peptone medium and simplified extraction of poly-β-hydroxybutyrate. *Appl. Environ. Microbiol.* **59**: 4236-4244.
3. Lemos, P. C., Viana, C., Salgueiro, E. N., Ramos, A. M., Crespo, J. P. S. G. and Reis, M. A. M., 1998, Effect of carbon source on the formation of polyhydroxyalkanoates (PHA) by a phosphate-accumulating mixed culture. *Enzyme and Microb. Technol.* **22**: 662-671.
4. Van Loosdrecht, M. C. M., Pot, M. A. and Heijen, J. J., 1997, Importance of bacterial storage polymers in bioprocesses. *Wat. Sci. Technol.* **35**: 41-47.
5. Pereira, H., Lemos, P. C., Carrondo, M. J. T., Crespo, J. P. S. G., Reis, M. A. M. and Santos, H., 1996, Model for carbon metabolism in biological phosphorus removal processes based on *in vivo* ^{13}C-NMR labelling experiments. *Water Res.* **30**: 2128-2138.
6. Mino, T., Liu, W. T., Satoh, H. and Matsuo, T., 1996, Possible metabolisms of polyphosphate accumulating (PAOs) and glycogen accumulating non-poly-P organisms (GAOs) in the enhanced biological phosphate removal process. *Med. Fac. Landbouww. Univ. Gent.* **61**: 1769-1775.
7. Wentzel, M. C., Lotter, L. H., Ekama, G. A., Loewenthal, R. E. and Marais, G. V., 1991, Evaluation of biochemical models for biological excess phosphorus removal. *Wat. Sci. Technol.* **23**: 567-576.
8. Satoh, H., Mino, T. and Matsuo, T., 1992, Uptake of organic substrates and accumulation of polyhydroxyalkanoates linked with glycolysis of intracellular carbohydrates under anaerobic conditions in the biological excess phosphate removal processes. *Wat. Sci. Technol.*, **26**: 933-942.
9. Rees, G. N., Vasilidis, G., May, J. W. and Bayly, R. C., 1992, Differentiation of polyphosphate and poly-β-hydroxybutyrate granules in an *Acinetobacter sp.* isolated from activated sludge. *Lett. Appl. Micro.*, **25**: 63-69.
10. Satoh, H., Iwamoto, Y., Mino, T. and Matsuo, T., 1998, Activated sludge as a possible source of biodegradable plastic. *Wat. Sci. Technol.*, **38**: 103-109.
11. Lee, S. Y., 1996, Bacterial polyhydroxyalkanoates. *Biotechnol. Bioeng.*, **49**: 1-14.
12. Akiyama, M., Taima, Y. and Doi, Y., 1992, Production of poly(3-hydroxyalkanoates) by a bacterium of the genus *Alcaligenes* utilising long-chain fatty acids. *Appl. Microbiol. Biotechnol.*, **37**: 698-701.
13. Shimizu, H., Tamura, S., Shioya, S. and Suga, K., 1993, Kinetic study of poly-D(-)-3-hydroxybutyric acid (PHB) production and its molecular weight distribution control in a fed-batch culture of *Alcaligenes eutrophus. J. Ferment. Bioeng.*, **76**: 465-469.
14. Taidi, B., Mansfield, D. and Anderson, A., 1995, Turnover of poly(3-hydroxybutyrate) (PHB) and its influence on the molecular mass of the polymer accumulated by *Alcaligenes eutrophus* during batch culture. *FEMS Microbiol. Lett.*, **120**: 201-206.
15. Taidi, B., Anderson, A. J., Dawes, E. A. and Byron, D., 1994, Effect of carbon source and concentration on the molecular mass of poly(3-hydroxybutyrate) produced by *Methylobacterium extorquens* and *Alcaligenes eutrophus. Appl. Environ. Microbiol.*, **40**: 786-790.
16. Hiramitsu, M., Koyama, N. and Doi, Y., 1993, Production of poly(3-hydroxybutyrare-co-4-hydroxybutyrate) by *Alcaligenes latus. Biotechnol. Lett.*, **15**: 461-464.
17. Chen, G. and Page, W., 1994, The effect of substrate on the molecular weight of poly-β-hydroxybutyrate produced by *Azotobacter vinelandii uwd. Biotechnol. Lett.* **16**: 155-160.

18. Daniel, M., Choi, J., Kim, J. and Lebeault, J., 1992, Effect of nutrient deficiency on accumulation and relative molecular weight of poly-β-hydroxybutyric acid by methylotrophic bacterium, *Pseudomonas 135. Appl. Microbiol. Biothecnol.*, **37**: 702-706.

19. Young, F., Kastner, J. and May, S., 1994, Microbial production of PHB from D-xylose and lactose by *Pseudomonas cepacia. Appl. Environ. Microbiol.*, **60**: 4195-4198.

20. Anderson, A., Williams, D., Taidi, B., Dawes, E. and Ewing, D., 1992, Studies on copolyester synthesis by *Rhodococus ruber* and factors influencing the molecular mass of polyhydroxybutyrate accumulated by *methylobacterium extorquens* and *Alcaligenes eutrophus. FEMS Microbiol. Rev.*, **103**: 93-102.

21. Beccari, M., Majone, M., Ramadori, R. and Tozzi, G., 1997, Biodegradable polymers from wastewater treatment by using selected mixed cultures. XII Convegno Div. Chim. Ind. E Cruppu Interdiv. Catalisi, SCI, Giarden, Italy, 22-25 July .

22. Majone, M., Dircks, K. and Beun, J. J., 1999, Aerobic storage under dynamic conditions in activated sludge processes. *The state of the art. Wat. Sci. Technol.*, **39**: 61-73.

23. Majone, M., Massanisso, P. and Ramadori, R., 1998, Comparison of carbon storage under aerobic and anoxic conditions. *Wat. Sci. Technol.*, **38**: 77-84.

24. Majone, M., Massanisso, P., Carucci, A., Lindrea, K. and Tandoi, V., 1996, Influence of storage on kinetic selection to control aerobic filamentous bulking. *Wat. Sci. Technol.*, **34**: 223-232.

Controlled Synthesis of Biodegradable Polyesters

PIETER J. DIJKSTRA, ZHIYUAN ZHONG, WIM M. STEVELS and JAN FEIJEN
Department of Chemical Technology and Institute of Biomedical Technology, University of Twente, P.O. Box 217, 7500 AE Enschede, The Netherlands

Abstract: The application of new initiators for the ring-opening polymerization of cyclic esters has markedly improved the macromolecular engineering of biodegradable polyesters. A wide variety of complex macromolecular architectures can nowadays be prepared using these initiators. Decisive factors in achieving this are the range of monomers that can be polymerized, and the favourable rates of polymerization obtained for these monomers. The number of papers dealing with the exciting synthetic aspects of these initiators is rapidly increasing. A historic overview of the developments in this area is presented in this paper.

1. INTRODUCTION

Synthetic polymer chemistry has evolved over the past decades to the point where both structure and properties can be controlled very accurately. This holds for bulk as well as for specialty polymers. The properties of polymeric materials are primarily determined by their chemical structure.

Considerable research effort has been directed towards the synthesis of macromolecules with complex structures. Examples include block copolymers, graft copolymers, star copolymers, dendrimers and other macromolecular architectures. The combination of advanced polymerization methods and efficient coupling methods has resulted in a broad array of chemical structures that can be prepared nowadays. In order to obtain a high molecular weight product, reactions are repeated many times, and thus a high selectivity, is required. The functionalities used in coupling methods

Biorelated Polymers: Sustainable Polymer Science and Technology
Edited by Chiellini *et al.*, Kluwer Academic/Plenum Publishers, 2001

need to have a high reactivity, because the concentration of reacting groups is usually low.

Synthesis of polymers with special features that are *stable* for extended periods of time is a challenge, but the synthesis of polymers with special features that are *intentionally unstable* for some purpose is even a bigger challenge.

2.   DEGRADABLE POLYMERS

A degradable polymer is a polymer designed to undergo a significant change in its chemical structure under specific environmental conditions resulting in a loss of some properties[1]. An important class of degradable materials are naturally occurring biodegradable polymers, such as proteins and polysaccharides. These are polymers which are degraded hydrolytically, by the action of enzymes, or by microorganisms such as bacteria, fungi and algae. A second class of degradable materials are synthetic biodegradable polymers. These materials that are degraded hydrolytically with or without the action of enzymes have been designed for biomedical applications and thus intended to degrade *in vivo*. Despite the large number of biodegradable polymers synthesized, only a limited number of polymers has been commercialized or are in the pre-marketing phase. Considerable progress is currently made in the design and manufacturing of synthetic biodegradable polymers for environmental applications. These materials have to compete for their price with synthetic materials like poly (ethylene) and commercialization of these polymers can be expected in the near future. Materials for biomedical and environmental applications have been summarized in Table 1.

Table 1. Commercially available biodegradable synthetic polymers

Polymer	Producer(s)
Poly(ε-caprolactone) and copolymers	Union Carbide, Solvay
Poly(lactic acid) and copolymers	Cargill, Mitsui Toatsu, PURAC, Neste Oy, Boehringer
Poly(glycolic acid) and copolymers	Boehringer, Ethicon
(Co)polyesters derived from poly(butylene succinate)	Showa (Bionolle)
Co-poly(amide-ester) derived from nylon-6	Bayer (BAK1095)
Aromatic-aliphatic copolyesters derived from poly(butylene terephtalate)	BASF, Eastman Chemical
Poly(propylene terephalate)	Du Pont

Blends or copolymers are often prepared to alter the properties of the homopolymers. In many applications, fillers, stabilizers, pigments, or other additives are added to the polymer, in which sometimes residues of chain extenders, catalysts and solvents are still present. The ultimate biodegradability of a product can be influenced by all these factors.

When *in vivo* use of a degradable polymer is considered, the presence of such impurities constitute a major concern. It must be shown that all contaminations can and are completely removed or approval of the medical authorities for their presence must be obtained. If at all possible, it is prohibitively expensive and time-consuming in most cases and one is usually restricted to conventional synthetic routes or to suppliers of biomedical grade polymers. For general purpose biodegradable polymers, the requirements are less strict and the presence of more substances can be tolerated. The market volume of these, e.g. in packaging, is potentially much higher than for biomedical applications, creating opportunities for plastic manufacturers. This is an important driving force for the development of new materials, production processes and synthetic methods.

3.   LIVING POLYMERIZATIONS

The synthesis of special macromolecular structures imposes severe restrictions on the polymerization process. Ultimately, the synthesis of polymers with control of molecular weight, molecular weight distribution and end group identity can only be achieved by a polymerization process which has a high selectivity in initiation and propagation and termination. Also, the relative rates of these processes must be favorable. If these conditions are fulfilled, the polymerization is called a living polymerization. The following practical definition of a living polymerization originates from Swarc:[2] "Polymerizations can be considered living if their end-groups retain the propensity of growth for at least as long a period as needed for the completion of an intended synthesis, or any other desired task." Molecular weights are determined by the ratio of monomer to initiator. The molecular weight distributions are small, ideally Poisson distributions with a polydispersity of (approximately) $M_w/M_n = 1+ 1/$degree of polymerization. Living polymerizations are almost invariable carried out in solution to allow polymerization to proceed in a homogeneous environment at low temperature.

Many examples of living polymerizations are now available. Commonly they are classified by the nature of the propagating centre or polymerization mechanism. An overview of the most prominent living polymerizations is given in Table 2.

In many of these polymerizations, livingness is only obtained after addition of a co-catalyst, electron donor(s) and/or common salt(s). The field of living radical polymerization has only recently been discovered and many improvements can still be expected, as well as a broadening of the scope of suitable monomers and initiators. Furthermore, it should be noted that there are examples of living polymerizations for which the requirements of livingness are fulfilled only by a very limited number of monomers or catalysts, or by specific combinations of monomer and catalyst. Examples are group-transfer polymerization of methacrylates[9] and ring-opening metathesis polymerization of strained cyclic olefins[10].

Table 2. Typical living polymerizations

Anionic polymerization:[2-4]	
Initiators:	alkaline earth alkyls, amides and alcoholates radical anions
Monomers:	styrenes, butadiene, isoprene, vinylpyridines, methacrylates, ethylene oxide
Cationic polymerization:[5-7]	
Initiators:	proton acids (HI, HClO$_4$, HBF$_4$, CF$_3$SO$_3$H)
	Lewis acids/co-catalyst combinations (TiCl$_4$, AlCl$_3$, BF$_3$/H$_2$O)
	stable carbenium cations generated by the inifer method
Monomers:	1,1-substituted alkenes, styrenes, vinyl-ethers, N-vinylcarbazoles, oxazolines
Radical polymerization:[8]	
Initiators:	nitroso radical/organic peroxide or /aluminium alkyl/Lewis base
Monomers:	styrenes, vinyl acetate, methacrylates

4.   POLYESTERS

The large scale production of the polyesters is usually by *poly(condensation)* in bulk. Block copolymers can not be prepared by this synthetic method and control of molecular weight, molecular weight distribution and end group identity is difficult to achieve. Side reactions frequently occur, because poly(condensation) is carried out at elevated temperatures.

In some cases polyesters can be prepared by *ring-opening polymerization*, instead of polycondensation. This is a method which is in principle suitable for macromolecular engineering. Hydroxy acids can react by condensation intermolecularly to give linear polymer (Eq. 1), but alternatively can react intramolecularly to give a cyclic product. This can be a cyclic oligomer, or the cyclic ester of the parent hydroxy acid (Eq. 2). Such a molecule is called a lactone, which by consecutive ring-opening reactions can in principle be polymerized to give linear polymer (Eq. 3).

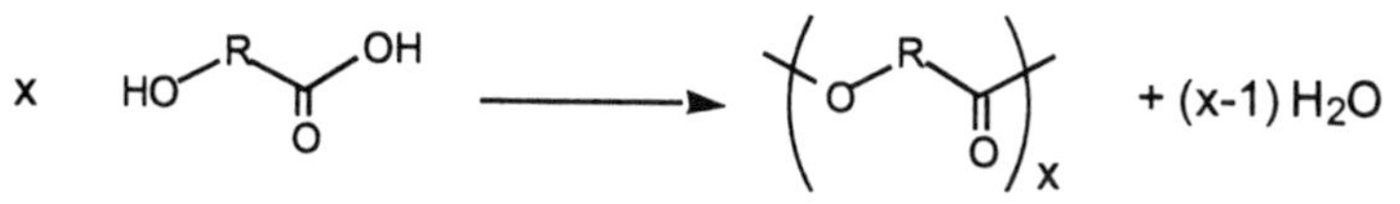

$$x \quad \text{HO} \overset{R}{\underset{O}{\diagup}} \text{OH} \longrightarrow \left(O \overset{R}{\underset{O}{\diagup}} \right)_x + (x-1)\,H_2O \quad \textbf{(1)}$$

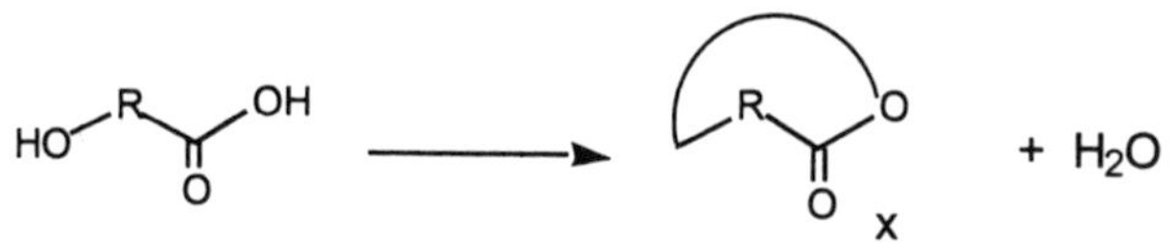

$$\text{HO} \overset{R}{\underset{O}{\diagup}} \text{OH} \longrightarrow \underset{x}{\diagup} + H_2O \quad \textbf{(2)}$$

$$x \quad \diagup \longrightarrow \left(O \overset{R}{\underset{O}{\diagup}} \right)_x \quad \textbf{(3)}$$

The lactones most frequently encountered in literature related to polymerizations are lactide, glycolide and, ε-caprolactone the first two being dilactones (Fig 1). Although polymerizations of substituted four-membered ring lactones are also of some importance, the resulting polymers are more conveniently prepared by microbiological methods.

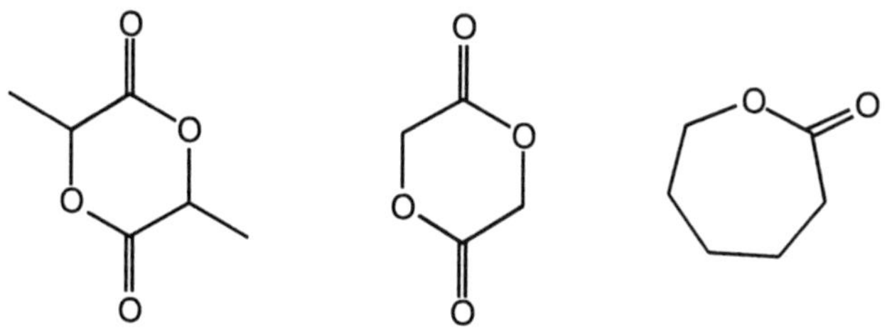

Figure 1. Stuctures of lactone monomers

ε-Caprolactone and L-lactide are medium strained cyclic esters and the thermodynamic parameters for polymerization of these have been determined and are listed in Table 3.

Table 3. Polymerizability of lactone monomers

	state[a]	T °C	-ΔH kJ/mol	-ΔS J/K.mol	ΔG(100 °C) kJ/mol	refs.
β-propiolactone	lc	127	74	51	-55	11
γ-butyrolactone	lc	77	-6	29	17	11
δ-valerolactone	lc	77	10	13	-5	11
ε-caprolactone	lc	77	14	8	-11	11
DL-lactide	lc	127	27	13	-22	12
LL-lactide	ss	100	23	25	-9	13

a) state refers to the state of respectively monomer and polymer: c = condensed amorphous state; l = liquid; s = solution

Poly(lactide) is a polymer which is known to have excellent properties for use as a biomedical material.[14] *Random* copolymers of lactide and other monomers such as glycolide are widely applied.[15]. Polymers based on L-lactide and ε-caprolactone are interesting materials for use in biomedical applications such as drug delivery devices. In the following sections attention is focussed on lactide and ε-caprolactone, because they can be applied in macromolecular engineering. For glycolide this is difficult due to the very limited solubility of monomer and polymer in acceptable solvents.

4.1   Tin compounds

In the vast majority of articles dealing with poly(lactone) synthesis, stannous di(ethyl-hexanoate), Sn(Oct)$_2$, is used for bulk polymerization of lactones. The mechanism of this reaction has been the centre of considerable controversy.[16-19] Polymerization in the presence of hydroxyl groups is nowadays proposed to occur by the reactions outlined in Eq. 4-6.

$$\text{(4)}$$

$$\text{(5)}$$

$$\text{(6)}$$

L = 2-ethyl-hexanoate

Initially, hydroxyl functionalities coordinate to stannous octoate. In the next step a lactone coordinates. Both the lactone ester and hydroxyl group are now activated for a nucleophilic attack of the latter on the ester. Polymerization has now been initiated and propagation proceeds by repetitions of the reaction sequence. Whether and how lactones are polymerized by stannous octoate in the absence of impurities remains unsolved, and hypothetical.

Typically, polymers with molecular weight distributions of 1.7 are isolated. Control over end group identity and molecular weights can be obtained up to M_n 's of $30 \cdot 10^3$ by the addition of alcohols acting as initiators or chain transfer agents. At higher molecular weights, initiation by impurities becomes important. High molecular weight poly(lactides) ($M_n >$ 100.000) are often prepared without co-initiators and at very low stannous octoate concentrations, indicating the presence of impurities in the stannous octoate. Also the poor reproducibility of synthesises carried out in this way is a further indication that these polymerizations are not really under control. Room temperature polymerization of L-lactide in solution can be performed, but is too slow for practical purposes. Well controlled synthesis of block copolymers is difficult to achieve, due to the higher temperatures used in polymerizations. However, some useful tapered block copolymers of ε-caprolactone and L-lactide were prepared using $Sn(Oct)_2$.[20]

4.2 Aluminium alkoxides

There is an extensive literature on ε-caprolactone polymerization using homogeneous aluminium alkoxides. Most publications deal with the reactivity of aluminium isopropoxide $Al(O^iPr)_3$. Although the earliest report on this polymerization dates back to 1975[21], it took twenty years before a proper understanding of the reactivity of this compound was gained[22].

Already from the first study it was apparent that the polymerization of ε-caprolactone initiated by $Al(O^iPr)_3$ is first order in monomer and that polymerization proceeds by alkoxide attack on the acyl carbon atom, followed by acyl oxygen cleavage (Eq. 7)[21].

$$Al(O^iPr)_3 \;+\; \text{[R} \rangle\text{O]} \longrightarrow {}^iPrO\overset{O}{\overset{\|}{C}}\text{-R-OAl}\!\!< \qquad (7)$$

$$R = (CH_2)_5$$

Polymers with isopropoxy carbonyl and hydroxyl groups were isolated after quenching the reactions with some proton donor. The kinetic order in initiator was less straightforward to determine, and also "the number of active alkoxide groups per aluminium" was a point of controversy[23, 24].

It has long been known that in solution, $Al(O^iPr)_3$ exists in two different aggregated forms, a trimer and a tetramer (Eq. 8)[25].

$$(Al(O^iPr)_3)_3 \qquad \overset{\longrightarrow}{\longleftarrow} \qquad (Al(O^iPr)_3)_4 \qquad (8)$$

However, it was only recently recognized that these species are in slow equilibrium and have an entirely different reactivity in initiating ε-caprolactone polymerization[26]. The trimer is much more reactive than the tetramer. Thus, when a mixture of trimer and tetramer is used to initiate the polymerization of ε-caprolactone, the trimer reacts entirely with conversion of all its isopropoxy groups into polymer chain ends, whereas the tetramer remains unreacted in time needed to complete ε-caprolactone conversion. If pure tetramer is used as initiator, a long induction period is observed and only part of the tetramers eventually participate in the polymerization.

Another class of aluminium compounds used in ε-caprolactone polymerization are aluminium porphyrins. In two studies the use of (5,10,15,20-tetraphenylporphinato)-aluminium alkoxides (TPPAlOR) was evaluated[27, 28]. Living polymerization of ε-caprolactone and δ-valerolactone could be achieved. It was established that a major improvement in polymerization rate was obtained if an inert Lewis acid was added, e.g. TPPAlCl[29]. The polymerization proceeds then by the mechanism shown in Eq. 9-10.

$$\text{TPPAlCl} + \underset{O}{\overset{R}{\bigcirc}}O \longrightarrow \text{TPPAl}\underset{O}{\overset{Cl}{\diagdown}}\underset{O}{\diagup}R \qquad (9)$$

$$R = (CH_2)_5$$

$$\underset{\text{TPPAlOMe}}{\text{TPPAl}}\overset{Cl}{\diagdown}R \longrightarrow \text{TPPAlOR}\overset{O}{\overset{\|}{C}}\text{OMe} + \text{TPPAlCl} \quad (10)$$

$$R = (CH_2)_5$$

This method of monomer activation later proved of prime importance in the polymerization of other monomers as well, such as methylmethacrylate[30,31] and methacrylonitrile[32]. Living chain transfer in the presence of alcohols is observed with the aluminium porphyrins, as with other aluminium alkoxide systems. A general drawback of the use of aluminium porphyrins is their difficult preparation and their strong colour difficult to remove from the polymer. Sometimes, activities are very low due to the large steric bulk generated by the porphyrin ligand and/or light irradiation is necessary to keep polymerization going.

Aluminium isopropoxide was used to initiate L- and D,L-lactide homopolymerization at 70 °C[33]. All alkoxide groups were found to initiate polymerization. The presence of the different aggregates does not influence the polymerization to a large extent as in the case of ε-caprolactone, probably because reactions are carried out at higher temperatures and the reactivity of lactide in polymerization is lower than that of ε-caprolactone.

Again, tri(alkoxide)s are more active than alkyl aluminium alkoxides. Even at the elevated temperatures used the propagation rate constants for L-lactide polymerization are low compared to ε-caprolactone polymerization using similar initiators. As for ε-caprolactone, TPPAlOR is an effective initiator for L-lactide polymerization[34]

Macromolecular engineering with aluminium alkoxides.
A first group of specialty polymers prepared by aluminium alkoxides are end-functionalized polylactones. Many examples are available, which can be divided in two groups. A first method is to prepare a functionalized polymer by using a functionalized aluminium alkoxide (Eq. 11).

$$\text{>A1OX} + x \;\;\bigodot_{R}^{O} \longrightarrow \text{XO}\left(\!\!\begin{array}{c}O\\\|\\C-R-O\end{array}\!\!\right)_x\!\!\text{-Al<} \tag{11}$$

The second method is to prepare a functionalized polymer by reacting some reagent with the living aluminium alkoxide, with the terminal hydroxyl group generated after quenching the polymerization with a proton donor, or with the functionality incorporated by the first method (Eq. 12-14).

$$\text{R'O}\left(\!\!\begin{array}{c}O\\\|\\C-R-O\end{array}\!\!\right)_x\!\!\text{-Al<} + X \longrightarrow \text{R'O}\left(\!\!\begin{array}{c}O\\\|\\C-R-O\end{array}\!\!\right)_x\!\!\text{-X} \tag{12}$$

$$\text{R'O}\left(\!\!\begin{array}{c}O\\\|\\C-R-O\end{array}\!\!\right)_x\!\!\text{-H} + X \longrightarrow \text{R'O}\left(\!\!\begin{array}{c}O\\\|\\C-R-O\end{array}\!\!\right)_x\!\!\text{-X} \tag{13}$$

$$\text{XO}\left(\!\!\begin{array}{c}O\\\|\\C-R-O\end{array}\!\!\right)_x\!\!\text{-H} + X' \longrightarrow \text{X'X}\left(\!\!\begin{array}{c}O\\\|\\C-R-O\end{array}\!\!\right)_x\!\!\text{-H} \tag{14}$$

Three main groups of functionalized polymers can be identified, independent of the preparation method. Vinyl- or methacroyl-functionalized macromonomers have been prepared for copolymerization with other vinyl monomers or for cross-linking reactions[35-38] or for coupling by hydrosilylation[39]. Amine terminated polyesters have been prepared for coupling with other functionalities[40, 41] or possibly as macroinitiators for N-carboxy anhydride (NCA) polymerization, to obtain block copolyester-amino acids[42]. Also, functionalities can be introduced which are easily detected by spectroscopic techniques (NMR, UV), allowing molecular weight determination by end group analysis[43]. Coupling reactions can be used to prepare special structural features, such as α,ω-functionalized- and star-shaped polymers[44,45]. In a recent review the chemistry and possibilities of alumininum alkoxides in macromolecular engineering has been described[46].

4.3   Compounds of the rare earth elements

The rare earth elements here referred to include the group IIIB metals (Sc, Y and La) and the lanthanide elements (Ce to Lu) and are sometimes loosely referred to as lanthanide elements, as the main characteristics and reactivity of these elements are similar in many cases.

The first examples of lactone polymerization were given in a Du Pont patent entitled "yttrium and rare earth compounds catalyzed lactone polymerization" written by McLain and Drysdale, filed in 1989 and assigned in 1991[47]. In two first publications, based on this patent, the use of yttrium alkoxides in ε-caprolactone and lactide polymerization was described and some first examples of block copolymerization were given. In general, the activity of these catalysts appeared much higher than determined for aluminium alkoxides, especially in lactide polymerization, without extensive macrocycle formation, transesterification or racemization as in anionic polymerization. No examples of melt polymerization were given. Control over molecular weight, end group identity and molecular weight distribution was also obtained. Initially, commercially available yttrium isopropoxide was used, but it appeared that this initiator has a very low activity in lactide polymerization.

Thus, although the yttrium isopropoxide certainly has its merits, the limited availability of different yttrium alkoxides also limits the array of macromolecular structures that can be prepared. An approach to more flexible systems has been developed, in which the active initiator is generated in situ[48] The concept is based on sterically very crowded tri(2,6-di-t-butylphenoxy) lanthanides that can not initiate lactone polymerization, but do exchange with sterically less crowded alcohols to generate alkoxides that turn into active initiators for e.g. lactide, ε-caprolactone or δ-valerolactone polymerization. In this way, any alcohol can effectively be used to functionalize a polyester chain as shown in Fig 2 (vide supra). In a similar way the in situ formation of such a catalyst/initiator system from the commercially available [tris(hexamethyldisilyl)amide]Yttrium with isopropanol can be performed[49].

4.4   Compounds of miscellaneous elements

Solution polymerizations of ε-caprolactone and lactide have been carried out using other homogeneous alkoxides, notably Zn, Mg, Ti and Zr alkoxides.50 Interesting examples include living polymerization of ε-caprolactone by zinc alkoxides[51], monocyclo-pentadienyl titanium complexes[52,53] and cationic zirconium and hafnium complexes of the Cp2MMe+ B(C6F5)4- type (M = Zr, Hf; Cp = cyclopentadienyl, C_5H_5).[54] Lactide polymerization using Mg compounds has been reported[55], including one with a very defined ligand environment, of tri[(3-*tert*-butyl)pyrazolyl]borate magnesium ethoxide[56].

4.5   Calcium alkoxides

It is highly preferable to use nontoxic catalysts in the synthesis of polyesters intended for biomedical and pharmaceutical applications, since complete removal of catalyst residues from the polymer is often impossible.

Like aluminium compounds calcium-based catalysts may serve this purpose. However, calcium compounds, such as calcium oxide, carbonate and carboxylate, revealed a low catalytic activity towards the ring-opening polymerization of L-lactide in the bulk at temperatures ranging from 120 to 180°C.[57,58] Calcium hydride appeared more effective and has been used to prepare PLA/PEO/PLA triblock copolymers, but racemization was observed.[59,60] Polyglycolide homopolymers of high molecular weights and random copolymers of glycolide and ε-caprolactone were synthesized using calcium acetylacetonate as a catalyst, but the molecular weight of the obtained polymers could not be controlled.[61] The calcium compounds so far investigated do not allow the macromolecular engineering of polyesters.

Based on the concept described above a novel and efficient calcium alkoxide initiating system, generated *in situ* from bis(tetrahydrofuran)calcium, bis[bis(trimethylsilyl)amide] and an alcohol, for the ring-opening polymerization of cyclic esters has been developed (Fig 2). The solution polymerization in THF using mild conditions is living, yielding polyesters of controlled molecular weight and tailored macromolecular architecture. The polymerizations initiated with the 2-propanol-Ca[N(SiMe₃)₂]₂(THF)₂ system are first-order in monomer with no induction period. At high 2-propanol/Ca[N(SiMe₃)₂]₂(THF)₂ ratios, complete conversion of the 2-propanol occurs due to the fast and reversible transfer between dormant and active species.

Initiation:

$$Ca(N(SiMe_3)_2)_2 + 2\ ROH \longrightarrow Ca(OR)_2 + 2\ HN(SiMe_3)_2$$

Propagation:

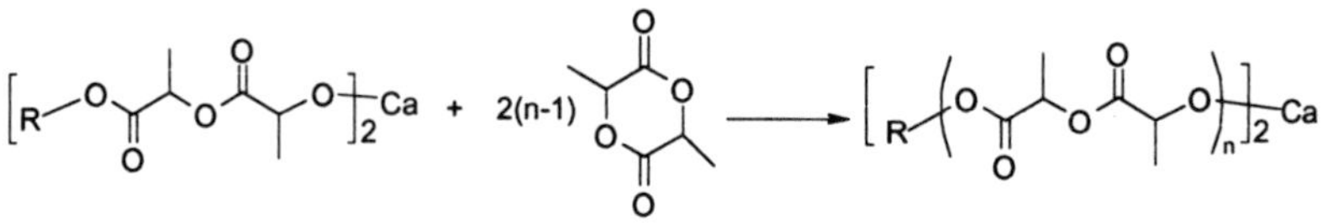

Termination:

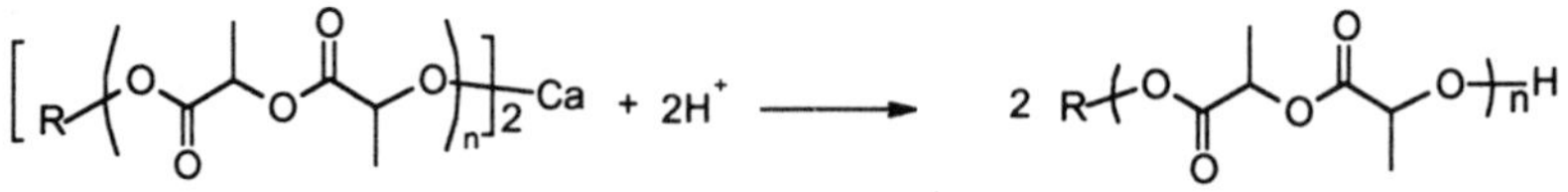

Reversible transfer:

Figure 2. In situ generation of a calcium alkoxide from bis(tetrahydrofuran)calcium, bis[bis(trimethylsilyl)amide] and an alcohol, for the ring-opening polymerization of cyclic esters

4.6   Perspectives

The range of polymers prepared using lanthanides and earth alkaline alkoxides can be expected to increase, especially bearing in mind that the polymerization activity of these alkoxides is not restricted to the polymerization of lactone-type monomers and that monomers such as cyclic anhydrides, carbonates, oxides and others can be (co)polymerized satisfactorily. Furthermore, a broader range of ligand environments of the metal elements must be evaluated for a full appreciation of the potential of these initiators. Recent examples on the stereoselective polymerization of lactides reveal the possibilities of these new catalyst-initiators in macromolecular engineering. Eventually, it may be expected that for every specific polymer an optimal initiator will be available.

REFERENCES

1. Stapert, H.R., Dijkstra, P.J., and Feijen, J., 1996, *Polymer Products and Waste Management*, (Smits, M. ed.), International Books: Utrecht, The Netherlands, pp 71-106
2. Swarc, M., and Beylen, van M., 1993, *Ionic Polymerization and Living Polymers*, Chapman & Hall, New York
3. Hsieh, H.L., and Quirk, R.P, 1996, *Anionic Polymerization: Principles and Practical Applications*, Marcel Dekker, New York
4. A number of reviews by various authors can be found 1989, *in Comprehensive Polymer Science*, (Eastmond, G.C., Ledwith, A., Russo, S., Sigwalt, P. eds), Pergamon Press: New York, Vol. 3, pp. 365-578
5. Kennedy, J.P., and Ivan, B. 1991, *Designed Polymers by Carbocationic Macromolecular Engineering*, Oxford University Press: New York
6. Matyjaszewski, K. 1996, *Cationic Polymerizations: Mechanism, Synthesis and Applications*, Marcel Dekker: New York
7. A number of reviews by various authors can be found 1989, In *Comprehensive Polymer Science*, (Eastmond, G.C., Ledwith, A., Russo, S., Sigwalt, P.eds), Pergamon Press: New York, pp. 579-851
8. Veregin, R.P.N., Odell, P.G., Michalak, L.M., and Georges, M.K., 1996, Molecular Weight Distributions in Nitroxide-Mediated Living Free Radical Polymerization: Kinetics of the Slow Equilibria between Growing and Dormant Chains, *Macromolecules*, **29**: 3346
9. Sogah, D.Y., Hertler, W.R., Webster, O.W., and Cohen, G.M., 1987, Group Transer Polymerization. Polymerization of Acrylic Monomers, *Macromolecules*, **20**: 1473
10. Grubbs, R.H., and Tumas, W., Polymer Synthesis and Organotranstion Metal Chemistry. 1989, *Science*, **243**: 907
11. Lebedev, B.V., and Estropov, A.A., 1982, Thermodynamics of the Polymerization of Lactones. *Dokl. Phys. Chem.*, **264**: 334
12. Lebedev, B.V., Estropov, A., Kiparisova, G., and Belov, V.I., Thermodynamics of the Glycolide, Polyglycolide and Glycolide Polymerization process in the 0-550 °K.1978 *Vysokomol. Soedin.*, **A20**: 1297
13. Duda, A., and Penczek, S., 1990, Thermodynamics of L-Lactide Polymerization. Equilibrium Monomer Concentration. *Macromolecules*, 23: 1636
14. Vert, M., Schwarch, G., and Coudane, J., 1995, *J. Mater. Sci., Pure Appl. Chem.*, **A32**: 787
15. Vert, M., Li, S.M., Spenlehauer, G., and Guerin, P., Bioresotbability and Biocompatability of Aliphatc Polyesters. 1992, *J. Mater. Sci.: Mater. in Med.*, 3: 432
16. Kricheldorf, H.R., and Kreiser-Saunders, I., Polylactones 31. Sn (II)octoate-initiated Polymerization of L-Lactide. A Mechanistic Study. 1995, Boettcher, *C. Polymer*, **36**: 1253
17. Veld, in't P.J.A., Velner, E.M., Witte, van P., Hamhuis, J., Dijkstra, P.J., and Feijen, J., 1997, Melt Block Copolymerization of ε-Caprolactone and L-Lactide. *J. Pol. Sci. A: Pol. Chem.*, **A 35**: 219
18. Zhang, X., Macdonald, D.A., and Goossen, M.F.A., Mcauley, K.B. Mechanism of Lactide Polymerization in the Presence of Stannous octoate: The Effect of Hydroxy and Carboxylic Acid Substances. 1994, *J. Pol. Sci. A: Pol. Chem.*, **32**: 2965
19. Nijenhuis, A.J., Grijpma, D.W., and Pennings, A.J., 1992, Electrochemical and Chemical Syntheses of Poly(thiophenes) Containing Oligo(oxyethylene) Substituents, *Macromolecules*, **25**: 6419
20. Grijpma, D.W., and Pennings, A.J., Polymerization Temperature Effects on the Properties of L-Lactide and ε-Caprolactone copolymers. 1991, *Pol. Bull.*, **25**: 335

21. Ouhadi, T., Stevens, C., and Teyssié, P., Mechanism of ε-Caprolactone Polymerization of Aluminum Alkoxides. 1975, *Makromol. Chem., Suppl.* **1**: 191

22. Duda, A., and Penczek, S., Of the Difference of Reactivities of Various Aggregated Forms of Aluminum Triisopropoxide in Initiating Ring-Opening Polymerizations. 1995, *Macromol. Rapid Commun.*, **16**: 67

23. Jacobs, C., Dubois, P., Jérôme, R., and Teyssié, P, 1991, Macromolecular Engineering of Polylactones and Polylactides. 5. Synthesis and Characterization of Diblock Copolymers based on Poly-ε-caprolactone and Poly(l,l or dl)lactide by Aluminium Alkoxides. *Macromolecules*, **24**: 3027

24. Kricheldorf, H.R., Berl, M., and Scharnagl, N., 1988, Poly(lactones). 9. Polymerization Mechanism of Metal Alkoc\xide Initiated Polymerizations of Lactide and Various Lactones. *Macromolecules*, **21**: 286

25. Shiner, V.J., Whittaker, D., and Fernandez, V.P., 1963, *J. Am. Chem. Soc.*, **85**: 2318.

26. Duda, A., and Penczek, S, 1995, Polymerization of ε-Caprolactone Initiated by Aluminum Isopropoxide Trimer and-or Tetramer. *Macromolecules*, **28**, 5981

27. Yasuda, T., Aida, T., and Inoue, S,. Reactivity of (Porphinato)Aluminum Phenoxide and Alkoxide as Active Initiators for Polymerization of Epoxide and Lactone. 1986, *Bull. Chem. Soc. Jpn.*, **59**: 3931

28. Endo, M., Aida, T., and Inoue, S., 1987, "Immortal" Polymerization of ε-Caprolactone Initiated by Aluminium Pophyrin in the presence of Alcohol. *Macromolecules*, **20**: 2982

29. Shimasaki, K., Aida, T., and Inoue, S., 1987, Living Polymerization of δ-Valerolactone with Aluminium Porphyrin. Trimolecular Mechanism by the Participation of Two Aluminium Porphyrin Molecules. *Macromolecules*, **20**: 3076

30. Sugimoto, H., Aida, T., and Inoue, S., 1993, Organoboron Compounds as Lewis Acid Accelerators for the Aluminum Porphyrin-Mediated Living Anionic Polymerization of Methyl Methacrylate. *Macromolecules*, **26**: 4751

31. Sugimoto, H., Aida, T., and Inoue, S,. 1994, Lewis Acid-Promoted Living Anionic Polymerization of Alkyl Methacrylates Initiated with Aluminum Porphyrins. Importance of Steric Balance between a Nucleophile and a Lewis Acid, *Macromolecules*, **27**: 3672

32. Sugimoto, H., Saika, M., Hosokawa, Y., Aida, T., and Inoue, S., 1996, Accelerated Living Polymerization of Methacrylonitrile with Aluminum Porphyrin Initiators by Activation of Monomer or Growing Species. Controlled Synthesis and Properties of Poly(methyl methacrylate-b-methacrylonitrile)s, *Macromolecules*, **29**: 3359

33. Dubois, P., Jacobs, C., Jérôme, R., and Teyssié, P., 1991, Macromolecular engineering of Polylactones and Polylactides. 4. Machanism and Kinetics of Lactide Homopolymerization by Aluminium Isopropoxide. *Macromolecules*, **24**: 226634. Tromifoff, L., Aida, T., and Inoue, S., Formation of Poly(lactide) with controlled Molecular Weight. Polymerization of Lactide by Aluminum Porhyrin. 1987, *Chem. Lett.*, **991**: pp

35. Dubois, P., Jérôme, R., and Teyssié, P., 1991, Macromolecular Engineering of Polylactones and Polylactides. 3. Synthesis, Characterizaion, and Application of Poly(ε-caprolactone) Macromonomers. *Macromolecules*, **24**: 977

36. Barakat, I., Dubois, P., Jérôme, R., Teyssié, P., and Goethals, E., Macromolecular Engineering of Polyesters and Polylactones XV. Poly(D,L-lactideMacromonomers and Precursors of Biocompatible Graft Copolymer and Bioerodible Gels. 1994, *J. Pol. Sci.: A: Pol. Chem.*, **32**: 2099

37. Barakat, I., Dubois, P., Grandfils, C., and Jérome, R., Macromolecular Engineering of Polyesters and Polylactones XXI. Controlled Synthesis of Low Molecular Weight Polylactide Macromonomers. 1996, *J. Pol. Sci.: A: Pol. Chem.*, **34**: 497

38. Sawhney, A.S., Pathak, C.P., and Hubell, J.A., 1993, Bioerodible Hydrogels Based on Photopolymerized Poly(ethylene glycol)-co-poly(a-hydroxy acid) Diacrylate Macromers. *Macromolecules*, **26**: 581

39. Bachari, A., Bélorgy, G., Hélary, G., and Sauvet, G., Synthesis and Characterization of Multiblock Copolymers poly[poly(L-lactide)-block-poly-dimethylsiloxane]. 1995, *Macromol. Chem. Phys.*, **196**: 411

40. Degée, P., Dubois, P., Jérôme, R., and Teyssié, P., 1992, Macromolecular Engineering of Polylactones and Polylactides. 9. Synthesis, Characterization, and Application of w-Primary Amine Poly(ε-caprolactone), *Macromolecules*, 25, 4242

41. Stassen, S., Archambeau, S., Dubois, P., Jérome, R., and Teyssié, P., Macromolecular Engineering of Polyesters and Polylactones XVI. On the Way to the Synthesis of w-Aliphatic Primary Amine Poly(ε-caprolactone) and Polylactides. 1994, *J. Pol. Sci.: A: Pol. Chem.*, **32**: 2443

42. Gotsche, M., Keul, H., and Höcker, H., Amino-Terminated poly(L-lactide) as Initiators for the Polymerization of N-carboxyanhydrides: Synthesis of Poly(L-lactide)-block-Poly(α-Amino Acid). 1995, *Macromol. Chem. Phys.*, **196**: 3891

43. Sosnowski, S., Slomkowski, S., Penczek, S., and Florjanczyk, Z., Telechelic Poly(ε-caprolactone Terminated at Both Ends with OH Groups and its Derivatization. 1991, *Makromol. Chem.*, **192**: 1457

44. Dubois, P., Zhang, J.X., Jérôme, R., and Teyssié, P., Macromolecular Engineering of Polyesters and Polylactones: 13. Synthesis of Telechelic Polyesters by Coupling Reactions. 1994, *Polymer*, **35**: 4998

45. Dong, T., Dubois, P., Jérôme, R., and Teyssié, P., 1994, Macromolecular Engineering of Polylactones and Polylactides. 18. Synthesis of Star-Branched Aliphatic Polyesters Bearing Various Functional End Groups. *Macromolecules*, **27**: 4134

46. Mecerreyes, D., P., Jérôme, R., and Dubois, P., Novel Macromolecular Architectures Based on Aliphatic Polyesters: Relevance of the "Coordination-Insertion" Ring-Opening Polymerization. 1999, *Advances in Polymer Science*, **147**: 1

47. McLain, S.J., and Drysdale, N.E., 1991, U.S. Patent, No. 5.028.667

48. Stevels, W.M., Ankoné, M.J.K., Dijkstra, P.J., and Feijen, J., 1996, Kinetics and Mechanism of ε-Caprolactone Polymerization Using Yttrium Alkoxides as Initiators, *Macromolecules*, **29**: 8296

49. Martin, E., Dubois, P., and Jerome, R., 2000, Controlled Ring-Opening Polymerization of ε-Caprolactone Promoted by "in Situ" Formed Yttrium Alkoxides, *Macromolecules*, **33**: 1530

50. Kricheldorf, H.R., Berl, M., and Scharnagl, N., 1988, Poly(lactones). 9. Polymerization Mechanism of Metal Alkoxide Initiated Polymerizations of Lactide and Various Lactones. *Macromolecules*, **21**: 286

51. Barakat, I., Dubois, P., Jérome, R., and Teyssié, P., 1991, Living Polymerization and Selective End Functionalization of ε-Caprolactone using Zinc Alkoxides as Initiators. *Macromolecules* **24**: 6542

52. Okuda, J., and Rushkin, I.L., 1993, Mono(cyclopentadienyl)titanium Complexes as Initiators for the Living Ring-Opening Polymerization of ε-Caprolactone. *Macromolecules* **26**: 5530

53. Okuda, J., Konig, P., Rushkin, I.L., Kang, H-.C., and Massa, W., 1995, Indenyl effect in d0-transition metal complexes: synthesis, molecular structure and lactone polymerization activity of (Ti(h5-C9H7)Cl2(OMe)). *J. Organomet. Chem.* **501**: 37

54. Mukaiyama, M., Hayakawa, M., Oouchi, K., Mitani, M., and Yamada, T., Preparation of Narrow Polydispersity Polycaprolactone Catalyzed by Cationic Zirconocene Complexes. 1995, *Chem. Lett.* **737**: pp

55. Kricheldorf, H.R., and Lee, S-.R., Polylactones 32. High-Molecular-Weight Polylactides by Ring-Opening Polymerization with Dibutyl Magnesium or Butylmagnesium Chloride. 1995, *Polymer* **36**: 2995

56. Chisholm, M.H., and Eilerts, N.W., Single Site Metal Alkoxide Catalysts for Ring-Opening Polymerizations. Poly(dilactide) Synthesis Employing {HB93-Bu^{t}pz)$_3$}Mg(Oet). 1996, *J. Chem. Soc., Chem. Commun.* 853

57. Wood, R. J., Suter, P.M., and Russell, R.M., 1995, Mineral requirements of elderly people. *Am. J. Clin. Nutr.* **62**: 493

58. Turnlund, J.R., Betschart, A.A., Liebman, M., Kretsch, M.J., and Sauberlich, H.E., 1992, Vitamin B-6 depletion followed by repletion with animal- or plant-source diets and calcium and magnesium metabolism in young women. *Am. J. Clin. Nutr.* **56**: 905

59. Koo, W.W.K., and Tsang, R C., 1991, Mineral Requirements Of Low-Birth-Weight Infants. *J. Am. Coll. Nutr.* **10**: 474

60. Li, S.M., Rashkov, I., Espartero, J.L., Manolova, N., and Vert, M., 1996, Synthesis, Characterization, and Hydrolytic Degradation of PLA-PEO-PLA Triblock Copolymers with Long Poly(L-lactic acid) Blocks, *Macromolecules* **29**: 57

61. Dobrzynski, P., Kasperczyk, J., and Bero, M., 1999, Application of Calcium Acetylacetonate to the Polymerization of Glycolide and Copolymerization of Glycolide with e Caprolactone and L Lactide, *Macromolecules* **32**: 4735

Transglucosylation and Hydrolysis Activity of *Gluconobacter oxydans* Dextran Dextrinase with Several Donor and Acceptor Substrates

MYRIAM NAESSENS and ERICK J. VANDAMME
Department of Biochemical and Microbial Technology, Faculty of Agricultural and Applied Biological Sciences, University of Gent, Coupure links 653, B-9000 Gent, Belgium

Abstract: In this paper the reactivity of *Gluconobacter oxydans* dextran dextrinase (DDase) towards several glucosyl donor and acceptor molecules was studied. The donor/acceptor assay reflecting most accurately the DDase transglucosylation activity, could then be used as an evaluation tool for the optimization of the DDase production by *G. oxydans*. Different combinations of glucosyl donors (maltodextrin or maltose) and glucosyl acceptors (glucose, maltose or cellobiose) were incubated with a crude *G. oxydans* cell extract as a biocatalyst. The synthesis of panose revealed to be the most reliable indicator for DDase transglucosylation activity. Measurements of increasing glucose concentrations or decreasing maltose concentrations were less suitable for the quantification of DDase activity, since maltose seemed to be subjected to some hydrolysis during the enzymatic assays. In order to determine whether the hydrolytic activity, revealed in the donor/acceptor assays, originated from the DDase enzyme itself or from another enzyme present in the DDase preparation, the *G. oxydans* cell extract was subjected to native PAGE and zymogram analysis. The cell extract contained a number of proteins, the majority being smaller than 140 kDa. One protein band could be detected between the MW markers of catalase and ferritin (with a MW of 232 kDa and 440 kDa resp.), corresponding to DDase, which has a MW of 300 kDa, according to Yamamoto et al[1]. The zymogram showed an uncoloured zone with the same Rf value as DDase, leading to the conclusion that DDase itself was most probably the enzyme displaying the hydrolytic activity observed in the donor/acceptor assays.

1.   INTRODUCTION

Dextran dextrinase is the enzyme elaborated by two strains of *Gluconobacter oxydans* (previously assigned to the genus *Acetobacter*), which were isolated from ropy beer by Shimwell[2] in 1947. Hehre and Hamilton[3] showed that the ropiness formation originated from the conversion of maltodextrins, natural constituents of beer, into dextran and that this conversion was catalysed by dextran dextrinase. Yamamoto et al.[4] reported three transglucosylation modes of the DDase enzyme, as illustrated in Fig 1.

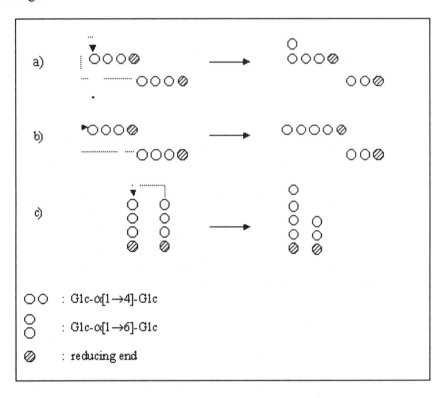

Figure 1. DDase transglucosylation modes: transfer of glucosyl residue from a) $\alpha(1\rightarrow4)$ linked donor to $\alpha(1\rightarrow6)$ linked acceptor; b) $\alpha(1\rightarrow4)$ linked donor to $\alpha(1\rightarrow4)$ linked acceptor; c) $\alpha(1\rightarrow6)$ linked donor to $\alpha(1\rightarrow6)$ linked acceptor.

The main action mode consists of transferring an $\alpha(1\rightarrow4)$ linked glucosyl residue forming an $\alpha(1\rightarrow6)$ linkage. Two secondary modes, namely the transfer of an $\alpha(1\rightarrow4)$ linked glucosyl residue forming an $\alpha(1\rightarrow4)$ linkage and the transfer of an $\alpha(1\rightarrow6)$ linked glucosyl residue forming an $\alpha(1\rightarrow6)$ linkage, are outweighed by the main action mode, resulting in the accumulation of α(1→6) linked glucosyl residues, which is dextran.

Dextran dextrinase can be used for the transglucosylation of food additives such as stevioside, palatinose, and soybean glycoside, in order to eliminate undesirable flavors[5,6,7]. The aim of this work was to monitor the transglucosylation activity of DDase towards several glucosyl donor and acceptor substrates. The most suitable donor/acceptor combination, yielding an easy quantifiable transglucosylation product, could thereafter be used for DDase activity measurements. This work forms part of an ongoing research project aiming at the optimisation of the dextran dextrinase production by *Gluconobacter oxydans*.

2.   MATERIALS AND METHODS

2.1   Enzyme preparation

G. oxydans ATCC 11894 was cultured in 4 l of sorbitol medium containing 20 g/l D-sorbitol (Vel) and 5 g/l yeast extract (Oxoid), at pH 6.0 during 18h at 25°C and 200 rpm. Cells were collected by centrifugation (10 min at 10 000 rpm, 4°C) and washed with 10 mM Na-acetate buffer, pH 4.8. The cells were resuspended in 50 ml of the same buffer, mixed with 50 ml of water saturated n-butanol and left to rest for 30 min at 4°C. The mixture was centrifuged (10 min at 10 000 rpm, 4°C) and DDase was extracted in the water layer. The solution obtained after dialysis against the same buffer, was lyophilised and dissolved in 40 ml of buffer for use in the activity assays.

2.2   DDase assays

All solutions were prepared with 10 mM Na-acetate buffer, pH 4.8. Reaction mixtures containing 250 µl of donor (10 g/l), 250 µl of acceptor (10 g/l) and 500 µl of enzyme solution, were incubated at 30°C for appropriate times. In the case of maltose acting as donor as well as acceptor, the acceptor solution consisted of buffer to avoid doubling of the maltose concentration in the reaction mixture. The enzymatic reactions were stopped by heating at 100°C for 10 min. For the preparation of blank solutions, reagents were kept on ice before mixing, after which they were immediately inactivated. All assays were performed in two-fold.

2.3   Analysis of transfer products

Reaction products were identified and quantified by Dionex® high performance anion exchange chromatography (CarboPac™ PA-100 column), coupled with pulsed amperometric detection. Samples were diluted if necessary and supplemented with an internal standard of fucose (30 ppm) prior to analysis.

2.4   Gel electrophoresis and zymogram analysis

Proteins present in the *Gluconobacter* cell extract were precipitated with 5.3 M $(NH_4)_2SO_4$, dialysed, lyophilised and redissolved in 1 ml of Tris/HCl 10 mM, pH 4.8. If necessary, the sample was diluted with the same buffer. Any insoluble material was removed by centrifugation (5 min at 10 000 rpm, 4°C). One µl of prepared sample was applied to the gel.

Native gel electrophoresis was carried out with the PhastSystem™ of Amersham Pharmacia Biotech, with PhastGel gradient 8-25, PhastGel native buffer strips and a HMW calibration kit (Amersham Pharmacia Biotech). Proteins were visualised with coomassie staining.

The zymogram was prepared by submerging the native gel in 2% (w/v) soluble starch (Merck) in Na-acetate 10 mM, pH 4.8 for 20h at 30°C. Undegraded starch was stained blue with a lugol solution (1 g I_2 and 2 g KI in 300 ml of distilled water).

3.   RESULTS AND DISCUSSION

3.1   Maltodextrins as donor and glucose as acceptor

The reaction mixture in the first donor/acceptor assay consisted of maltodextrin and glucose, incubated with *G. oxydans* cell extract. It was expected that glucosyl units, originating from maltodextrin, would be transferred to glucose by DDase, resulting in the formation of maltose and isomaltose, and that this would be reflected in a decrease in glucose concentration. Such a decrease could not be observed in the assay with maltodextrins as glucosyl donor and glucose as glucosyl acceptor (Fig 2). On the contrary, the amount of glucose in the reaction mixture increased with 3 mM.

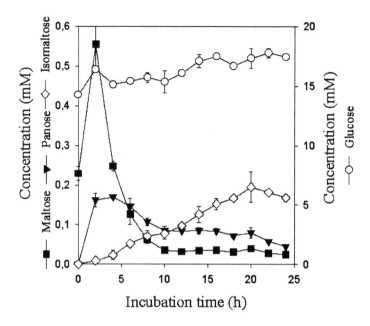

Figure 2. Maltodextrin/glucose assay: Concentration of reaction products in function of incubation time.

However, maltose and isomaltose (transglucosylation products of glucose) and panose (transglucosylation product of maltose) appeared in the reaction mixture, indicating the presence of DDase activity. The fast disappearance of maltose from the reaction mixture could not solely be caused by its transglucosylation, since it was not reflected in a parallel increase in panose or maltotriose concentration. Maltose was probably subjected to hydrolysis, which would also explain the increase in glucose concentration. Panose hydrolysis could account for the gradual decrease in panose concentration. However, it is equally probable that panose was used as a substrate for DDase transglucosylation, resulting in non-identified higher oligosaccharides, which would also cause panose disappearance from the reaction mixture.

3.2 Maltose as donor and as acceptor

Yamamoto et al.[8] reported the ability of DDase to transfer a glucosyl unit from maltose to another maltose molecule, resulting in the formation of panose (α-D-Glc-[1$\rightarrow$6]-α-D-Glc-[1$\rightarrow$4]-D-Glc) and glucose. At first sight, the fluctuations in maltose, glucose and panose concentration in the assay with maltose acting as glucosyl donor as well as acceptor, seemed to confirm the transglucosylation activity of the *G. oxydans* cell extract (Fig 3).

However, the concentration fluctuations of the three sugars should have been of the same order of magnitude if they were all involved in the same transglucosylation reactions. This was hardly the case. Whereas the panose concentration varied from 0 to 0.30 ± 0.01 mM, the concentration of glucose and maltose ranged from 0 to 11 ± 1 mM and from 6.1 ± 0.3 to 0 mM, respectively, indicating the presence of some maltose hydrolysing activity, in addition to the DDase transglucosylation activity. Panose was formed in amounts comparable to those in the maltodextrin/glucose assay, despite the presence of larger amounts of maltose in the maltose/maltose assay.

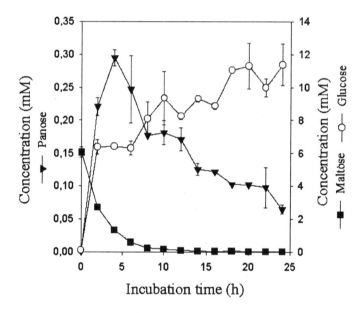

Figure 3. Maltose/maltose assay: Concentration of reaction products in function of incubation time.

3.3   Maltose as donor and cellobiose as acceptor

The results of the assay with maltose as glucosyl donor and cellobiose as acceptor were very similar to those of the maltodextrin and glucose assay. DDase showed only minor reactivity towards cellobiose (Fig 4). The enzyme activity seemed focused on maltose acting as glucosyl donor as well as acceptor.

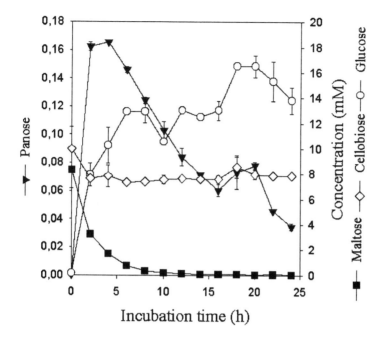

Figure 4. Maltose/cellobiose assay: Concentration of reaction products in function of incubation time.

The evolutions in maltose and glucose concentrations were not in accordance with the amount of panose synthesised, pointing again towards maltose hydrolysis in addition to maltose transglucosylation.

The concentration in panose showed a similar evolution as observed in the maltodextrin/glucose assay and the maltose/maltose assay.

3.4   Native PAGE and zymogram analysis

Figure 5 demonstrates that the *Gluconobacter* cell extract contained a number of proteins with the majority smaller than 140 kDa. One protein band could be detected between the MW markers of catalase and ferritin (with a MW of 232 kDa and 440 kDa resp.). This corresponds with DDase, which has a MW of 300 kDa according to Yamamoto et al[1].

The zymogram in Figure 5 shows a starch degradation zone with the same Rf value as the protein band corresponding to DDase, leading to the conclusion that DDase itself was the enzyme displaying the hydrolytic activity observed in the donor/acceptor assays.

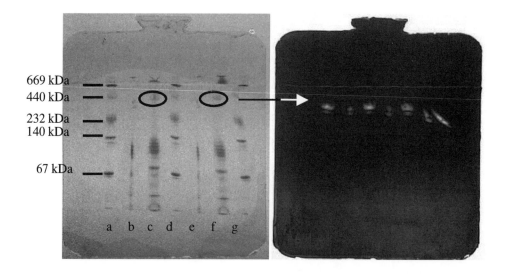

Figure 5. Left: Native PAGE on cell extract. Lane a,d,g: MW markers; Lane b,e: sample, 5 times diluted; Lane c,f: sample, undiluted; Right: Zymogram on cell extract.

4.   CONCLUSION

All donor/acceptor DDase assays demonstrated the presence of DDase activity in the *Gluconobacter* cell extract, but the transglucosylation reactions were easiest monitored in the assay with maltose as donor and acceptor.

The synthesis of panose revealed to be the most reliable indicator for DDase transglucosylation activity. One unit of DDase activity could then be defined as:

1U= *the amount of enzyme which causes the formation of 1 μmol of panose in 1 min at 30°C and pH 4.8*

Measurements of increasing glucose concentrations or decreasing maltose concentrations were less suitable for the quantification of DDase activity, since maltose seemed to be subjected to some hydrolysis during the enzymatic assays. Native PAGE and zymogram analysis of the *Gluconobacter* cell extract revealed that it was most probable that this hydrolysis was caused by the dextran dextrinase itself.

REFERENCES

1. Yamamoto, K., Yoshikawa, K., Kitahata, S., Okada, S. , 1992, Purification and some properties of dextrin dextranase from *Acetobacter capsulatus* ATCC 11894. *Bioscience, Biotechnology and Biochemistry*, **56**: 169-173
2. Shimwell, J.L., 1947, A study of ropiness in beer. *Journal of the Institute of Brewing*, **53**: 280-294
3. Hehre, H., Hamilton, D.M., 1949, Bacterial conversion of dextrin into a polysaccharide with the serological properties of dextran. *Proceedings of the Society for Experimental Biology and Medicine*, **71**: 336-339
4. Yamamoto, K., Yoshikawa, K., Okada, S., 1993, Detailed action mechanism of dextrin dextranase from *Acetobacter capsulatus* ATCC 11894. *Bioscience, Biotechnology and Biochemistry*, **57**: 47-50
5. Lobov, S.V., Ohtani, R.K.K., Tanaka, O., Yamasaki, K., 1991, Enzymic production of sweet stevioside derivatives: Transglucosylation by glucosidases. *Agricultural and Biological Chemistry*, **55**: 2959-2965
6. Japanese patent: Preparation of food additives palatinose derivatives with sugars. JP 92-123945 920515
7. Japanese patent: Food product based on soybean glycosides. JP 83-64183 830411
8. Yamamoto, K., Yoshikawa, K., Okada, S., 1994, Substrate specificity of dextrin dextranase from *Acetobacter capsulatus* ATCC 11894. *Bioscience, Biotechnology and Biochemistry*, **58**: 330-333

New Highly Functionalised Starch Derivatives

UTE HEINZE, VERA HAACK and THOMAS HEINZE
Institute of Organic Chemistry and Macromolecular Chemistry, Friedrich Schiller University of Jena, Humboldstraße 10, D-07743 Jena, Germany

Abstract: Pure, well soluble *p*-toluenesulfonyl (tosyl) starch samples with a DS_{Tos} range from 0.4 to 2.0 were prepared by reacting starch with tosyl chloride in the presence of triethylamine dissolved in the solvent *N,N*-dimethyl acetamide in combination with LiCl. The thermal degradation starts at a temperature of 166 °C for a sample with DS_{Tos} of 0.61 which is sufficiently high for subsequent modifications of the remaining OH groups, for instance by acylation reactions. The total DS_{Tos} can be determined using the signal of the methyl protons of the tosyl, acetyl or propionyl moieties of peracylated samples. Moreover, the NMR characterisation reveals that a predominant functionalisation at position 2 occurs. On the other hand, 6-*O*-tosyl starch products can be synthesised via 2-*O*-acetyl starch which is accessible by a new acylation procedure. The functionalisation patterns of new synthesised polymers with an unusual distribution of functional groups and high DS values were unambiguously characterised by various NMR techniques.

1.   INTRODUCTION

In contrast to the recent progress in cellulose chemistry, which is in particular stimulated by the use of non-aqueous cellulose solvents and regioselectively protected or activated derivatives[1], a fully satisfying solution to design high-performance materials based on starch is still missing. Most of the starch derivatives, which are available commercially, have low degree of substitution (DS = 0.01 – 0.20). Examples are the preparation of starch acetate or succinate, the reaction with ethylene or propylene oxide to form hydroxyalkyl starch, with sodium tripolyphosphate to synthesise starch phosphate, and with sodium monochloroacetate to obtain carboxymethyl starch[2]. Many other types of starch derivatives have been prepared, including products with high DS values (DS > 0.5), but most of these compounds have not been commercialised. Therefore, in the field of starch functionalisation, products of high degree of substitution (DS) and with controlled distribution of the functional groups are presently in the centre of our interest, which includes the development of methods for the determination of the functionalisation patterns. Moreover, specifically designed starch materials, which can be easily prepared and used for subsequent reactions via different synthesis paths, are a challenge of the recent polysaccharide chemistry and of great importance for future applications of the renewable and unique polymer starch.

In this context our interest was focused on sulfonic acid esters of starch, derived from the reaction of organic sulfonyl chloride (e.g. benzene, methane, or toluene sulfonyl chloride). These derivatives can be employed as partially protected and reactive intermediates. Up to now, especially the reaction of starch with *p*-toluenesulfonyl chloride in pyridine was studied, i.e. under heterogeneous reaction conditions, which may be accompanied by several side reactions[3-6]. Alternatively, a homogeneous procedure was published using dimethyl sulfoxide as solvent for starch. However, the sulphur contents of the products were very low and an extensive degradation of the polymer occured[7]. The extent of *p*-toluenesulfonylation of primary and secondary groups was determined by the iodination method[8]. It was revealed that the reaction proceeds faster at *O*-6 than at *O*-2 and *O*-3 [9].

With regard to characterisation of the polysaccharide structures obtained by chemical functionalisation, a very powerful method is the NMR spectroscopy. Especially the ^{1}H-NMR spectroscopy of polysaccharide acetates or of derivatives which are subsequently modified by complete acetylation reaching a total DS of 3 [10-13] is extensively used.

In the present paper we wish to report about an effective homogeneous synthesis of pure tosyl starch samples having a wide range of DS using the solvent system *N,N*-dimethyl acetamide (DMA) in combination with LiCl. Moreover, the synthesis of regioselective *O*-6 starch tosylates via 2-*O*-protected starch acetate is discussed yielding products useful for subsequent displacement reactions. The products are characterised by elemental analysis and FTIR spectroscopy. The DS values and the functionalisation pattern of the new polymers are determined by means of ^{13}C- and ^{1}H-NMR spectroscopy including two-dimensional methods using also the subsequently acetylated and propionylated samples.

2.   EXPERIMENTAL

2.1   Materials

The starch material used was HYLON VII (70 % amylose, *National Starch & Chemical GmbH*, Neustadt, Germany) after drying at 100°C for 24 h under vacuum. The solvents were dried and distilled prior to use according to conventional methods. Anhydrous LiCl was used after drying at 130°C for 2 h under vacuum. Other reagent grade chemicals were used without further purification.

2.2   Measurements

Elemental analysis was carried out by means of a LECO CHNS/932 ultimate analyser. The FTIR spectra were recorded on a Nicolet Impact 400 spectrometer using KBr pellets. ^{1}H- and ^{13}C-NMR spectra were acquired on a Bruker AMX 400 spectrometer in DMSO-d_6 or CDCl$_3$ with an accumulation number between 10000–75000 scans. The solubility of the tosyl starch samples were determined at a concentration of 1 g/100 ml in various organic solvents. The thermogravimetric analyses were performed by means of a thermogravimetric analysing system (home made) using 10 mg samples. The curves were run in air at a heating rate of 10 K/min from ambient temperature up to 600°C.

2.3   Synthesis of 2-*O*-acetyl starch[14]

2.0 g starch were dissolved in 40 ml DMSO at 80 °C. After cooling 2.3 mol vinyl acetate/mol AGU and 40 mg (2 %, w/w) NaCl were added. The mixture were allowed to react for 70 h at 40 °C under stirring. The NaCl was removed by centrifugation, the product precipitated in 400 ml isopropyl alcohol, filtrated off, washed intensively with isopropyl alcohol and dried at 50 °C under vacuum.

2.4   Dissolution of starch in *N,N*-dimethyl acetamide (DMA)/LiCl

Dried starch (2.5 g, 15.4 mmol, actual weight of water-free polymer) was suspended in 60 ml of DMA and kept at 160°C for 1 h under stirring. After the slurry had been allowed to cool to 100°C, 5.0 g of anhydrous LiCl were added. The starch dissolves completely during cooling to room temperature under stirring.

2.5   Synthesis of *p*-toluenesulfonyl (tosyl) starch, and tosyl starch acetate

For a typical preparation, to the 4.3 % (w/v) solution of 2.5 g starch (15.4 mmol) in DMA/LiCl a mixture of 8.5 ml (61.4 mmol) triethylamine and 5.0 ml of DMA was added under stirring. After cooling down to about 8°C, a solution of 5.9 g (30.8 mmol) *p*-toluenesulfonyl chloride in 7.4 ml DMA was dropwisely added under stirring within 30 min. The homogeneous reaction mixture was additionally stirred 24 h at 8°C and then slowly poured into 700 ml of ice water. The precipitate was filtrated off, carefully washed with about 2.3 l of distilled water and 250 ml of ethanol, dissolved in 120 ml of acetone and reprecipitated into 370 ml of distilled water. After filtration and washing with ethanol, the product was dried at 60°C for 8 h under vacuum. The tosyl starch acetates were prepared under comparable conditions using 2-*O*-acetyl starch as starting material.

2.6   Acetylation of tosyl starch

For a typical preparation, to a stirred solution of tosyl starch (1.0 g, DS_{Tos} = 1.02) in 20 ml pyridine 0.3 g *N,N*-dimethylaminopyridine (DMAP) and 10 ml (106.0 mmol) acetic anhydride were added. The mixture was kept at 20°C for 20 h and at 80 °C for 5 h. After cooling down to room temperature, the product was isolated by precipitation in 150 ml ethanol, filtered off, washed with ethanol and dried at 50°C under vacuum.

2.7   Propionylation of tosyl starch

Tosyl starch sample (DS_{Tos} = 1.02; 0.3 g) was allowed to react with 5 ml (39 mmol) propionic acid anhydride/0.1 g DMAP in 5 ml anhydrous pyridine. The mixture was kept at 20 °C for 20 h and at 80 °C for 5 h. After cooling down to room temperature, the product was isolated by precipitation in 100 ml of ethanol, filtered off, thoroughly washed with ethanol and dried at 50 °C under vacuum.

2.8   Reaction of tosyl starch and tosyl starch acetate with iodine

1.0 g of tosyl starch (see 2.5, DS_{Tos} = 1.35) or 1.0 g tosyl starch acetate (see 2.6, DS_{Ac} = 0,70, DS_{Tos} = 1,37) was dissolved in 50 ml of acetylacetone. Then 1.7 g of anhydrous NaI was added and the reaction mixture was kept for 2 h at 130°C. After cooling down to room temperature, the product was isolated by precipitation into 200 ml of ethanol, filtrated off and carefully washed with 3.2 l of distilled water. The sample was soaked overnight in 170 ml ethanol and then in 170 ml 0.1 M $Na_2S_2O_3$ solution for 1 h. After filtration and washing with water and ethanol, the product was dried at 50°C for 6 h under vacuum.

3. RESULTS AND DISCUSSION

In order to guarantee an even distribution of the functional groups within the polymer chain, a homogeneous procedure for the tosylation reaction was preferred. It was already shown that homogeneous tosylation using the well-known starch solvent dimethyl sulfoxide does not yield products of sufficient degree of substitution[7]. Therefore, we carried out the reaction in the solvent system *N,N*-dimethyl acetamide (DMA)/LiCl which is usually applied for homogeneous functionalisation of cellulose.

3.1 Homogeneous synthesis of tosyl starch

The tosylation of starch was carried out with *p*-toluenesulfonyl chloride (Tos-Cl) and triethylamine as a base within 24 h at 8°C in DMA/LiCl solution.

To evaluate the accessible range of the degree of substitution (DS_{Tos}) the molar ratio Tos-Cl/anhydroglucose unit (AGU) was varied from 1.0 to 6.0 mol/mol. Regarding the results of tosylation (Tab. 1) it has to be underlined that the DS_{Tos} increases with increasing amount of Tos-Cl per AGU. That means that the DS_{Tos} can be simply controlled by adjusting the molar ratio. In comparison to the reaction at 8°C, the tosylation of starch at ambient temperature leads to products with lower DS_{Tos} values and higher chlorine content by using the same molar ratio Tos-Cl/AGU. This indicates that a nucleophilic displacement reaction of the tosylate groups by chloride especially at higher temperature can occur as a side reaction.

Table 1. Conditions and results of homogeneous tosylation of starch with *p*-toluenesulfonyl chloride (Tos-Cl) at 8 °C

Molar ratio Tos-Cl/AGU	Reaction time (h)	DS_{Tos}	Cl (%)	IDT^a (°C)
1.0	24	0.61	0.19	165.9
1.5	1	0.81	0.00	-
1.5	5	1.02	0.27	-
1.5	24	1.02	0.20	198.3
1.5^b	24	0.87	0.43	-
2.0	24	1.35	0.30	192.6
2.0	1	1.09	0.14	196.1
2.0	3	1.08	0.25	202.7
2.0	5	1.15	0.36	201.9
3.0	24	1.43	0.42	-
3.0	1	1.14	0.00	200.9
6.0	24	2.02	0.32	214.0

[a] Initial decomposition temperature (IDT) of starch = 303.4 °C
[b] at room temperature

3.2   Spectroscopic characterisation of tosyl starch

3.2.1   FTIR spectroscopy

The structure of the tosyl starch was confirmed by means of FTIR spectroscopy. The spectra show the characteristic absorption bands of the starch backbone as well as signals at 3064, 1597, 1495 and 816 cm^{-1} of the aromatics. Furthermore, two bands with a high intensity at 1360 and 1176 cm^{-1} (v SO$_2$) confirm the presence of the tosyl groups.

3.2.2 ^{13}C NMR spectroscopy

Fig 1 shows a standard ^{13}C-NMR spectrum of tosyl starch (DS_{Tos} = 1.02) in DMSO-d_6 and the assignment of the signals. The peak for the C-6 atom of the AGU influenced by *O*-6 tosylation appears at δ = 68.4 ppm (C-6$_s$), i.e. it exhibits a down field shift compared to the corresponding carbon of unmodified starch (C-6, δ = 59,8 ppm). From the signal intensities it may be concluded that the 6 position is not or only slight tosylated. On the other hand, the signal at δ = 93.5 ppm which is assigned to C-1' (C-1 atom influenced by *O*-2 toslyation) indicates an extensive reaction at the position 2. Moreover, a signal for C-1 (unmodified C-2) is not visible which would appear with a high field shift of 3 - 5 ppm. The tosylated secondary hydroxyl groups itself give a new signal at $\delta = 79.4$ ppm. From the spectra it may be concluded that a distribution of tosyl groups within the AGU in the order *O*-2 >>*O*-6 and *O*-3 occurred.

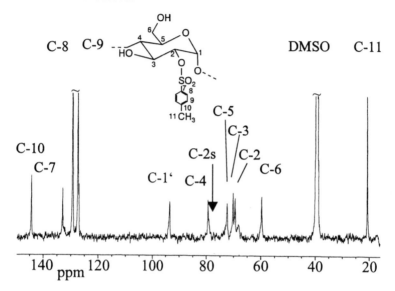

Figure 1. ^{13}C-NMR spectrum of tosyl starch (DS$_{Tos}$ = 1.02) recorded in DMSO-d$_6$ at 60 °C

3.3   Some selected properties

3.3.1   Solubility

An important property of the tosyl starch synthesised is their solubility in common organic solvents. The derivatives with a DS$_{Tos}$ ≥ 0.61 are soluble in DMSO and additionally in different dipolar aprotic solvents like DMA, *N,N*-dimethyl formamide (DMF) as demonstrated in Fig 2. Samples with a slightly higher DS$_{Tos}$ are soluble in dioxane (≥ 0.98), tetrahydrofuran and acetone (≥ 1.15). At a DS$_{Tos}$ of 2.02 the tosyl starch samples dissolve even in the non-polar solvents dichloromethane and chloroform.

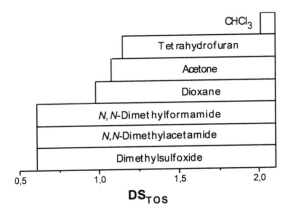

Figure 2. Solubility of tosyl starch in organic solvents in dependence on DS_{Tos}

3.3.2   Thermal behaviour

Thermal analyses of starch and homogeneously synthesised tosyl starch were studied by means of thermogravimetry (TG) under air at a heating rate of 10 K/min. The initial degradation temperatures (IDT) show that the thermal decomposition of the tosyl starch was initiated at lower temperature compared to starch (see Tab. 1). With increasing content of tosyl groups (DS_{Tos} = 0.61 - 2.02) an increase in stability from 166 to 214 °C occurs. The thermal stability is sufficiently high for processing and subsequent chemical modifications.

3.4   Subsequent modifications of tosyl starch

3.4.1   Iodination

In contrast to the generally accepted idea, the determination of the extend of tosylation of primary hydroxy groups by the analysis of the corresponding iododeoxy derivatives represents a semiquantitative method. Deviations from a perfect analysis were already published[15-17], and the results should be, therefore, confirmed by alternative methods. For the determination of the partial DS_{Tos} of tosyl starch at the primary position the samples were converted with NaI (in acetyl acetone) under conditions for a selective substitution of the 6 tosylate groups. A maximal DS_I of 0.7 could be obtained starting from tosyl starch with $DS_{Tos} = 2.02$ (Tab. 2).

The result indicates that even at this high DS_{Tos} no complete tosylation of the primary function occurred. That means that a remarkable reaction at the secondary OH groups appeared. These results are in good agreement with those of ^{13}C-NMR spectroscopic investigations.

Table 2. Iodination of tosyl starch with NaI in acetylacetone for 2 h at 130 °C

Tosyl starch DS_{Tos}	6-Iodo-6-deoxy tosyl starch DS_I
1.08	0.20
1.15	0.23
2.02	0.73

3.4.2   Acetylation and Propionylation

Subsequently acetylated, respectively propionylated starch derivatives are suitable for the determination of the substituent distribution by using ^{13}C- and ^{1}H-NMR spectroscopy[14]. In order to obtain well resolved spectra suitable for quantitative assessment it is necessary that all free OH groups are completely acylated. This complete esterification could be achieved by reacting the polymers with a 40 molar excess of acetic acid anhydride for 20 h at room temperature and additionally 5 h at 80 °C applying *N,N*-dimethylaminopyridine (DMAP) as a catalyst. The synthesized tosyl starch acetates as well as propionates show the same solubility in common organic solvents as the starting tosyl starch samples, however, they dissolve additionally in $CHCl_3$ independent of DS_{Tos}.

Standard ^{13}C-NMR spectra of the tosyl starch acetates and propionates are appropriate to gain information about the distribution of the functional groups by using the carbonyl signals. In contrast to a starch triacetate which possesses three separate signals at 170.5, 170.1 and 169.2 ppm in agreement to a C=O moiety at position 2,6 and 3, the ^{13}C-NMR spectrum of tosyl starch acetate (DS_{Tos} = 1.02) shows only two carbonyl peaks at 169.9 and 169.4 ppm indicating that the position 6 and 3 are acetylated. Comparable results were obtained for tosyl starch propionates[18].

The ^{1}H-NMR spectra of tosyl starch acetate respectively tosyl starch propionate are suitable to calculate the DS_{Tos} values (Eq. 1) from the protons both of the methyl groups or the aromatic rings of the tosylate moieties as well as from the methyl groups of the acetyl or propionyl ester function.

$$DS = \frac{I_{Signal}/n}{\Sigma I_{AGU}/7}$$

Equation 1, n = H atoms of the signal of aromatic (n = 4) or methyl protons (n = 3) of tosyl moieties or methyl protons (n = 3) of acetyl and propionyl ester moieties, AGU = anhydroglucose unit

In some cases even the determination of the distribution of the functional groups within the AGU is possible with regard to the primary versus secondary positions. It becomes obvious that with increasing DS_{Tos} the resolution of the spectra decreases. This might be a result of the increasing shielding of the rather bulky aromatic groups. As a consequence, samples of high DS_{Tos} should preferably investigated after perpropionylation since the distance of the corresponding methyl groups of the propionyl esters and the polymer backbone is longer compared to acetate. For an reliable assignment of the signals of the methyl protons concerning their position within the AGU the HMBC (heteronuclear multibond coherence) technique was used indicating that the propionyl ester function is located at position 3.

From these results and in accordance with the above described findings of iodination of tosyl starch, it becomes obvious that the tosylation occurs faster at *O*-2 compared to *O*-6 and *O*-3.

3.5   Regioselective starch tosylates via 2-*O*-acetyl starch

It is well known that the nucleophilic displacement reactions at tosylated polysaccharides are limited or at least mainly directed towards the primary positions[19]. Therefore, our interest was focused on 6-*O*-tosyl starch samples with $DS_{Tos} \leq 1$. One suitable synthesis path is the protection of *O*-2 and the subsequent tosylation. A useful protecting group may be the acetyl ester function. It was recently found that in contrast to conventional esterification processes of starch with acetic anhydride, which leads to a statistic distribution of the ester groups, an acetylation of starch dissolved in DMSO with acetic acid vinyl ester in the presence of sodium chloride yields 2-*O*-acetyl starch of varying DS_{Ac} from 0.1 to 1.0. The functionalisation patterns of these new starch products were unambiguously proved by means of various NMR measurements including two dimensional methods[14].

R = H or Tos

However, up to now no subsequent modification of the 2-*O*-acetyl starch is known. We have found that the 2-*O*-acetyl starch samples are not soluble in common organic solvents but they swell to a certain extend. Heterogeneous conversions of 2-*O*-acetyl starch swollen in DMA with Tos-Cl/N(C$_2$H$_5$)$_3$ leads to a small DS_{Tos} only (Tab. 3). On the other hand, 2-*O*-acetyl starch can be dissolved in DMA in combination with LiCl which is an appropriate reaction medium for tosylation as discussed for unmodified starch in detail above. As shown in Tab. 3, the homogeneous procedure in DMA/LiCl is suitable to prepare products of high DS_{Tos}. At a molar ratio Tos-Cl : modified AGU of 2:1 respectively 5:1 using a 2-*O*-acetyl starch with $DS_{Ac} = 0.7$ a DS_{Tos} of 0.53 respectively 1.37 is reached. From ^{13}C-NMR spectroscopy a preferred tosylation of *O*-6 can be concluded.

Table 3. Tosylation of 2-O-acetyl starch with Tos-Cl and triethylamine as a base heterogeneously in DMA and dissolved in DMA/LiCl

DS_{Ac}	LiCl	Tos-Cl: modified AGU	T [°C]	Elemental analysis S (%)	DS_{Tos}
1,0	—	1 : 1	8	1.58	0.11
1,0	—	1.5 : 1	8	2.80	0.21
1,0	—	2 : 1	8	3.57	0.27
1,0	+	2 : 1	8	4.79	0.40
1,0	+	2 : 1	Room temp.	6.21	0.56
1,0	+	6 : 1	8	9.83	1.19
0,7	—	2 : 1	8	3.72	0.27
0,7	+	2 : 1	8	6.24	0.53
0,7	+	5 : 1	8	10.89	1.37

Preliminary experiments of conversions of the 2-O-acetyl-6-O-tosyl starch samples with NaI reveal that a nucleophilic displacement reaction occurs. For instance, using a sample with DS_{Ac} of 0.7 and DS_{Tos} of 1.37 a treatment with a 3 molar excess of NaI in acetylacetone for 2 h at 130 °C yields a product with DS_I of 0.6 (remaining $DS_{Tos} = 0.5$, $DS_{Ac} = 0.6$).

4. CONCLUSION

The homogeneous tosylation of starch with Tos-Cl and triethylamine as a base dissolved in DMA/LiCl represents a suitable and effective method for the preparation of organo-soluble tosyl starch samples. These products are well soluble in various organic solvents. From ^{13}C- and ^{1}H-NMR measurements as well as by analysis of the corresponding iododeoxy tosyl starch samples can be concluded that the tosylation preferentially occurred at the 2 position. Consequently, tosyl starch samples with a new pattern of functionalisation are accessible. The quantitative determination of the distribution of functional groups was studied by means of two dimensional ^{1}H-NMR spectroscopy after peracetylation respectively perpropionylation of the tosyl starch samples. It appears that the tosylation occurs with high regioselectivity at position 2 with regard to samples with a $DS_{Tos} \leq 1$. A regioselective functionalisation of the primary OH-groups was achieved by subsequent tosylation of 2-*O*-protected acetyl starch which can be synthesised by transesterification using starch dissolved in DMSO, vinyl carbonic acid and an inhomogeneous salt (e.g. NaCl). The new 6-*O*-tosyl-2-*O*-acetyl starch products are included in a program to design advanced polysaccharide products by nucleophilic displacement reactions.

REFERENCES

1. Heinze, Th., Glasser, W.G., 1998, The role of novel solvents and solution complexes for the preparation of highly engineered cellulose derivatives. *In Cellulose derivatives: modification, characterization, and nanostructures*, (Thomas J. Heinze, Wolfgang G. Glasser, eds.) ACS Symposium Series 688. Washington, DC: American Chemical Society
2. Wurzburg, O.W., 1986, *Modified starches: properties and uses*. Boca Raton, USA: CRC Press
3. Clode, D.M., Horton, D., 1971, Preparation and characterization of the 6-aldehydo derivatives of amylose and whole starch. *Carbohydr. Res.*; **17**: 365-373.
4. Horton, D., Meshreki, M.H., 1975, Syntheses of 2,3-unsaturated polysaccharides from amylose and xylan. *Carbohydr. Res.* **40**: 345-352.
5. Teshirogi, T., Yamamoto, H., Sakamoto, M., Tonami, H., 1978, Syntheses and reactions of Aminodeoxycelluloses. *Sen-i Gakkaishi* **34**: T510-T515.
6. Teshirogi, T., Yamamoto, H., Sakamoto, M., Tonami, H., 1979, Synthesis of mono- and di-aminated starches. *Sen-i Gakkaishi* **35**: T479-T485.
7. Weill, C.E., Kaminsky, M., Hardenbergh, J., 1980, Random substitution of amylose. *Carbohydr. Res.* **84**: 307-313.
8. Mahoney, J.F., Purves, C.B., New methods for investigating the distribution of ethoxyl groups in a technical ethylcellulose. *J. Amer. Chem. Soc.* **1942**; 64: 9-15.
9. Horton, D., Luetzow, A.E., Theander, O., 1973, Preparation of 6-chloro-6-deoxyamylose of various degrees of substitution; an alternative route to 6-aldehydoamylose. *Carbohydr. Res.* **27**: 268-272.
10. Heinze, Th., Schaller, J., 2000, New water soluble cellulose esters synthesized by an effective acylation procedure. Macromol. Chem. Phys. 201: 1214-1218.
11. Klemm, D., Stein, A., 1995, Silylated cellulose materials in design of supramolecular structures of ultrathin cellulose films. J.M.S.-Pure Appl. Chem. A32: 899-904.
12. Matulova, M., Toffanin, R., Navarini, L., Gilli, R., Paoletti, S., Cesaro, A., 1994, NMR analysis of succinoglycans from different microbial sources: partial assignment of their 1H and 13C NMR spectra and location of the succinate and the acetate groups. *Carbohydr. Res.* **65**: 167-179.
13. Deus, C., Friebolin, H., Siefert, E., 1991, Partiell acetylierte Cellulose-Synthese und Bestimmung der Substituentenverteilung. *Makromol. Chem.* **192**: 75-83.
14. Dicke, R., 1999, Ph.D. Thesis, Friedrich Schiller University of Jena, Germany.
15. Hall, D.M., Horne, J.R., 1973, Model compounds of cellulose: trityl ethers substituted exclusively at C-6 primary hydroxyls. *J. Appl. Polym. Sci.* **17**: 2891-2896.
16. Takahashi, S.-I., Fujimoto, T., Bama, B.M., Miyamoto, T., Inogaki, H., 1986, 13C-NMR spectral studies on the distribution of substituents in some cellulose derivative. *J. Polym. Sci., Part A: Polym. Chem.* **24**: 2981-2993.

17. Rahn, K., Diamantoglou, M., Klemm, D., Bergmans, H., Heinze, T., 1996, Homogeneous synthesis of cellulose p-toluenesulfonates in N,N-dimethyl acetamide/LiCl solvent system. *Angew. Makromol. Chem.* **238**: 143-163.
18. Dicke, R., Rahn, K., Haack, V., Heinze, T., Starch derivatives of high degree of functionalization 2. Determination of the functionalization pattern of p-toluenesulfonyl starch by peracylation and NMR spectroscopy. *Carbohydr. Polym.*, in press.
19. Heinze, T., 1998, *Ionische Funktionspolymere aus Cellulose: Neue Synthesekonzepte, Strukturaufklärung und Eigenschaften.* Shaker Verlag, Aachen

Preparation of Dextran-Based Macromolecular Chelates for Magnetic Resonance Angiography

MARIA G. DUARTE[1], CARLOS F.G.C. GERALDES[1], JOOP A. PETERS[2], and MARIA H. GIL[3]
[1]*Department of Biochemistry, University of Coimbra, Portugal, [2]Laboratory of Applied Organic Chemistry and Catalysis, Delft University of Technology, The Netherlands, [3]Department of Chemical Engineering, University of Coimbra, Portugal*

Abstract: The preparation of macromolecular conjugates of dextran, which could be used in the design of in perfusion contrast agents for Magnetic Resonance Angiography (MRA), is described. In a first step, an amino group was introduced in the diethylenetriamine pentaacetic acid (DTPA) structure, in order to obtain a complex with one linking group. The derivatized DTPA was characterized by NMR, and subsequently covalently linked to dextran using carbonyldiimidazole (CDI) as an activating agent. The degree of substitution of macromolecular conjugates of dextran can be tuned by the molar ratio of CDI/dextran applied.

1. INTRODUCTION

In the last decade there has been great interest in the development of macromolecular contrast agents to be used in Magnetic Resonance Imaging (MRI), a new diagnostic technique which provides a better contrast resolution of the tissues than others techniques, such as X-ray, ultrasound and scintigraphy.

Low molecular weight chelates of the paramagnetic ion Gd^{3+}, such as Gd-DTPA or Gd-DOTA, are generally used in MRI as contrast agents due to their high magnetic moment and long electronic relaxation time[1-4]. However, they present low vascular residence and need repeated doses for prolonged clinical examination. Also, because they rapidly diffuse into the interstitial

Biorelated Polymers: Sustainable Polymer Science and Technology
Edited by Chiellini *et al.*, Kluwer Academic/Plenum Publishers, 2001

space, they are not useful as blood-pool agents for Magnetic Resonance Angiography (MRA). To overcome these problems, macromolecular complexes have been developed by conjugation of low molecular weight complexes to natural or synthetic polymeric materials, like polysaccharides[5-12], albumin[13,14] and polyaminoacids like polylysine and polyornithine[15-20].

Dextran has been studied as a carrier for covalent coupling of these Gd^{3+} chelates since it shows a good biocompatibility, good solubility and availability of this polymer with different molecular weights. DTPA and other chelates have been linked to dextran through the bisanhydride derivative either directly [7] or by using a spacer arm[6,12]. This kind of linkage leads to the crosslinking of the polysaccharide.

In this work we prepared a dextran-Gd^{3+}-DTPA complex by linking to the polysaccharide a DTPA moiety, previously chemically modified to contain one single amine group. This kind of coupling will prevent the crosslinking of the polysaccharide and does not decrease the number of binding sites of the chelate to the paramagnetic ion.

2.   MATERIALS AND METHODS

2.1   Materials

Diethylenetriamine pentaacetic acid (DTPA) 97% purity, ethylenediamine and carbonyldiimidazole were purchased from Aldrich and used as supplied. Acetic anhydride, pyridine, dimethylsulfoxide and dimethylformamide, also from Aldrich, were of Analar purity, and were dried over molecular sieves 3Å before use. Dextran (MW 40000), di-t-butyl dicarbonate of 97% purity and Dowex 50W×8×200 ion exchange resin, were supplied by Sigma.

Dialyses were performed with a MWCO 12-14000 Membrane purchased from Medicel, London, U.K.

2.2   Synthesis and characterization of dextran conjugates of DTPA

Preparation of DTPA-bis-anhydride (1): A modification of the procedure described by Krejacarek and Tucker was applied[21]. Diethylenetriaminepentaacetic acid (0.125 mol) was mixed with 0.558 mol of dried acetic anhydride and 0.770 mol of dried pyridine. The resulting suspension was stirred for 24 h at 65°C. Then the solid obtained was filtered, washed with dried diethyl ether and dried under vacuum, until constant weight at 25°C. (yield: 97,5%). ^{1}H-NMR (DMSO-d_6) δ 3.7 (s,8H); δ 3.29 (s,2H); δ 2.43 (t,4H); δ 58 (t,4H). ^{13}C-NMR (DMSO-d_6) δ 171.9, 165.7, 54.5, 52.5, 51.7, 50.7

Preparation of Ethylenediamine-BOC (2): The method was described by Carvalho *et al.* [22] was applied. A solution of 0.65 mol of ethylenediamine in 50 ml of chloroform was cooled in an ice bath. Then a solution of di-t-butyl dicarbonate (0.1 mol) in 15ml of chloroform was added dropwise with stirring. The reaction mixture was stirred for another 18 h at room temperature. After filtration, the solution was concentrated by rotary evaporation and then washed 3 times with 50 ml of toluene. The final product was purified by distillation under reduced pressure at 120°C and analyzed by NMR (yield:73,6%). ^{1}H-NMR (DMSO-d_6) δ 6.76 (s,1H); δ 2.92 (q,2H); δ 2.53 (q,2H); δ 1.60 (s,2H); δ 1.38 (s,9H). ^{13}C-NMR (DMSO-d_6) δ 155.6, 77.2, 43.6, 41.6, 28.2.

Preparation of DTPA-NH$_2$(3): A solution of 50 mmol of DTPA-bis-anhydride (1), 50 mmol of ethylenediamine-BOC (2) and 100 ml of triethylamine in 200 ml of DMF was stirred for 3 h at room temperature. The DMF was then removed by rotary evaporation. The obtained dark oil was diluted with 100 ml of distilled water. The solution was cooled in an ice bath and 30 ml of concentrated HCl was added, with stirring. After 2 hours the solution was neutralized with NaOH and then was concentrated by rotary evaporation. Distilled water (50 ml) was added and the mixture obtained was then filtered to remove salts. The filtrate was acidified to pH 2.5 with 5M HCl and then evaporated to adjust the volume to 50ml.

This solution was loaded on a column with Dowex 50W×8×200 ion-exchange resin. The column was eluted with water, 1M and 2M ammonium hydroxide. The fractions obtained were lyophilized and analyzed by NMR. The pure fractions were combined and dissolved in a minimal amount of water. After addition of 75ml of acetic acid, the solution was concentrated by rotary evaporation. Ammonium acetate was removed by repeated co-evaporation with water. The final solution was lyophilized.

Preparation of Dextran-DTPA-NH$_2$ (4): To 0.1g of dextran dissolved in 50 ml of dried DMSO, carbonyldiimidazole (CDI) was added while stirring. After 15 min of reaction, at room temperature, DTPA-NH$_2$ (3) was added. After stirring for 24 h, and the product was precipitated in methanol:ether (6:4). The precipitate was dissolved in 10 ml of water and purified by dialysis for 3 days and then lyophilized.

2.3   Activation Studies of Dextran

After being activated with different amounts of CDI, the dextran (2 g of dextran in 40 ml of dried DMSO) was treated with ethylenediamine. After reaction with CDI during 15 min at room temperature, 5 ml of ethylenediamine were added. The reaction was carried out with stirring for 60 min and the product was precipitated in 100 ml methanol:ether (6:4), dissolved in 10 ml of water and purified by dialysis for 3 days and then lyophilized.

2.4   NMR Measurements

The NMR spectra were recorded on a Varian Unity 500 spectrometer (at an external field of 11.8 T). ^{13}C-NMR and ^{1}H-NMR spectra were recorded at 125.697 MHz and 499.824 MHz, respectively, in D_2O (99.8% D from Sigma Chem. Co.) solutions. The pH of the solutions was adjusted with DCl and CO_2-free NaOD (from Sigma Chem. Co.) using a Crison MicropH 2002 pH-meter with an Ingold 405-M5 combined electrode. For the ^{1}H-NMR spectra, the water signal was used as internal reference, set at δ 4.75 ppm (at 298 K) and was suppressed by a pre-saturation pulse. ^{13}C NMR spectra were measured with broad-band proton decoupling. The methyl resonance of TSP (sodium-3-trimethylsilylpropionate-2,2-3,3-d$_4$) was used as internal reference for the ^{13}C-NMR spectra and was set at δ 0 ppm. Assignments of ^{1}H and ^{13}C NMR were based in the results of two-dimensional heteronuclear correlation spectra (HMQC) and on literature data for similar systems [23]. The temperature precision of the experiments was ± 0.5°C.

3.   RESULTS AND DISCUSSION

3.1   Preparation of DTPA-NH$_2$

The polysaccharide selected for linking the DTPA chelate was dextran with a molecular weight of 40000 D. The dextran polymer was chosen due the fact that it is biocompatible, biodegradable and has a high molecular weight which prevents its diffusion into the interstitial space of the cells, and consequently would be maintained for longer times in the blood pool system. Nevertheless, the DTPA complex should be linked by a biodegradable linkage, like an amide bond, to the polysaccharide to prevent too long retention of the metal complexes in the body.

The preparation of the DTPA derivative for coupling to dextran was carried out by three steps, as shown in Fig. 1.

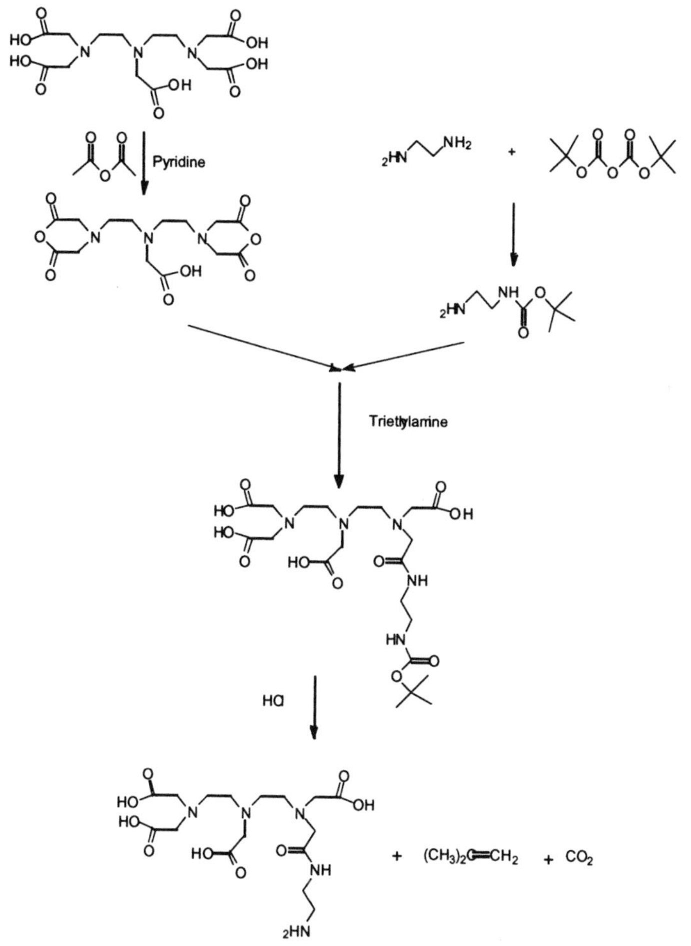

Figure 1. Reaction steps for the preparation of DTPA-NH2

The first step was the preparation of the bis-anhydride. In the second step a protective group, BOC, was linked to ethylenediamine and in the third step both products were reacted and the BOC group was cleaved.

Fig. 2 represents the ^{1}H-NMR and ^{13}C-NMR spectra of DTPA-NH$_2$. Only the methylene carbons could be assigned on the basis of the results of the HMQC spectra obtained for the region between δ 0 and 70 ppm. In these spectra we can observe the presence of a signal at δ 55.2 ppm in the ^{13}C spectrum and a broad signal at δ 3.8 ppm in the proton spectrum, that can be assigned to the methylene groups (mainly of the carboxylate arms) of unmodified DTPA. These shift values are in agreement with the proton NMR titration curves of DTPA observed by Geraldes et al. [23].

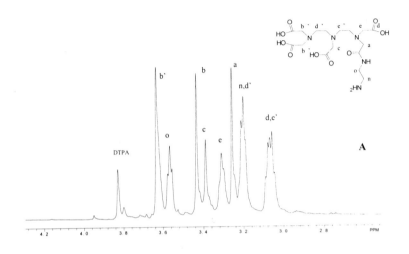

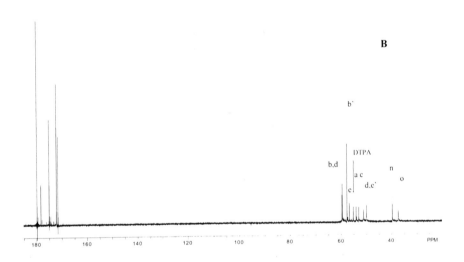

Figure 2. NMR spectra of DTPA-NH$_2$: (A) 1H NMR spectrum obtained in D$_2$O+NaOD (0.01M) at 25°C, pH 8.01; (A) ^{13}C NMR spectrum obtained in D$_2$O+NaOD (0.01M) at 25°C, pH 8.05

From the integration of this signal in the ^{1}H-NMR spectrum we estimate the purity of DTPA-NH$_2$ to be 91%.

3.2   Activation Studies of Dextran

The capacity of the activating agent, CDI, to react with the OH groups of dextran and subsequently link to NH2 groups of the chelating was studied using ethylenediamine as a model compound. The degree of derivatization of dextran was calculated from the ^{1}H-NMR spectra by comparison of the integrals of the signals of the anomeric protons of the dextran moiety with those of the ethylene protons of ethylenediamine moiety. The results (Fig. 3) show a linear relationship between the CDI concentration and the degree of substitution. The NMR spectra indicate that no side products are formed during the coupling.

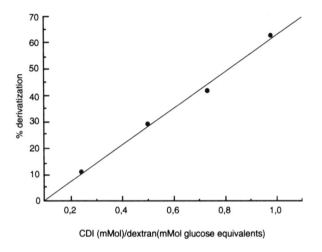

CDI (mMol)/dextran(mMol glucose equivalents)

Figure 3. Degree of derivatization of dextran by activation with CDI

3.3   Preparation of Dextran-DTPA-NH$_2$

After activation of dextran with different concentrations of CDI, DTPA-NH$_2$ was coupled to the activated dextran, as shown in the scheme of Fig 4. Four samples were prepared and, in order to obtain higher degrees of derivatization of the dextran with DTPA-NH$_2$, the CDI:dextran ratio was extended to 1.5.

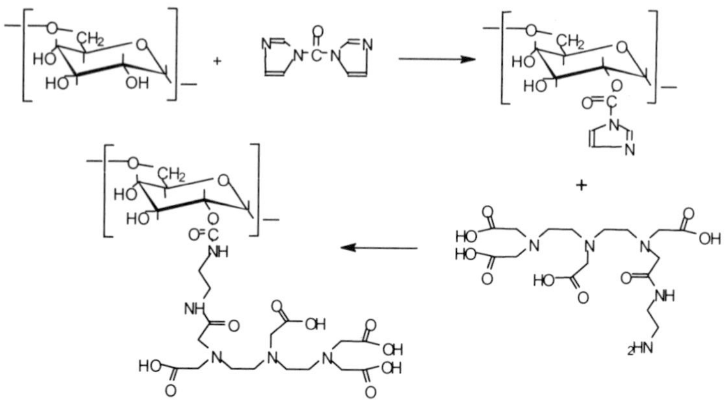

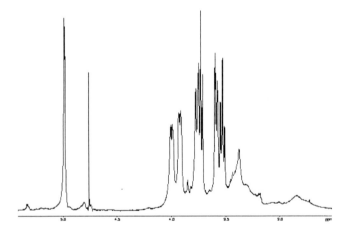

Figure 4. Reaction of the dextran with DTPA-NH$_2$ by activation with CDI

Fig. 5 shows the ^{1}H NMR spectrum of the dextran derivative with 71% degree of substitution, defined as DTPANH$_2$ units/glucose units. The derivatization degree was calculated from the ^{1}H-NMR spectra, by integration of the anomeric protons of dextran, at δ 4.9 ppm, and the integration of the other signals of dextran and those of DTPA (2-4.5 ppm), which overlap, as can be seen in the spectrum. The value of the integral of the anomeric proton signal can be used to calculate the integral of the other dextran protons, which can be subtracted from the total integral of the signals between 2 and 4.5 ppm to give the value of the integral of DTPA proton resonances present in the conjugated polymer.

Figure 5. 1H NMR spectrum of Dextran-DTPA-NH$_2$ with a degree of derivatization of 71% in D$_2$O+NaOD at 25°C, pH 7.0

Fig 6 shows that the percentages of DTPA-NH$_2$ linked to the dextran obtained in the different experiments are proportional to the quantity of CDI in the reaction mixture. We can also observe that they are lower then expected if we take into account the preliminary studies of the activation of dextran with CDI, where ethylenediamine was the coupling agent. This could be due to the higher molecular weight of the DTPA-NH$_2$, where steric effects can be present.

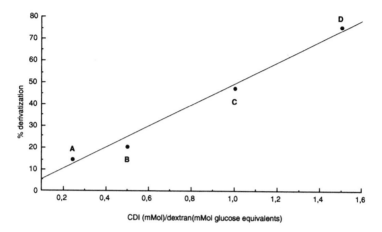

Figure 6. Degree of derivatization of Dextran-DTPA-NH$_2$ by reacting with different molar ratios of CDI/dextran (glucose equivalents):A – 15%, B – 20%, C – 45% and D – 71%

Nevertheless, these high molecular weight complexes show high propensity to be complexed with Gd^{3+} ions and be used as MRA contrast agents, since they have high solubility and are biodegradable and biocompatible.

4.   CONCLUSION

The results obtained in this work suggest that it is possible to modify the DTPA ligand and prepare DTPA-NH$_2$, which is a chelating agent with one active group. This chelating agent can be linked covalently to dextran with good yield of coupling by using CDI as an activating agent. With this work we have shown that different macromolecular chelates can be prepared, which can be used to prepare contrast agents for Magnetic Resonance Angiography.

MARIA G. DUARTE et al.

ACKNOWLEDGMENTS

The authors thank F.C.T., Portugal, for the financial support from the Praxis XXI project 2/2.2/SAU/1194/95 and for a Ph.D. grant to M. G. Duarte (BD/5056795). Financial support from the Biomed II (MACE) project of the E.U. is also acknowledged. This project was carried out within the E.U. COST Chemistry D8 and COST D18 Actions of the E.U..

REFERENCES

1. Lauffer, R.B., 1987, Paramagnetic metal complexes as water proton relaxation agents for NMR imaging. *Chem. Rev.*, **87(5)**: 901-927
2. Peters, J.A., Huskens, J. and Raber, D.J., 1996, Lanthanide induced shifts and relaxation rate enhacements. *Progr. Magn. Reson. Spectrosc.*, **28(3/4)**: 283-350
3. Aime, S., Botta, M., Fasano, M. and Terreno, E., 1998, Lanthanide (III) chelates for NMR biomedical applications. *Chem. Soc. Rev.*, **27(1)**: 19-29
4. Caravan, P., Ellison, J. J., McMurry, T. J. and Lauffer, R.B., 1999, Gadolinium (III) chelates as MRI contrast agents: Structure, dynamics and applications. *Chem. Rev.*, **99(9)**: 2293-2352
5. Armitage, F. E., Richardson, D. E. and Li, K. C. P., 1990, Polymeric contrast agents for magnetic resonance imaging. Synthesis and characterization of gadolinium diethylenetriaminepentaacetic acid conjugated to polysaccharides. *Bioconjugate Chem.*, **1(6)**: 365-374
6. Meyer, D., Schaefer, M., Bouillot, A., Beaute, S. and Chambon, C., 1991, Paramagnetic dextrans as magnetic resonance contrast agents. *Invest. Radiol.*, **26**: S50-52
7. Rongved, P. and Klaveness, J., 1991, Water-soluble polysaccharides as carriers of paramagnetic contrast agents for magnetic resonance imaging: Synthesis and relaxation properties. *Carbohydr. Res.*, **214**: 315-323
8. Rongved, P., Lindberg, B. and Klaveness, J., 1991, Cross-linked, degradable starch microspheres as carriers of paramagnetic contrast agents for magnetic resonance imaging: Synthesis, degradation and relaxation properties. *Carbohydr. Res.*, **214**: 325-330
9. Rongved, P., Fritzell, T. H., Strande, P. and Klaveness, J., 1996, Polysaccharides as carriers for magnetic resonance imaging contrast agents: synthesis and stability of a new amino acid linker derivative. *Carbohydr. Res.*, **287**: 77-89
10. Corot, C., Schaefer, M., Beauté, S., Bourrinet, P., Zehaf, S., Bénizé, V., Sabatou, M. and Meyer, D., 1997, Physical, chemical and biological evaluations of CMD-A2-Gd-DOTA. *Acta Radiol.*, 38(S412): 91-99.C. Corot, M. Schaefer, S. Beauté, P. Bourrinet, S. Zehaf, V. Bénizé, M. Sabatou, and D. Meyer, *Acta Radiol.*, **38** (1997) S412, 91.
11. Rebizak, R., Schaefer, M. and Dellacherie, E., 1997, Polymeric conjugates of Gd^{3+} diethylenetriamine pentaacetic acid and dextran 1. Synthesis, characterization and paramagnetic properties. *Bioconjug. Chem.*, **8(4)**: 605-610
12. Rebizak, R., Schaefer, M. and Dellacherie, E., 1998, Polymeric conjugates of Gd^{3+} diethylenetriamine pentaacetic acid and dextran 2. Influence of spacer arm length and conjugate molecular mass on the paramagnetic properties and some biological parameters. *Bioconjug. Chem.*, **9**: 94-99
13. Lauffer, R. B. and Brady, T. J., 1985, Preparation and water relaxation properties of proteins labeled with paramagnetic metal chelates. *Magn. Reson. Imaging* **3(1)**: 11-16

14. Schmiedel, U., Ogan, M., Paajanen, H., Marotti, M., Crooks, L. E., Brito, A. C. and Brash, R. C., 1987, Albumin labeled with gadolinium-DTPA as an intravascular blood pool enhancing agent for MR imaging: biodistribution and imaging studies. *Radiology,* **162(1)**: 205-210

15. Fossheim, S. L., Kellar, K.E., Månsson, S., Collet, J.M., Rongved, P., Fahlvik, A.K. and Klaveness, J., 1999, Investigation of lanthanide-based starch particles as a model system for liver contrast agents. *J. Magn. Reson. Imaging,* **9**: 295-303

16. Sieving, P. F., Watson, A. D., Rocklage, S. M., 1990, Preparation and characterization of paramagnetic polychelates and their protein conjugates. *Bioconjug. Chem.,* **1**: 65-71

17. Schuhmann-Giampieri, G., Schmitt-Willich, H., Frenzel, T., Press, W. R. and Weinmann H. J., 1991, In vivo and in vitro evaluation do Gd-DTPA-polylysine as a macromolecular contrast agent for magnetic resonance imaging. *Invest. Radiol.,* **26(11)**: 969-974

18. Spanoghe, M., Lanens, D., Dommisse, R., Van der Linden, A. and Alderweireldt, F., 1992, Proton relaxation enhancement by means of serum albumin and poly-lysine labeled with DTPA-Gd^{3+}: relaxivities as a function of molecular weight and conjugation efficiency. *Magn. Reson. Imaging* **10(6)**: 913-917

19. Desser, T. S., Rubin, D. L., Muller, H. H., Qing, F., Khodor, S., Zanazzi, G., Young, S. W., Ladd, D.L., Wellons, J. A., Kellar, K. E., Toner, J., and Snow, R., 1994, Dynamics of tumor imaging with Gd-DTPA-polyethylene glycol polymers: dependence on molecular weight. *J. Magn. Reson. Imaging,* **4(3)**: 467-472

20. Aime, S., Botta, M., Crich, S.G., Giovenzana, G., Palmisano, G. and Sisti, M., 1999, Novel paramagnetic macromolecular complexes derived from the linkage of a macrocyclic Gd(III) complex to polyamino acids through a squaric acid moiety. *Bioconj. Chem.,* **10**: 192-199

21. Krejcarek, G. E. and Tucker, K. L., 1977, Covalent attachment of chelating groups to macromolecules *Biochem. Biophys. Res. Commun.,* **77**: 581-585

22. J. F. Carvalho, S. P. Crofts, J. Vadarojan, 1992, Peptide linkage-containing macrocyclic compounds methods for their preparation, and their use as chelating agents in pharmaceuticals or diagnostic agents containing said chelating agents, in detoxification agents or tomagraphic imaging agents. PCT Int. Appl. WO 92 08,707

23. Geraldes, C. F. G. C., Urbano, A. M., Alpoim, M. C., Sherry, A. D., Kuan, K.-T., Rajagopalan, R., Matos, F. and Muller, R. N., 1995, Preparation, physico-chemical characterization, and relaxometry studies of various gadolinium(III)-DTPA-bis(amide) derivatives as potential magnetic resonance contrast agents. *Magn. Reson. Imaging* **13(3)**: 401-420

Fatty Esterification of Plant Proteins

FABIOLA AYHLLON-MEIXUEIRO, VÉRONIQUE TROPINI, and
FRANÇOISE SILVESTRE
*Laboratoire de Chimie Agro-industrielle, UMR 1010 INRA/INP-ENSCT - 118, route de
Narbonne, 31077 Toulouse Cedex 04, France*

Abstract: The present work opens new outlets for the valorization of plant proteins by
              chemical modification. Sunflower and wheat gluten proteins were esterified by
              n-octanol in the presence of an acid catalyst. By varying reaction parameters
              (temperature, catalyst concentration, reaction time), the optimum reaction
              conditions of esterification were defined. The esterification yield for each type
              of proteins was obtained with a maximum for sunflower proteins. The
              difference was attributed to their own structure and composition. Furthermore,
              an increase of amino groups was observed indicating a hydrolysis
              phenomenon of proteins. The esterification was accompanied by a loss of
              weight due to a loss of small peptides in solution. However, the reaction leads
              to chemically-modified proteins with new functional properties. In both cases,
              the esterified proteins were less soluble at basic pH compared to native
              proteins, indicating thus a hydrophobation effect.

1.   INTRODUCTION

Over the last two decades, there has been an increasing interest in the
industrial use of plant proteins for non-food applications because of their
renewability or biodegradability[1]. Plant proteins have thus been used for the
fabrication of materials such as films and coatings, adhesives, thermoplastics
and surfactants[2]. However, for many applications, it is necessary to confer
and/or to improve some specific properties by chemical modification of the
native proteins[3]. Particularly, the esterification of their carboxyl and amide
groups by a fatty alcohol (Fig 1) could lead to a protein-derivative with
improved functional properties. Such modification would result into a lower
water sensitivity of the protein-based products and it would therefore offer

new opportunities for the extension of their application in different fields. The present work intends to demonstrate some of these aspects.

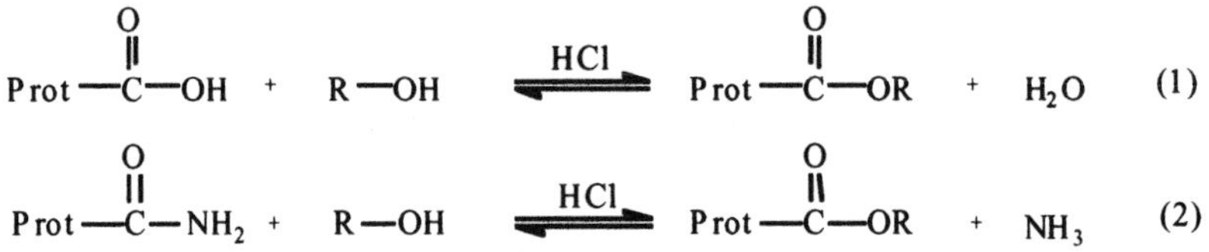

$$Prot\text{—}\overset{\overset{\displaystyle O}{\|}}{C}\text{—}OH \;+\; R\text{—}OH \;\underset{\xleftarrow{\hspace{1cm}}}{\overset{HCl}{\xrightarrow{\hspace{1cm}}}}\; Prot\text{—}\overset{\overset{\displaystyle O}{\|}}{C}\text{—}OR \;+\; H_2O \qquad (1)$$

$$Prot\text{—}\overset{\overset{\displaystyle O}{\|}}{C}\text{—}NH_2 \;+\; R\text{—}OH \;\underset{\xleftarrow{\hspace{1cm}}}{\overset{HCl}{\xrightarrow{\hspace{1cm}}}}\; Prot\text{—}\overset{\overset{\displaystyle O}{\|}}{C}\text{—}OR \;+\; NH_3 \qquad (2)$$

Figure 1. Esterification of carboxylic acid (1) and amide groups (2) of proteins by an aliphatic alcohol

2.   MATERIALS AND METHODS

Alkali-extracted proteins from sunflower oil cake (89% proteins, Nx6.25) and wheat gluten (76.5% proteins, Nx5.7) were reacted with n-octanol in the presence of an acid catalyst. Temperature, reaction time and catalyst concentration were varied according to an experimental design to maximize the esterification yield. The latter was determined by alkaline hydrolysis and subsequent analysis by gas chromatography. The hydrolysis of the peptide chain was traced by the determination of the amount of free amino groups in esterified proteins using 2,4,6-trinitrobenzene-sulfonic acid (TNBS) assays[4]. The solubility curves of modified proteins in water as a function of pH were obtained by Kjeldahl analysis[5] to determine the composition of the soluble and insoluble parts.

3.   RESULTS

Wheat gluten and sunflower proteins, containing a large amount of carboxylic groups (high amount of aspartic and glutamic acid), were successfully esterified by n-octanol and the degree of esterification was determined. We observed that the extent of esterification depended largely on temperature, acid concentration and reaction time. By increasing the reaction time, the extent of esterification increased but after 2 hours this value decreased (Fig 2). This diminution may be caused by the following phenomena.

Concurrently to an increase of esterification, a loss of weight was observed (Fig 2). This occurred even if fatty alkali chains were grafted. It was then suggested that small esterified peptides were lost in solution during the isolation of the final product.

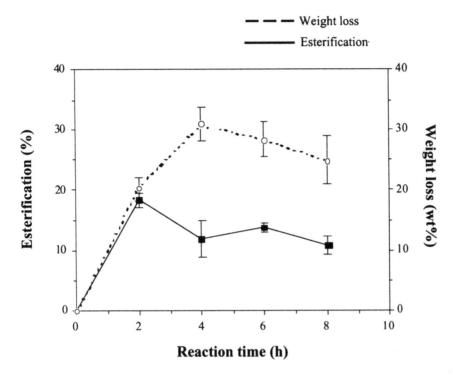

Figure 2. Degree of esterification and weight loss of wheat gluten as a function of reaction time (reactions carried out at 110°C, 3.5 meq/g protein HCl 5N).

Furthermore, during esterification, hydrolysis of proteins occurred. This phenomenon was evaluated by measuring the amount of amino groups after reaction (Fig 3). An increase of amino groups indicates the hydrolysis of the peptide bonds. Indeed, the latter is the consequence of the acid attack of the protein at relatively high temperatures yielding lower molecular weight peptides.

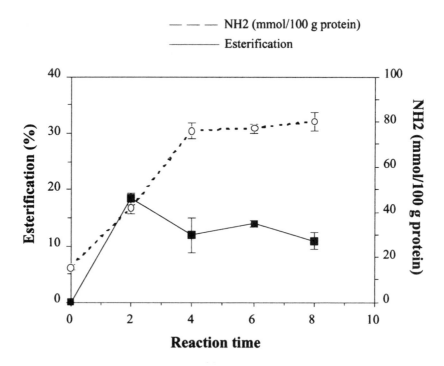

Figure 3. Degree of esterification and amount of free amino groups as a function of reaction time (reactions carried out at 110°C, 3.5 meq/g protein).

By varying reaction conditions, the optimum extent of esterification was obtained for each kind of proteins. Table 1 shows the highest esterification values obtained from an experimental design. Sunflower proteins underwent esterification in a larger extent than wheat gluten proteins. Only 25 % of the carboxyl groups of wheat gluten could be esterified with this experimental procedure. The different structure of proteins may account for this difference.

Table 1. Reaction conditions leading to maximum esterification yields

Type protein	Reaction conditions	Esterification yield
Wheat gluten	110°C, 4h, HCl 5N (4.5 meq/g protein)	25 %
Sunflower	90°C, 6h, HCl 5N (4 meq/g protein)	53 %

Esterification with a fatty alcohol modified the functional properties of proteins, particularly their solubility in alkaline water. For wheat gluten and sunflower esterified proteins, solubility at basic pH decreased compared to native proteins, indicating thus a hydrophobation effect. Fig 4 shows the effect of esterification on solubility of sunflower proteins. The solubility of unmodified sunflower proteins increased significantly in alkaline media and at pH 12 reaches 70%. On the contrary, the esterified product showed the same solubility in acid medium and did not varied significantly when pH was increased. The chemically-modified proteins showed a marked hydrophobic character.

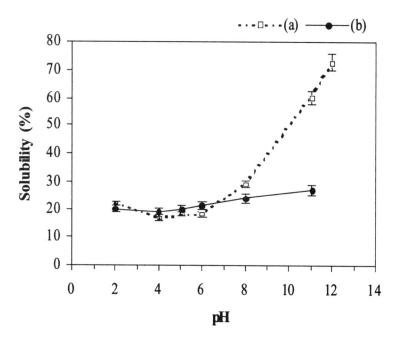

Figure 4. Solubility curves of native (a) and esterified (b) sunflower proteins (reactions carried out at 90°C, 6h, 4 meq/g protein HCl 5N).

4.   CONCLUSION AND PERSPECTIVES

Sunflower and wheat gluten proteins were successfully esterified by n-octanol. The grafting of hydrophobic groups to gluten and sunflower proteins decreased the solubility of the final product obtained therefrom.

The modification of their solvent affinity obtained by this method can open new outlets for the non-food valorization of modified proteins: elaboration of agromaterials, water-resistant protein-based films, etc.

REFERENCES

1. Sanchez, A.C.; Popineau, Y.; Mangavel, C.; Larré, C.; Guéguen, J., 1998, Effect of different plasticizers on the mechanical and surface properties of wheat gliadin films. *J. Agric. Food Chem.* **4**: 4539-4544.
2. De Graaf, L. A.; Kolster, P.; Vereijken, J. M., 1998, *Plant proteins from European Crops, Food and non-food applications*; Springer Verlag Berlin, Heidelberg, N.Y., p 335-339.
3. Means, G. E.; Feeney, R. E., 1990, Chemical modifications of proteins: history and applications. *Bioconjugate chem.* **1**: 2-12.
4. Lens, J.P. ; Dietz, C.H.J.T. ; Verhelst, K.C.S. ; Vereijken, J.M. ; De Graaf, L.A. 1999, Hydrofobization of wheat gluten using acylation and alkylation reactions. *Cereal Chem.* Submitted.
5. Ayhllon-Meixueiro, F., 2000, Ph. D., Films biodégradables à base de protéines de tournesol : Mise au point et étude des propriétés. Institut National Polytechnique de Toulouse

Chemical Modification of Wheat Gluten

[1]VÉRONIQUE TROPINI, [2]JAN-PLEUN LENS, [2]WIM J. MULDER, and [1]FRANÇOISE SILVESTRE
[1]*Laboratoire de Chimie Agro-industrielle, ENSCT-INPT, UMR-INRA, 118 route de Narbonne, 31077 Toulouse Cedex 4, France;* [2]*Agrotechnological Research Institute ATO, Subdivision Industrial Proteins, PO Box 17, 6700 A A Wageningen, The Netherlands.*

Abstract: Wheat gluten was cross-linked using water soluble 1-ethyl-3-(3-dimethylaminopropyl) carbodiimide HCl (EDC) together with N-hydroxysuccinimide (NHS) in order to improve its functional properties in non-food applications. By varying reaction parameters, it was demonstrated that cross-linking of the proteins occurred through different mechanisms. The pH-dependent formation of two types of cross-links is suggested. At pH 5-7, intermolecular cross-linking took place, while at pH 11 cross-links were preferentially formed within one protein chain (intramolecular cross-linking). However at pH 3 no reaction was observed.

1.   INTRODUCTION

Due to its very specific physiochemical properties such as elasticity, extensibility, insolubility in water, and viscoelasticity, wheat gluten is a protein that has a large potential to be used in food, feed, and non-food applications. To valorise wheat gluten in the non-food sector, chemical modifications can be performed to render the protein with specific properties[1,2]. For example, different applications demand a high water resistance. Especially this property of wheat gluten is subjected to improvement. One way to increase the water resistance is to cross-link the protein chains[3]. In the present study, the reaction of wheat gluten with the water soluble cross-linker 1-ethyl-3-(3-dimethylaminopropyl) carbodiimide HCl (EDC) in the presence of N-hydroxysuccinimide (NHS) was investigated. The extent and character of the cross-linking reaction was determined by measuring the amount of amino groups of modified wheat gluten and its solubility in aqueous solutions of different pH.

2.   MATERIALS AND METHODS

Wheat gluten (Amylum, Belgium) was dispersed in aqueous solution of different pH . The pH level was adjusted by adding aqueous solution of 0.5 M sodium hydroxide (NaOH) or 0.1 M H_2SO_4. Subsequently, NHS and EDC (Sigma, USA) were added using a molar ratio of NHS to EDC to COOH of 5:5:1. Prior to the reaction, the amount of COOH groups from the end groups, acid aspartic, and acid glutamic within these proteins was measured by tritration. Keeping the pH constant, the reaction was stirring for 4 hours at room temperature, and then stopped by adding acetic acid (Riedel-de-Haën AG, Germany). Later the dispersion was dialyzed and lyophilised giving a solid product.

The efficiency of cross-linking wheat gluten with EDC/NHS was evaluated by measuring:

- The amount of amino groups using a trinitrobenzenesulfonic acid (TNBS, Sigma, USA) assay[4].

- The solubility of the protein in aqueous solutions of different pH :

Wheat gluten (0.05 g) was dispersed in 45 g of deionized water. The pH was adjusted by adding an aqueous solution of 0.5 M NaOH or 50% (v/v) acetic acid. The dispersion was stirred at RT during 1 hr using a magnetic stirring device. The weight was adjusted to 50 g and the dispersion was stirred for another 5 min. Subsequently, 25 mL of the dispersion was centrifuged (15 min, 11,500×g). The supernatant was filtered and the nitrogen content of the supernatant was determined by Kjeldahl analysis. The nitrogen content of the original sample was determined by completely combusting about 30 mg of the protein. Subsequently, the amount of nitrogen was obtained using a rapid-N apparatus, Foss Electric (Hoorn, The Netherlands). Finally, the protein solubility was calculated by the yield of these two values obtained, multiplied by 100.

3.   RESULTS AND DISCUSSION

Cross-linking of proteins using EDC occurs through activation of the carboxylic acids groups, followed by a reaction with the free amino groups of the same or another protein chain (Fig 1). To enhance cross-linking, the catalytic reagent NHS was added to the reaction mixture. Its action is supposed to reduce the possibility of hydrolysis of activated species and to suppress side reactions.

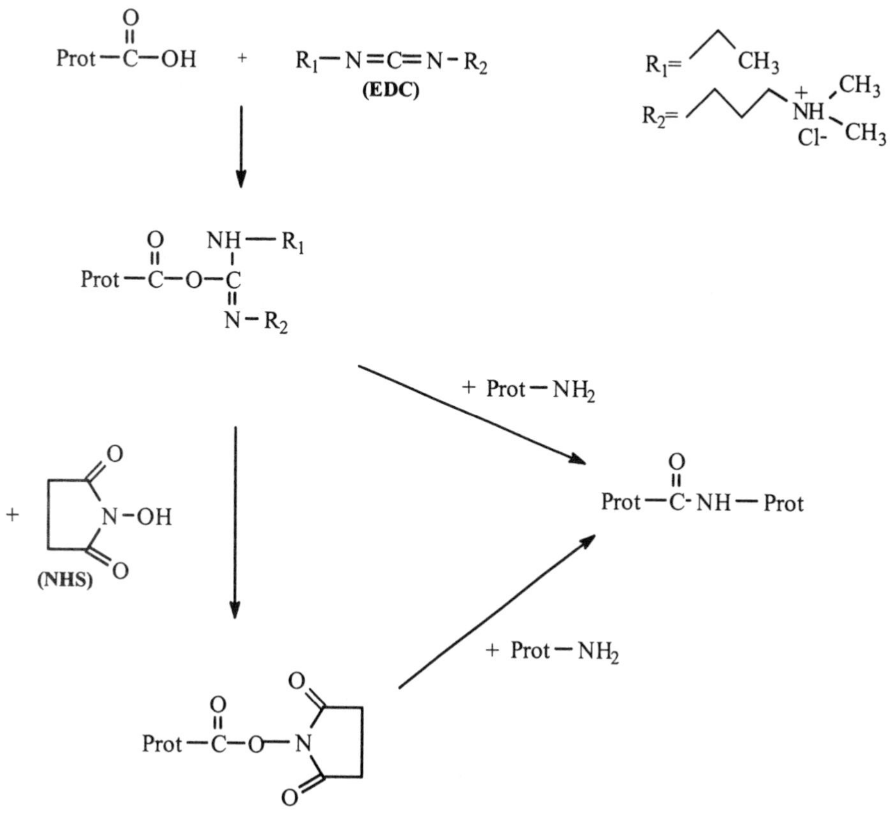

Figure 1. Cross-linking of proteins with EDC and NHS.

Cross-linking was dependent on the reaction time, the molar ratio of added reagents and the pH of the reaction mixture. By increasing the pH of the reaction mixture, the amount of amino groups of modified wheat gluten decreased. Furthermore, cross-linking affected the solubility of wheat gluten in aqueous solutions of pH 3 and 11 (Fig 2).

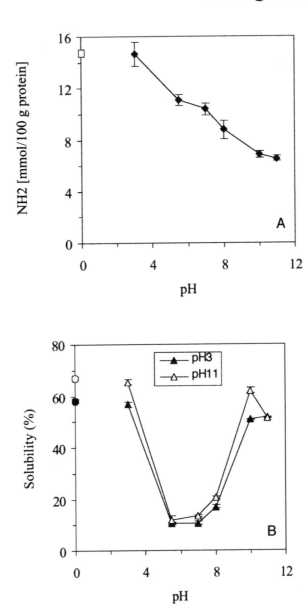

Figure 2. Amount of free amino groups (A) and solubility at pH 3 and 11 (B) of wheat gluten as a function of pH (reaction carried out at RT, 4 hrs, COOH:NHS:EDC=1:5:5). The amount of amino groups in native gluten was 14,8 mmol/100 g protein (□). The solubility of native gluten was 58% at pH 3 (●) and 67% at pH 11 (O).

If the reaction was carried out at pH 3, a low conversion of amino groups was accompanied by a minor decrease in solubility. However, wheat gluten that was modified at pH 5-7 showed a very low solubility (11%). Finally, despite the large decrease in the amount of amino groups, cross-linking at pH 11 resulted in only a slight decrease of the water solubility of wheat gluten. These observations are explained by the reactivity of the free amino groups and the accessibility of both the carboxylic acid and the amino groups.

At low pH, the protein is well soluble. However, the amino groups of the lysine residues are protonated, which eliminates the nucleophilic character of this group and prevents cross-linking. At pH 5-7, the protein has a zero net charge. Protein-protein interactions are favoured, creating aggregates. Reagents may only reach the surface of the protein. These conditions favour intermolecular cross-linking between carboxylic acid and amino groups of different protein chains. Finally, at pH 11, the nucleophilic deprotonated amino groups readily react with the activated carboxylic acid groups. Additionally, at this high pH, the protein chains are highly unfolded and EDC and NHS can penetrate the proteins. Consequently, preferentially intramolecular cross-links are formed. Thus, two different cross-link mechanisms are suggested (Fig 3). By controlling these mechanisms, the specific properties of the final product can be adjusted.

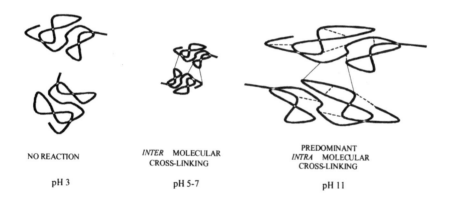

NO REACTION   *INTER* MOLECULAR CROSS-LINKING   PREDOMINANT *INTRA* MOLECULAR CROSS-LINKING

pH 3   pH 5-7   pH 11

Figure 3. Proposed mechanism of cross-linking proteins with EDC/NHS at different pH.

4. CONCLUSION

Chemical modification is a powerful tool to improve protein properties. There are different ways to modify proteins. Cross-linking, which is the covalent linking of protein chains, decreases the water sensitivity of wheat gluten. This may lead to improved water stability and to an increased strength of protein based products. The present work opens new outlets for the valorisation of proteins. In industrial products, proteins, which are biodegradable, may replace synthetic polymers; wheat gluten may be used in technical applications such as coatings, adhesives, paints, cosmetics, and detergents.

REFERENCES

1. Lens, J.-P., Mulder, W. J., and Kolster, P., 1999, Modification of wheat gluten for nonfood applications. *Cereal Foods World* **44**: 5-9.
2. Means, G. E., and Feeney, R. E., 1990, Chemical modifications of proteins: history and applications. *Bioconjugate Chem.* **1**: 2-12.
3. Lundblad, R. L., and Noyes, C. M., 1984, *Chemical reagents for protein modification*; CRC Press: Boca Raton, FL, Vol. 2.
4. Lens, J.-P., Dietz, C. H. J. T., Verhelst, K.C. S, Vereijken, J. M., and De Graaf, L. A. . 1999, Hydrofobization of wheat gluten using acylation and alkylation reactions. *Cereal Chemistry* submitted.

Enzymatic Crosslinking Enhance Film Properties of Deamidated Gluten

COLETTE LARRÉ, CLAUDE DESSERME, JACKY BARBOT, CÉCILE MANGAVEL and JACQUES GUÉGUEN
Institut National de la Recherche Agronomique, Unité de Biochimie et Technologie des Protéines, B. P. 71627, 44316 Nantes Cedex 3, France

Abstract: Covalent bonds were introduced in deamidated gluten films by transglutaminase. The type of cross-links was modulated by adding diamines in the film-forming solution. Biochemical and mechanical properties of the resulting films were investigated. Film properties of deamidated gluten were compared to those of native gluten. Plasticizer content modified considerably the film properties and a ratio of 0.35 was chosen for the following experiments. At this ratio, the addition of increasing diamine concentration in the film solution was shown to have a slight positive plasticizer effect. As shown by SDS PAGE of the film proteins, transglutaminase was efficient in catalysing the formation of polymers in the film solution. The loss of solubility was related to the formation of high molecular weight polymers in the film. In all cases, the action of transglutaminase induced simultaneously an increase in strain and stress of the films. The addition of diamines in the film solution affected more the strain than the stress properties. These diamines, able to react at their two extremities, probably acted as spacers between gluten proteins. The IR study didn't show any modification of the secondary structure of the proteins in the film. Transglutaminase was efficient in introducing new cross-links in gluten films. This modification enhanced their mechanical properties and modified their solubility properties.

1. INTRODUCTION

Considerable research work has been done these forty last years to develop edible films and coatings[2,4,8]. Compared with the numerous applications of traditional polymeric films, only few of these edible films have been applied commercially. However, even if the mechanical and barrier properties of traditional films are generally better, edible films offered advantages including the renewable nature of the raw material and their reduced environmental impact[9].

Wheat gluten films have been extensively studied[5,6,7,15]. When prepared by casting they are treated in acidic or basic conditions in order to help dispersion and added with a plasticizing agent to overcome the brittleness of the films and improve their flexibility. Krull and Inglett[10] casted films from gluten hydrolysates alone or in mixture with chemically modified polypeptides. Good tensile strength properties, comparable with those of polyvinyl chloride, were obtained with a mixture of polypeptides and ethylene-treated derivatives (6 :1, w :w).

Only few studies have been reported on the effect of the addition of cross-links on film properties. Chemical cross-linkers such as formaldehyde, glyoxal or glutaraldehyde were efficient in enhancing the maximum puncture force of films made from cotton seed proteins[13]. Brault et al[1] introduced cross-links between aromatic compounds (i.e. dityrosine) in caseinate films by exposing them to γ-ionisation. Puncture strength and puncture deformation were both increased but this effect was very dependant of the ratio glycerol/protein.

Two types of enzymes were used to introduce cross-links in protein films : peroxidase and transglutaminase. A treatment by horseradish peroxidase in the preparation of films of thermally denatured soy proteins reduced the elongation at elastic limit and increased the tensile strength[16].

The effect of cross-links established through the action of a transglutaminase on film properties was also studied in films casted from various proteins. After enzymatic treatment, the α_{s1} casein in the film became less water soluble but were still susceptible to proteolysis, moreover the tensile strength and strain of films were increased[14]. Similar results were obtained by Yildirim and Hettiarachchy[17] on films obtained with whey proteins, 11S globulins or a mixture, they exhibited higher mechanical properties and higher water vapour permeability. Mahmoud and Savello[12] showed that the mechanical properties of whey proteins films obtained after crosslinking were dependent on the level of plasticizer.

Transglutaminase catalyses the self polymerisation of proteins through ε-(γ-glutamyl)-lysine bonds but can also introduce cross-links between external primary amines and glutamine residues of proteins. In the case of gluten in doughy state, this enzyme increased the proportion of high molecular weight polymers and strengthened the network rheological properties[11]. Despite its very low content in lysine, some intermolecular ε-(γ-glutamyl)lysine bonds were generated in the gluten.

The objective of this study was to produce films by cross-linking gluten or deamidated gluten using transglutaminase and to characterise the effect of the introduction of intermolecular covalent bonds. Adding various diamines in the film-forming solution was used to modulate the length of cross-links. Mechanical properties of the resulting films were investigated as well as their solubility in water.

2.   MATERIAL AND METHODS

2.1   Materials

The experiments were done with industrial glutens, native and deamidated at 20 % provided by Amylum. Putrescine, cadaverine, hexyldiamine, 1-8 diaminooctane, and glycerol were purchased from Sigma Chemical Co. Transglutaminase (1 U /mg) was supplied by Ajinomoto. Transglutaminase (Tgase) activity was determined according to the colorimetric hydroxamate procedure of Folk[3].

2.2   Preparation of gluten films

Deamidated glutens were dispersed in 0.05 M Tris HCl pH 8. Film forming solution were prepared at 10% protein with ratios in glycerol/proteins varying from 0.2 to 0.5, mixed using a polytron at 25 000g then centrifuged at 1000 g for 15 min in order to remove foam. The resulting solution was then casted on glass plate before drying (70°C during 60 min). The same procedure was followed for native gluten except that the dispersion was done in 0.1 N NaOH and the film forming solution was prepared at 12.5 % proteins.

When alkyldiamines were used, they were directly added to the film-forming solution. For enzymatic assays, the procedure was slightly modified. The enzyme was added immediately prior casting. The casted film-solution was then incubated for 3 hours at 37 °C. In order to avoid drying during the incubation, the glass plate was put into a tray which was covered. At the end of the incubation time, the drying procedure previously described was applied. This procedure was followed for enzyme treated films and their references without enzyme.

2.3   Mechanical characterisation of the films

Mechanical properties were analysed on five replicate for each film tested. The shape and size of the tested pieces of film were made according to ISO 527-2 standards. Elongation at break and tensile strength were measured at 20°C and at 57° ± 2 of hygrometry on an Adamel-Lhomargy Testing Instrument (model DY34). Tensile strength (σ) was calculated by dividing the maximum stress at break by the cross sectional area of the film, elongation (Δl) was expressed in percentage of the initial length of the piece tested ($l_0 = 20$ mm).

$$\sigma = E_0 \left(\Delta l / l_0 \right)$$

2.4   Statistical analysis

Statistical analyses were performed using Statgraphics Plus. The results were compared by analysis of variance and Fisher's least significance difference procedure.

2.5   Film protein solubility

Film proteins were fractionated into soluble and insoluble fraction in water. In the solubilization procedure, 40 mg of each film was added with 1 ml of water, stirred 20 hours at 20°C then centrifuged at 20 000g. The supernatant was recovered and diluted in order to obtain a final concentration of 12.5 mM borate buffer pH 8.5, 2% SDS. One millilitre of 12.5 mM borate buffer pH 8.5, 2% SDS was added to the pellet and put under stirring 20 hours at 20°C. The resulting solution was centrifuged at 20 000 g before further analysis. The absence or presence of pellet was controlled. Following this procedure, supernatants and resolubilized pellets were analysed by size exclusion chromatography using a superose 6 (Pharmacia) eluted with 12.5 mM borate buffer pH 8.5, 0.2 % SDS. The proportion of soluble and insoluble proteins was calculated from the peak areas.

3. RESULTS AND DISCUSSION

3.1 Effect of glycerol concentration on the properties of films obtained with deamidated gluten

Because of its insolubility at neutral pHs, the film-forming solution of native gluten was prepared at pH 10 while that of deamidated gluten was prepared at pH 8. Whatever the substrate used, the global effect of glycerol addition was very similar (Fig 1). In both cases, a minimum ratio of glycerol/ protein of 0.2 was necessary to overcome the brittleness of the film. The decrease of the maximum stress and the increase of the strain at break when the proportion of added glycerol increases reflected a classical plasticizer effect. The stress values were quite similar with both substrates, the main difference was noted on the elongation at break.

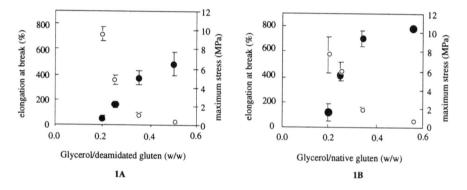

1A   **1B**

Figure 1. Effect of glycerol concentration in the film-forming solution on mechanical properties. 1A deamidated gluten at 10%.; 1B native gluten at 10 %
O   Maximum stress;   ● Elongation at break

A concentration of 10% in deamidated gluten and a ratio glycerol/protein of 0.35 was retained for further enzymatic experiments.

3.2 Enzymatic treatment

Preliminary assays of enzymatic treatment were performed on native gluten. In order to obtain homogeneous film-forming solutions, we prepared them in acidic (pH3) and basic (pH10) conditions but in these conditions transglutaminase was inefficient in generating any cross-linkages in the system. The preparation of gluten dispersion around neutral pHs led to a non homogeneous film-forming solution which could not be used for further characterisation. In order to overcome these difficulties and to be able to test the effect of the addition of intermolecular cross-links on film properties we chose to perform our experiments on a deamidated gluten commercially available. The ability of deamidated gluten to be substrate for transglutaminase was tested in solution. The electrophoretic analysis of the reaction products in reducing conditions showed the apparition of polymers of higher molecular weight which did not exist in non treated deamidated gluten (results not shown).

3.3   Effect of putrescine concentration in the film-forming solution

Putrescine as some other polyamines is a natural substrate for transglutaminase[3]. One or both amino groups of this molecule can react with glutaminyl residues leading either to N-(γ glutamyl)putrescine or to N^1,N^4-bis (γ glutamyl)putrescine. When putrescine is incorporated through both primary amino groups, a cross-link between two polypeptide chains occurs. As lysine content of gluten is very low, the addition of putrescine to the mixture potentially increases the number of intermolecular cross-linkings. The concentrations of added putrescine corresponded to ratios of putrescine added/glutamine from 0.025 to 0.3 mole/mole. Tensile strength (TS) and elongation at break (EB) obtained for each film with or without addition of transglutaminase are reported in Fig 2. TS and EB of films reference obtained in presence of increasing quantities of putrescine simultaneously decreased, indicating that putrescine didn't act as plasticizer. With or without addition of putrescine, transglutaminase action induced simultaneously an increase of both TS and EB. Variance analysis of the results obtained on treated or non treated films pointed out significant main effects of transglutaminase treatment and putrescine concentration on both tensile strength and elongation at break (p values < 0.05). A significant interaction between transglutaminase treatment and putrescine concentration was found for tensile strength (p value < 0.05) but not for elongation at break (p value $= 0.315$). This increase in TS and EB points to the formation of covalent linkages, which reinforced the film but were flexible enough to permit a gain in elongation. The addition of increasing concentrations of putrescine in transglutaminase treated films induced a slight and progressive decrease in TS. The effect on EB was not linear, at low putrescine concentrations EB increased then decreased at high concentrations. When only few putrescine molecules were present in the film solution, they could act as bifunctional cross-linkers between two glutaminyl residues but when their concentration increases some of them probably react only by one amino group and therefore did not create cross-linking between polypeptides. The highest elongation at break was obtained with ratios of putrescine/Gln around 0.18 meaning that a maximum of 36 % of glutamine had reacted.

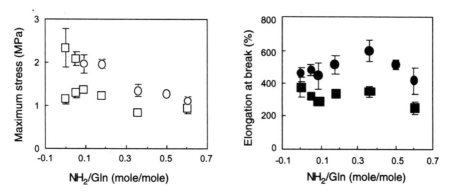

Figure 2. Tensile properties of deamidated gluten films cross-linked by transglutaminase. Treated film in circle, non treated film in square

The solubility of TG treated films was compared to that of non treated films (Fig 3). Films obtained with deamidated gluten were soluble at 78% in water, the addition of putrescine didn't modify the film solubility (Fig 3). The action of transglutaminase induced a decrease in the film water solubility. The highest insolubility was obtained for a ratio putrescine / glutamine of 0.18 mole/mole and could be related to the formation of polymers of high molecular weight.

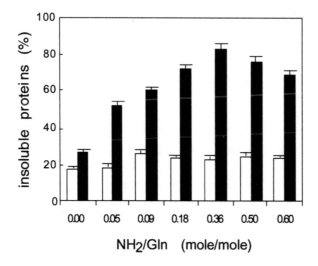

Figure 3. Evolution of water-insoluble proteins in films in presence of added putrescine and enzymatic treatment. In black: Film treated with transglutaminase. In white: Untreated films. Percentages are calculated from peak areas obtained by gel filtration.

In order to control the size of proteins in the water-insoluble fractions, they were solubilised in 2% SDS then analysed by gel filtration (results not shown). This analysis clearly showed the apparition of proteins of molecular weight higher than $2 \cdot 10^6$ which are generated by the transglutaminase reaction.

3.4   Effect of the addition of diamines with various length

Whatever the type of diamine used, transglutaminase treatment had a significant effect on films mechanical properties (Fig 4). Tensile strength and elongation at break were both increased (p value < 0.05). In the case of transglutaminase incubated films, differences between various diamines were studied using multiple range tests. For both properties two homogenous groups were identified but their meaning is not clear (Fig 4). In all cases the film water solubility decreased and polymers of high molecular weight were identified.

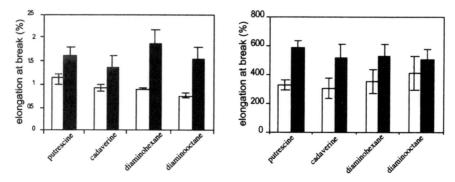

Figure 4. Mechanical properties of films prepared with various diamines added at the same concentration. Plasticizer/protein = 0.35; in white : Untreated films; in black: Films treated with transglutaminase

3.5   Effect of transglutaminase treatment on secondary structure of proteins in films

The protein conformation in films was characterised by infra red measurements. Spectra obtained by attenuated total reflection and data in the amide I region were compared in order to characterise the secondary structure of proteins in the film. The deconvolved spectra, as shown here, presented two main bands at 1650 and 1621 cm^{-1} (Fig 5). The band at 1650 is generally assigned to the helical conformation and the 1621 cm^{-1} bands reflects the presence of intermolecular hydrogen bonded beta sheet structure. The spectra recorded for gluten in film state are comparable to those obtained in doughy state. No difference was observed between the films tested indicating that protein secondary structures were similar for native or deamidated gluten and moreover that transglutaminase treatment didn't affect the secondary structure of proteins in the films. The formation of polymers in the film had no influence on the band at 1621 cm^{-1} which is usually related to the formation of aggregates.

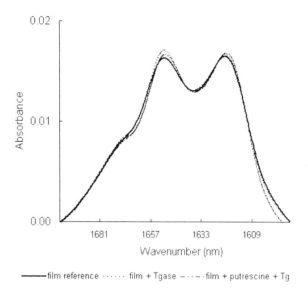

Figure 5 : Fourier deconvolved infrared spectra in the amide I region of deamidated gluten films

4. CONCLUSION

This study showed that transglutaminase was efficient in introducing covalent bonds in films obtained with slightly deamidated gluten. The formation of these covalent bonds has no influence on the secondary structure of proteins in the film. However they induced the formation of polymers of high molecular weight which are responsible for the higher insolubility of enzyme-treated films. On the basis of mechanical properties, we showed that the addition of covalent bonds by the way of transglutaminase increased the film integrity and heavy-duty capacity as well as its ability to stretch.

We also showed that the incorporation of diamine was efficient in increasing the elongation. This result could be attributed to the N^1,N^4-bis (γ glutamyl)diamine bond formed which could act like a spacer between the bound polypeptides and may allow a higher mobility of the polypeptides bound together.

ACKNOWLEDGMENTS

This research was funded by the European community in the frame of FAIR CT 96 1979.

REFERENCES

1. Brault D., D'Aprano G. and Lacroix M., 1997, Formation of free-standing sterilized films from irradiated caseinates.. *J. Agric. Food. Chem.*, **45**: 2964-2969.
2. Cuq B., Gontard N. and Guilbert S., 1998, Proteins as agricultural polymers for packaging production. *Cereal Chem*, **75(1)**: 1-9.
3. Folk J.E., Park M.H., Chung S.I., Schrode J., Lester E.P. and Cooper H.L., 1980, Polyamines as physiological substrates for transglutaminases. *J. Biol. Chem.*, **255**: 3695-3700.
4. Gennadios, A. and Weller, C.L., 1991, Edible films and coatings from soymilk and soy protein. *Cereal Foods World* **36**: 1004-1009.
5. Gennadios, A. and Weller, C.L., 1990, Edible films and coatings from wheat and corn proteins. *Food Technol.* **44**: 63-69.
6. Gontard, N. , Guilbert, S. and Cuq, J.L., 1992, Edible wheat gluten films : influence of the main process variables on film properties using surface response methodology. *J. Food Sci.* **57**: 190-195.
7. Herald, T.J. ; Ganasambandam, R. , McGuire, B.H. and Hachmeister, K.A. , 1995, Degradable wheat gluten films : preparation, properties and applications. *J. Food Sci.* **60**: 1147-1150.
8. Kester, J.J. and Fennema, O.R., 1986, Edible films and coatings : a review. *Food Technol.*, **40**: 47-59.
9. Krochta, J. M. and De Mulder- Johnston, C., 1997, Edible and biodegradable polymer films : challenges and opportunities. *Food Technol.* **51**: 61-74.
10. Krull L.H. and Inglett G.E., 1971, Industrial uses of gluten. Cereal Sci. Today **16(8)**: 232-261
11. Larré C., Deshaye G., Lefebvre J. and Popineau Y. 1998, Hydrated gluten modified by a transglutaminase. *Nahrung*, **42**: 155-157
12. Mahmoud R. and Savello P.A. , 1993, Solubility and hydrolyzability of films produced by transglutaminase catalytic crosslinking of whey protein, *J. Dairy Sci.*, **76**: 29-35.
13. Marquié, C. ; Aymard, C. ; Cuq, J. L. and Guilbert, S., 1995, Biodegradable packaging made from cottonseed flour: formation and improvement by chemical treatments with gossypol, formaldehyde and glutaraldehyde. *J. Agric. Food. Chem.*, **43**: 2762- 2767.
14. Motoki M., Aso H., Seguro K. and Nio N., 1987, α_{s1} casein film prepared using transglutaminase. *Agric. Biol. Chem.*, **51**: 993-996.
15. Roy S., Weller C.L., Gennadios A., Zeece M.G. and Testin R.F., 1999, Physical and molecular properties of wheat gluten films cast from heated film-forming solutions. *J. Food Sci.* **64**: 57-60.

16. Stuchell Y.M. and Krochta J.M., 1994, Enzymatic treatments and thermal effects on edible soy protein films *J. Food Sci.* **59,6**: 1332-1337.
17. Yildirim M. and Hettiarachchy N.S., 1997,Biopolymers produced by cross-linking soybean 11S globulin with whey proteins using transglutaminase. *J. Food Sci.*, **62, 2**: 270-275.

Laccase – a Useful Enzyme for Modification of Biopolymers

KRISTIINA KRUUS, MARJA-LEENA NIKU-PAAVOLA and LIISA
VIIKARI
VTT Biotechnology, Tietotie 2, Espoo, P.O.BOX 1500, FIN-02044 VTT, Finland

Abstract: Laccases are an interesting group of multi-copper enzymes, which have
              potential within various applications. They have surprisingly broad substrate
              specificities and can oxidize simple diphenols, polyphenols, diamines, and
              aromatic amines. Laccases oxidize their substrates by a one-electron transfer
              mechanism. Molecular oxygen is used as an electron acceptor. The substrate
              loses a single electron and usually forms a free radical. Laccases are widely
              distributed in nature. The best known laccase producers are from fungal origin.
              They have several functions in nature e.g. are involved in both polymerisation
              and de-polymerisation processes of lignin. Laccases have been used
              successfully in biglueing of lignocellulose material in order to produce
              lignocellulose based composites, like fibre or particle boards. Promising
              results with laccases have also been achieved in grafting reactions. Oxidation
              of biopolymer substrates such as starch or cellulose has been carried out with
              laccases combined with mediators.

1.   INTRODUCTION

Due to increasing environmental concerns, many industries have started
to develop clean and green technologies. The term clean technology refers to
processes, which have an improved environmental impact compared to
existing technologies. It may include both new waste-free or low-waste
processes. Ecological concerns have resulted in a steadily growing interest in
natural and renewable materials, and therefore issues, such as recyclability
and environmental safety are becoming more and more important, especially
when introducing new materials and products. Enzymatic processes have

Biorelated Polymers: Sustainable Polymer Science and Technology
Edited by Chiellini *et al.*, Kluwer Academic/Plenum Publishers, 2001

over the past years gained a steadily growing interest. They could offer an interesting alternative to conventional chemical processes.

Enzymes are powerful catalysts. They are proteins, which are completely degraded in nature. Enzyme reactions are performed in mild conditions: close to physiological temperature and pH. Because enzymes are specific, the side reactions are not very common. Enzymatic processes have gained an established position within various industrial applications in the detergent, starch processing, dairy, textile and pulp and paper industries.

2.   THE REACTIONS CATALYZED BY LACCASES

Laccases (benzenediol:oxygen oxidoreductases, EC 1.10.3.2) are a diverse group of multi-copper enzymes, which catalyze oxidation of a variety of aromatic compounds. Laccases oxidize their substrates by a one-electron transfer mechanism. They use molecular oxygen as the electron acceptor. The substrate loses a single electron and usually forms a free radical. The unstable radical may undergo further laccase-catalysed oxidation or non-enzymatic reactions including hydration, disproportionation, and polymerisation[1].

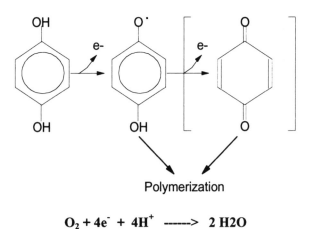

$$O_2 + 4e^- + 4H^+ \longrightarrow 2 H2O$$

Figure 1. Laccase catalyzed oxidation of diphenol.

Laccases have very broad substrate specificities. Simple diphenols like hydroquinone and catechol, polyphenols, diamines, and aromatic amines are good substrates for most laccases. The full range of compounds oxidised by laccases is not yet known. Due to the discovery of the mediator concept in early nineties[2-4], the range of substrates can even be increased. Mediators are small molecular weight compounds, which can be oxidized by laccase. The oxidized mediator will oxidize the actual substrate. A typical example of a mediator is hydroxybenzotriazole (HBT), which has been studied intensively for delignification together with laccase[5]. Laccase alone can only oxidize phenolic components in lignin. However, when combined with a mediator also non-phenolic groups will be oxidized. In principal, different types of monomers, as well as polymers, can be oxidized by a suitable enzyme mediator combination. Because of the broad substrate specificity of laccases, they possess great biotechnological potential. The most intensively studied applications for these enzymes include delignification, textile dye bleaching, effluent detoxification, detergent components, removal of phenolic components from wine, as well as biopolymer modification[6]. Although many possible applications for laccases have been reported recently, they are currently used in large scale only in finishing of denim fabrics.

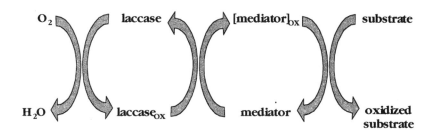

Figure 2. The laccase mediator concept.

3.   SOURCES OF LACCASES AND THEIR FUNCTIONS

It was as early as 1883 when laccase was first discovered from the extract of the Japanese laquer tree *Rhus vernicifera*. Laccase or laccase-like activity has also been demonstrated by other higher plants, including apricots, mango, rosemary, mung beans, poplar, maple, and tobacco. Some insects and a few bacteria have also been reported to produce laccase activity. However, most known laccases are of fungal origin, especially belonging to the class of white-rot fungi. Well known laccases have been isolated from *Coprinus, Myceliophthora, Phanerochaete, Phlebia, Pleurotus, Pycnoporus, Rhizoctonia, Schizophyllum, and Trametes*[7]. Laccase is commercially available as a commodity enzyme. Laccases are involved in several physiological functions, such as lignin biosynthesis, plant pathogenesis, insect sclerotization, and degradation of lignocellulosic materials. It is well recognised that laccases are involved in both polymerisation and de-polymerisation processes of lignin[1].

4.   POSSIBLE APPLICATIONS FOR LACCASES

4.1   Glueing of lignocellulose material

The ability of laccases to polymerize lignin can be exploited in lignin based adhesives. Promising results have been achieved, when laccase has been used in bonding of wood fibres by activating lignin. In conventional production of lignocellulose based composites, like fibre or particle boards, synthetic adhesives, usually urea- and phenol formaldehydes, are used in combination of hot pressing. It has been shown that with laccases the use of synthetic adhesives can be reduced or even omitted. The laccase catalyzed bonding can be achieved by activation of the lignin present in fibres (one component system) or using additional lignin (two component system), which is activated by the enzyme. Haars and Huttermann[8] used laccase treated lignosulphonates in their first experiments in order to bond wood material. In their later experiments[9], spent sulphite liquor was used with laccase for particle board and wood laminate production.*Trametes hirsuta* laccase was used with concentrated refining process water supplemented with additional Kraft lignin in order to prepare fibre boards and MDF boards[10]. The tensile strength measurements from the test fibre boards showed clearly that laccase treatment was comparable to a process where synthetic adhesives, urea formaldehyde resins, were used.

Kharazipour et al.[11] reported a one component system, where laccase treated wood fibres with no additional adhesive were used for MDF-boards preparation. The wood fibres were incubated with laccase for 2-7 days. After the incubation excess water was removed and the fibres were hot pressed to desired boards. The incubation time was later reduced to 12 hrs[12]. The results indicated that the technical values obtained from the enzymatically bound MDF-boards met the German Standards. Felby et al.[13] has also shown that production of MDF boards can be carried out without synthetic adhesives. They have studied the possible reaction mechanism involved in bonding. Although the mechanism is not completely understood, it presumably involves direct oxidation of fibre surface lignin and parallel radicalization of solubilized or colloidal lignin. These radicals will react further without enzymatic action. During hot pressing fibre to fibre bondings are formed between radicals and other reactive groups situated on separate fibres[14].

Besides laccases, peroxidases have been studied for activation of lignin in order to enhance bonding effect of wood fibres[12, 15].

4.2 Grafting reactions catalyzed by laccase

Copolymerization of organosolv lignin and acrylamide has been demonstrated by the action of laccase[16]. The reaction was carried out in organic solvent, dioxane, and it required also organic hydroperoxides. It is assumed that phenoxy radicals initiated by laccase oxidation of lignin do not *per se* cause genuine grafting. Instead, alkoxy radicals generated by decomposition of dioxane peroxide are supposed to initiate homopolymerization of acrylamide. The elongation of polyacrylamide chain is terminated by radical coupling to phenoxy radicals in lignin[17]. This reaction offers an attractive biotechnical alternative to utilize lignin, an abundant waste material from the pulp and paper industry. Lignin could thus be converted to added value synthetic polymer products.

Ikeda et al.[18] has used laccase in order to polymerize benzoic acid derivatives. The reaction was also carried out in organic solvents. This process was demonstrated to be an alternative for production of conventional phenol resins, however without the use of toxic formaldehyde.

Pedersen et al.[19] has modified the charge of lignocellulosic material by enzyme aided grafting of phenolic compounds onto lignocellulose material *e.g.* pulp. Lignocellulose material was treated with laccase and additional phenolic substance, such as ferulic acid. Laccase catalyzed oxidation of phenolic groups and formation of radicals in the aromatic moieties of both lignin and phenolic substance.

4.3 Oxidation of polymeric substrates

Starch is a naturally occurring polymer, whose derivatives are used as fillers for glueing and as coating agents in the paper and board industry. Oxidized starch as well as starch esters and cationized starch are used for surface sizing of paper, for improving strength and printing properties and as adhesives to bind the pigment particles in coating. Starch is presently oxidized industrially with the aid of inorganic agents, usually sodium hypochlorite and hydrogen peroxide[20].

We have introduced carbonyl and some carboxyl groups to native potato starch by treating the starch with laccase and the mediator Tempo (2,2,6,6-tetramethyl-1-piperidinyloxy). Native potato starch was incubated in the buffered solution with laccase and Tempo. After reaction the starch was collected and the carbonyl and carboxyl groups were determined by titration. The results showed clearly that neither Tempo nor laccase alone were able to oxidize the starch. However, as a result of a combined action of laccase and Tempo, carboxyl and especially carbonyl groups were introduced to starch[21].

Table. 1. Introduction of carboxyl and carbonyl groups to potato starch by laccase-Tempo-treatment.

Starch	COOH/100 Glc	CHO/100 Glc
Native potato starch	0.47-0.52	0
Laccase-Tempo oxidized starch	1.02	2.99

It has been shown that also cellulose can be oxidized similarly with the combined action of laccase and Tempo[22].

5.   CONCLUSION

This short review emphasizes the potental of laccase in the production of new biopolymers with interesting properties. Advances in biotechnology have already enabled more efficient production and design of new properties of enzymes. Previously, only hydrolytic enzymes have been used for the modification of biopolymers. Oxidation of polymer results either in polymerization through radical formation or in formation of carbonyl/ carboxylic groups. These reactions can be exploited for designing new functional polymers. Enzymes are fascinating alternatives for traditional technologies, because they are specific, powerful, and environmentally benign catalysts.

REFERENCES

1. Thurston, C.F. 1994. The structure and function of fungal laccases. *Microbiology* **140**: 19-26.
2. Bourbonnais, R., and Paice, M.1990. Oxidation of non-phenolic substrates: An expanded role for laccase in lignin biodegradation. *FEBS Lett.* **267**: 99-102.
3. Bourbonnais, R., and Paice, M.G. 1992. Demethylation and delignification of kraft pulp by Trametes versicolor laccase in the presence of 2,2'-azinobis-(3-ethylbenzthiazoline-6-sulphonate). *Appl. Microbiol. Biotechnol.* **36**: 823-827.
4. Call, H-P., Process for Modifying, Breaking down or Bleaching Lignin, Materials Containing Lignin or Like Substances. *PCT world patent application* WO 94/29510.
5. Call, H.P., and Mucke, I. 1997. History, owerview and applications of mediated lignolytic systems, especially laccase-mediator-system (Lignozyme-process). *J. Biotechnology* **53**: 163-202.
6. Gianfreda, L., Xu, F., and Bollag, J-M. 1999. Laccases: a useful group of oxidoreductive enzymes. *Bioremediation J.* **3(1)**: 1-25.

7. Xu, F., 1999. Recent process in laccase study: properties, enzymology, production, and applications. In: *The encyclopedia of bioprocessing technology: fermentation, biocatalysis and bioseparation* (Flickinger, M.C., and Drew, S.W. eds.) John Wiley&Sons, New York, pp. 1545-1554.

8. Haars, A., and Huttermann, A. 1983. Binder for wood materials. *German Patent* DE3037992.

9. Haars, A., Kharazipour, A., Zanker, H., and Huttermann, A. 1989. Room-temperature curing adhesives based on lignin and phenoloxidases. *ACS Symp.Ser.* **385**: 126-134.

10. Viikari, L., Hase, A., Qvintus-Leino, P., Kataja, K., Tuominen, S., and Gädda, L. A new adhesive for fibre boards. *PCT Word patent application* WO 98/31762.

11. Kharazipour, A., Huttermann, A., Kuhne, G., and Rong, M.1993. Verfahren zum Verkleben von Holzfragmenten und nach dem Verfahren hergestellte Formkörper. *European Patent* 0565109.

12. Kharazipour, A, Bergmann, K., Nonninger, K., and Huttermann, 1998. Properties of fibre boards obtained by activation of the middle lamella lignin of wood fibres with peroxidase and H2O2 before conventional pressing. *J.Adhesion Sci.Technol.* **12**: 1045-1053.

13. Felby, C., Pedersen, L.S., and Nielsen, B.R. 1997. Enhanced auto adhesion of wood fibres using phenol oxidases. *Holzforschung* **51**: 281-286.

14. Felby, C., Nielsen, B.R., Olesen, P.O., and Skibsted, L.H. 1997. Identification and quantification of radical reaction intermediates by electron spin resonance spectrometry of laccase-catalyzed oxidation of wood fibres from beech (Fagus sylvatica). *Appl. Microbiol Biotechnol* **48**: 459-464.

15. Nimz, H.H., Gurang, I., and Mogharab, I. 1976. Untersuchungen zur Vernezung technischer Sulfitablage. *Liebigs Ann. Chem.* 1421-1434.

16. Mai, C., Milstein, O., and Huttermann, A 1999. Fungal laccase grafts acrylamide onto lignin in presence of peroxides. *Appl. Microbiol. Biotechnol.* **51**: 527-531.

17. Mai, C., Milstein, O., and Huttermann, A., 2000. Chemoenzymatical grafting of acrylamide onto lignin. *J. Biotechnol.* **79**: 173-183.

18. Ikeda, R., Sugihara, J., Uyama, H., and Kobayashi, S. 1998. Enzymatic oxidative polymerization of 4-hydroxybenzoic acid derivatives to poly(phenylene oxide)s. *Polymer International* **47**: 295-301.

19. Pedersen, L.S., Felby, C., and Munk, N. (1997) Process for increasing the charge on a lignocellulosic material and products obtained thereby. *Patent WO* 9729237.

20. Röper, H. 1996. Applications of starch and its derivatives. *Carbohydrates in Europe* **14**: 22-30.

21. Viikari, L., Niku-Paavola, M-L., Buchert, J., Forssell, P., Teleman, A., and Kruus. K. 2000. Menetelmä hapetetun tärkkelyksen valmistamiseksi. *Finnish Patetnt* FI 105690.

22. Kierulff, J.V. Modification of polysaccharides by means of phenol oxidizing enzyme. *PCT world patent application* WO 99/32652.

PART 4

MATERIAL TESTING AND ANALYTICAL METHODS

Biodegradation of Polymeric Materials

An Overview of Available Testing Methods

MAARTEN VAN DER ZEE
ATO, BU Renewable Resources, Department Polymers, Composites and Additives, P.O. Box 17, NL-6700 AA Wageningen, The Netherlands

Abstract: This paper presents an overview of the current knowledge on the biodegradability of polymeric materials, in particular in relation to degradation under environmental conditions. The significance of defining 'biodegradation' and related terms, and complexities associated with the issue are discussed followed by the different aspects of assessing the potential, the rate, and the degree of biodegradation of polymeric materials. Particular attention is given to the ways for demonstrating complete mineralisation to gasses (such as carbon dioxide and methane), water and possibly microbial biomass. The presented overview of testing methods makes clear that there is no such thing as a single optimal method for determining biodegradation of polymeric materials. First of all, biodegradation of a material is not only determined by the chemical composition and corresponding physical properties; the degradation environment in which the material is exposed also affects the rate and degree of biodegradation. Furthermore, the method or test to be used depends on what information is requested. Therefore, one should always consider why a particular polymeric material should be (or not be) biodegradable when contemplating how to assess its biodegradability.

1. INTRODUCTION

There is a world-wide research effort to develop biodegradable polymers as a waste management option for polymers in the environment. Questions addressing the potential benefits of biodegradable polymers have become complex, because of the many disciplines and parties that are involved, including chemists, biochemists, microbiologists, environmentalists, legislators, and manufacturers, all trying to ensure that their perspectives and expectations are not ignored[1]. Until recently, most of the efforts were

Biorelated Polymers: Sustainable Polymer Science and Technology
Edited by Chiellini et al., Kluwer Academic/Plenum Publishers, 2001

synthesis oriented, and not much attention was paid to the identification of environmental requirements for, and testing of, biodegradable polymers. Consequently, many unsubstantiated claims to biodegradability were made, and this has damaged the general acceptance.

An important factor is that the term biodegradation has not been applied consistently. In the medical field of sutures, bone reconstruction and drug delivery, the term biodegradation has been used to indicate hydrolysis[2]. On the other hand, for environmentally degradable plastics, the term biodegradation may mean fragmentation, loss of mechanical properties, or sometimes degradation through the action of living organisms[3]. Deterioration or loss in physical integrity is also often mistaken for biodegradation[4]. Nevertheless, it is essential to have a universally acceptable definition of biodegradability to avoid confusion as to where biodegradable polymers fit into the overall plan of polymer waste management. Many groups and organisations have endeavoured to clearly define the terms 'degradation', 'biodegradation', and 'biodegradability'. But there are several reasons why establishing a single definition among the international community has not been straightforward, including:

- the variability of an intended definition given the different environments in which the material is to be introduced and its related impact on those environments,
- the differences of opinion with respect to the scientific approach or reference points used for determining biodegradability,
- the divergence of opinion concerning the policy implications of various definitions,
- challenges posed by language differences around the world.

As a result, many different definitions have officially been adopted, depending on the background of the defining organisation, and their particular interests.

Problems in defining biodegradation are related to basic principles such as whether the process needs to be irreversible, chemical bonds need to be broken, and whether physical destruction by living organisms such as chewing by mice or goats is also considered to be biodegradation. However, of more practical importance are the criteria for calling a material 'biodegradable'. A demonstrated potential of a material to biodegrade does not say anything about the time frame in which this occurs, nor the ultimate degree of degradation. The complexity of this issue is illustrated by the following common examples.

Low density polyethylene has been shown to biodegrade slowly (0.35% in 2.5 years)[5] and according to some definitions can thus be called a biodegradable polymer. However, the degradation process is so slow, that accumulation in the environment will occur. The same applies for polyolefin-starch blends which rapidly loose strength, disintegrate, and visually disappear if exposed to micro-organisms[6,7,8]. This is due to utilisation of the starch component, but the polyolefin fraction will nevertheless persist in the environment. Can these materials be called 'biodegradable'?

2. DEFINING BIODEGRADABILITY

In 1992, an international workshop on biodegradability was organised to bring together experts from around the world to achieve areas of agreement on definitions, standards and testing methodologies. Participants came from manufacturers, legislative authorities, testing laboratories, environmentalists and standardisation organisations in Europe, USA and Japan. Since this fruitful meeting, there is a general agreement concerning the following key points[9].

- For all practical purposes of applying a definition, material manufactured to be biodegradable must relate to a specific disposal pathway such as composting, sewage treatment, denitrification, and anaerobic sludge treatment.
- The rate of degradation of a material manufactured to be biodegradable has to be consistent with the disposal method and other components of the pathway into which it is introduced, such that accumulation is controlled.
- The ultimate end products of aerobic biodegradation of a material manufactured to be biodegradable are CO_2, water and minerals and that the intermediate products include biomass and humic materials. (Anaerobic biodegradation was discussed in less detail by the participants).
- Materials must biodegrade safely and not negatively impact the disposal process or the use of the end product of the disposal.

As a result, specified periods of time, specific disposal pathways, and standard test methodologies were incorporated into definitions. Standardisation organisations such as CEN, ISO, and ASTM were consequently encouraged to rapidly develop standard biodegradation tests so these could be determined. Society further demanded undebatable criteria for the evaluation of the suitability of polymeric materials for disposal in specific waste streams such as composting or anaerobic digestion. Biodegradability is usually just one of the essential criteria, besides ecotoxicity, effects on waste treatment processes, etc.

The used definitions for biodegradation remain rather broad since they have to be acceptable to all parties involved. In this paper, however, biodegradation is addressed from a more scientific point of view. The following definitions are therefore applied as a basis[10].

- *Degradation* is the irreversible process in which a material undergoes physical, chemical, and/or biochemical changes leading to an increase in entropy.
- *Biodegradation* is degradation catalysed by biological activity, ultimately leading to mineralisation and/or biomass.
- *Mineralisation* is the conversion of (organic) material to naturally occurring gasses and/or inorganic elements.
- *Biodegradability* is the potential of a material to be biodegraded. Biodegradability of a material shall be specified and measured by standard test methods in order to determine its classification with respect to specific environmental conditions.
- *Biodegradable* A material is called biodegradable with respect to specific environmental conditions if it undergoes biodegradation to a specified extent, within a given time, measured by standard test methods.

The chemistry of the key degradation process, is represented below by equations (1) and (2), where $C_{POLYMER}$ represents either a polymer or a fragment from any of the degradation processes defined earlier. For simplicity here, the polymer or fragment is considered to be composed only of carbon, hydrogen and oxygen; other elements may, of course, be incorporated in the polymer, and these would appear in an oxidised or reduced form after biodegradation depending on whether the conditions are aerobic or anaerobic, respectively.

Aerobic biodegradation:

$$C_{POLYMER} + O_2 \rightarrow CO_2 + H_2O + C_{RESIDUE} + C_{BIOMASS} + \text{salts} \qquad (Eq. 1)$$

Anaerobic biodegradation:

$$C_{POLYMER} \rightarrow CO_2 + CH_4 + H_2O + C_{RESIDUE} + C_{BIOMASS} + \text{salts} \qquad (Eq. 2)$$

Complete biodegradation occurs when no residue remains, and complete mineralisation is established when the original substrate, $C_{POLYMER}$ in this example, is completely converted into gaseous products and salts. However, mineralisation is a very slow process under natural conditions because some of the polymer undergoing biodegradation will initially be turned into biomass[11]. Therefore, complete biodegradation and not mineralisation is the measurable goal when assessing removal from the environment.

3.   MEASURING BIODEGRADATION

Biodegradation is very much affected by *(1)* the polymer chemistry, and *(2)* the environment. Although often research aims at investigating the biodegradability of a material, the effect of the environment cannot be neglected when measuring biodegradability. Microbial activity, and hence biodegradation, is influenced by:
– the presence of micro-organisms
– the availability of oxygen
– the amount of available water
– temperature
– chemical environment (pH, electrolytes, etc.)

In order to simplify the measurements, the environments in which biodegradation occurs are basically divided in two environments: *(a)* aerobic (with oxygen available) and *(b)* anaerobic (no oxygen present). These two, can in turn be subdivided into *(1)* aquatic and *(2)* high solids environments. Figure 1 schematically presents the different environments, with examples in which biodegradation may occur[10,12].

	aquatic	*high solids*
aerobic	• aerobic waste water treatment plants • surface waters; e.g. lakes and rivers • marine environments	• surface soils • organic waste composting plants • littering
anaerobic	• anaerobic waste water treatment • plants • rumen of herbivores	• deep sea sediments • anaerobic sludge • anaerobic digestion/ biogasification • landfill

Figure 1. Schematic classification of the different biodegradation environments

The high solids environments will be the most relevant for measuring biodegradation of polymeric materials, since they represent the conditions during biological municipal solid waste treatment, such as composting or anaerobic digestion (biogasification). However, possible applications of biodegradable materials other than in packaging and consumer products, e.g. in fishing nets at sea, or undesirable exposure in the environment due to littering, explain the necessity of aquatic biodegradation tests.

Numerous ways for the experimental assessment of polymer biodegradability have been described in the scientific literature. Because of slightly different definitions or interpretations of the term 'biodegradability', the different approaches are therefore not equivalent in terms of information they provide or the practical significance. Since the typical exposure environment involves incubation of a polymer substrate with micro-organisms or enzymes, only a limited number of measurements are possible: those pertaining to the substrates, to the micro-organisms, or to the reaction products. Four common approaches available for studying biodegradation processes have been reviewed in detail by Andrady[13].

1. Monitoring microbial growth
2. Monitoring the depletion of substrates
3. Monitoring reaction products
4. Monitoring changes in substrate properties

In the following paragraphs, different test methods for the assessment of polymer biodegradability are presented. Measurements are usually based on one of the four approaches given above, but combinations also occur. Before choosing an assay to simulate environmental effects in an accelerated manner, it is critical to consider the closeness of fit that the assay will provide between substrate, micro-organisms or enzymes, and the application or environment in which biodegradation should take place[14].

3.1   Enzyme assays

3.1.1   Principle

In enzyme assays, the polymer substrate is added to a buffered system, containing one or several types of purified enzymes. These assays are very useful in examining the kinetics of depolymerisation, or oligomer or monomer release from a polymer chain under different assay conditions. The method is very rapid (minutes to hours) and can give quantitative information. However, mineralisation rates cannot be determined with enzyme assays.

3.1.2   Applications

The type of enzyme to be used, and quantification of degradation, will depend on the polymer being screened. For example, Mochizuki et al.[119] studied the effects of draw ratio of polycaprolactone (PCL) fibres on enzymatic hydrolysis by lipase. Degradability of PCL fibres was monitored by dissolved organic carbon (DOC) formation and weight loss. Similar systems with lipases have been used for studying the hydrolysis of broad ranges of aliphatic polyesters[15-20], copolyesters with aromatic segments[16,21-23], and copolyesteramides[24,25]. Other enzymes such as α-chymotrypsin and α-trypsin have also been applied for these polymers[26,27]. Biodegradability of poly(vinyl alcohol) segments with respect to block length and stereo chemical configuration has been studied using isolated poly(vinyl alcohol)-dehydrogenase[28]. Cellulolytic enzymes have been used to study the biodegradability of cellulose ester derivatives as a function of degree of substitution and the substituent size[29]. Similar work has been performed with starch esters using amylolytic enzymes such as α-amylases, β-amylases, glucoamylases and amyloglucosidases[30]. Enzymatic methods have also been used to study the biodegradability of starch plastics or packaging materials containing cellulose[31-36].

3.1.3 Drawbacks

Caution must be used in extrapolating enzyme assays as a screening tool for different polymers since the enzymes have been paired to only one polymer. The initially selected enzymes may show significantly reduced activity towards modified polymers or different materials, even though more suitable enzymes may exist in the environment. Caution must also be used if the enzymes are not purified or appropriately stabilised or stored, since inhibitors and loss of enzyme activity can occur[14].

3.2   Plate tests

3.2.1   Principle

Plate tests have initially been developed in order to assess the resistance of plastics to microbial degradation. Several methods have been standardised by standardisation organisations such as the American Society for Testing and Materials (ASTM) and the International Organization for Standardization (ISO)[37-39]. They are now also used to see if a polymeric material will support growth[14,40]. The principle of the method involves placing the test material on the surface of a mineral salts agar in a petri dish containing no additional carbon source. The test material and agar surface are sprayed with a standardised mixed inoculum of known bacteria and/or fungi. The test material is examined after a predetermined incubation period at constant temperature for the amount of growth on its surface and a rating is given.

3.2.2   Applications

Potts[41] used the method in his screening of 31 commercially available polymers for biodegradability. Other studies, where the growth of either mixed or pure cultures of micro-organisms, is taken to be indicative for biodegradation, have been reported[3]. The validity of this type of test, and the use of visual assessment alone has been questioned by Seal and Pantke[42] for all plastics. They recommended that mechanical properties should be assessed to support visual observations. Microscopic examination of the surface can also give additional information.

A variation of the plate   test, is the 'clear zone' technique   [43] sometimes used to screen polymers for biodegradability. A fine suspension of polymer is placed in an agar gel as the sole carbon source, and the test inoculum is placed in wells bored in the agar. After incubation, a clear zone around the well, detected visually, or instrumentally is indicative of utilisation of the polymer. The method has for example been used in the case of starch plastics[44], various polyesters[45-47], and polyurethanes[48].

3.2.3   Drawbacks

A positive result in an agar plate test indicates that an organism can grow on the substrate, but does not mean that the polymer is biodegradable, since growth may be on contaminants, on plasticisers present, on oligomeric fractions still present in the polymer, and so on. Therefore, these tests should be treated with caution when extrapolating the data to field situation.

3.3   Respiration tests

3.3.1   Principle

Aerobic microbial activity is typically characterised by the utilisation of oxygen. Aerobic biodegradation requires oxygen for the oxidation of compounds to its mineral constituents, such as CO_2, H_2O, SO_2, P_2O_5, etc.. The amount of oxygen utilised during incubation, also called the biochemical (or biological) oxygen demand (BOD) is therefore a measure of the degree of biodegradation. Several test methods are based on measurement of the BOD, often expressed as a percentage of the theoretical oxygen demand (TOD) of the compound. The TOD, which is the theoretical amount of oxygen necessary for completely oxidising a substrate to its mineral constituents, can be calculated by considering the elemental composition and the stoichiometry of oxidation[13,49-52] or based on experimental determination of the chemical oxygen demand (COD)[13,53].

3.3.2 Applications

The closed bottle BOD tests were designed to determine the biodegradability of detergents[51,54]. These have stringent conditions due to the low level of inoculum (in the order of 10^5 micro-organisms/l) and the limited amount of test substance that can be added (normally between 2 and 4 mg/l). These limitations originate from the practical requirement that the oxygen demand should be not more than half the maximum dissolved oxygen level in water at the temperature of the test, to avoid the generation of anaerobic conditions during incubation.

For non-soluble materials, such as polymers, less stringent conditions are necessary and alternative ways for measuring BOD were developed. Two-phase (semi) closed bottle tests provide a higher oxygen content in the flasks and permit a higher inoculum level. Higher test concentrations are also possible, encouraging higher accuracy with directly weighing in of samples. The oxygen demand can alternatively be determined by periodically measuring the oxygen concentration in the aquatic phase by opening the flasks[50,55,56], by measuring the change in volume or pressure in incubation flasks containing CO_2-absorbing agents[49,57,58], or by measuring the quantity of oxygen produced (electrolytically) to maintain constant gas volume/pressure in specialised respirometers[49,52,55,57].

3.3.3 Suitability

BOD tests are relatively rapid (days to weeks), and sensitive, and are therefore often used as simple screening tests. However, the measurement of oxygen consumption is a non-specific, indirect measure for biodegradation, and it is not suitable for determining anaerobic degradation. The requirement for test materials to be the sole carbon/energy source for micro-organisms in the incubation media, eliminates the use of oxygen measurements in complex natural environments.

3.4 Gas (CO_2 or CH_4) evolution tests

3.4.1 Principle

The evolution of carbon dioxide or methane from a substrate represents a direct parameter for mineralisation. Therefore, gas evolution tests can be important tools in the determination of biodegradability of polymeric materials. A number of well known test methods have been standardised for aerobic biodegradation, such as the (modified) Sturm test[59-63] and the laboratory controlled composting test[64-66], as well as for anaerobic biodegradation, such as the anaerobic sludge test[66,67] and the anaerobic digestion test[68]. Although the principle of these test methods are the same, they may differ in medium composition, inoculum, the way substrates are introduced, and in the technique for measuring gas evolution.

3.4.2   Applications

Anaerobic tests generally follow biodegradation by measuring the increase in pressure and/or volume due to gas evolution, usually in combination with gas chromatographic analysis of the gas phase[69,70]. Most aerobic standard tests apply continuous aeration; the exit stream of air can be directly analysed continuously using a carbon dioxide monitor (usually infrared detectors) or titrimetrically after sorption in dilute alkali. The cumulative amount of carbon dioxide generated, expressed as a percentage of the theoretically expected value for total conversion to CO_2, is a measure for the extent of mineralisation achieved. A value of 60% carbon conversion, achieved within 28 days, is generally taken to indicate ready degradability. While this criterion is meant for water soluble substrates, it is probably applicable to very finely divided moderately-degradable polymeric materials as well[13]. Nevertheless, most standards for determining biodegradability of plastics consider a maximum test duration of 6 months.

Besides the continuously aerated systems, described above, several static respirometers have been described. Bartha and Pramer[71] describe a two flask system; one flask, containing a mixture of soil and the substrate, is connected to another chamber holding a quantity of carbon dioxide sorbant. Care must be taken to ensure that enough oxygen is available in the flask for biodegradation. Nevertheless, this experimental set-up and modified versions thereof have been successfully applied in the assessment of biodegradability of polymer films and food packaging materials[72-74].

The percentage of carbon converted to biomass instead of carbon dioxide depends on the type of polymer and the phase of degradation. Therefore, it recently has been suggested to regard the complete carbon balance to determine the degree of degradation[75]. This implies, that besides the detection of gaseous carbon, also the amount of carbon in soluble and solid products needs to be determined. Soluble products, oligomers of different molecular size, intermediates and proteins secreted from microbial cells can be measured as chemical oxygen demand (COD) or as dissolved organic carbon (DOC). Solid products, biomass, and polymer remnants require a combination of procedures to separate and detect different fractions. The protein content of the insoluble fraction is usually determined to estimate the amount of carbon converted to biomass, using the assumptions that dry biomass consists of 50% protein, and that the carbon content of dry biomass is 50%[75-77].

3.4.3 Suitability

Gas evolution tests are popular test methods because they are relatively simple, rapid (days to weeks) and sensitive. A direct measure for mineralisation is determined, and water-soluble or insoluble polymers can be tested as films, powders or objects. Furthermore, the test conditions and inoculum can be adjusted to fit the application or environment in which biodegradation should take place. Aquatic synthetic media are usually used, but also natural sea water[78,79] or soil samples[71,73,74,80] can be applied as biodegradation environments. A prerequisite for these media is that the background CO_2-evolution is limited, which excludes the application of real composting conditions. Biodegradation under composting conditions is therefore measured using an inoculum derived from matured compost with low respiration activity[64,65,81,82].

A drawback of using complex degradation environments such as mature compost is that simultaneous characterisation of intermediate degradation products of determination of the carbon balance is difficult due to the presence of a great number of interfering compounds. To overcome this, an alternative test is currently under development based on an inoculated mineral bed based matrix[83,84].

3.5 Radioactively labelled polymers

3.5.1 Principle and applications

Some materials tend to degrade very slowly under stringent test conditions without an additional source of carbon. However, if readily available sources of carbon are added, it becomes impossible to tell how much of the evolved carbon dioxide can be attributed to decomposition of the plastic. The incorporation of radioactive ^{14}C in synthetic polymers gives a means of distinguishing between CO_2 or CH_4 produced by the metabolism of the polymer, and that generated by other carbon sources in the test environment. By comparison of the amount of radioactive $^{14}CO_2$ or $^{14}CH_4$ to the original radioactivity of the labelled polymer, it is possible to determine the percent by weight of carbon in the polymer which was mineralised during the duration of the exposure[41]. Collection of radioactively labelled gasses or low molecular weight products can also provide extremely sensitive and reproducible methods to assess the degradation of polymers with low susceptibility to enzymes, such as polyethylene[5,85] and cellulose acetates[86,87].

3.5.2   Drawbacks

Problems with handling the radioactively labelled materials and their disposal are issues on the down side to this method. In addition, in some cases it is difficult to synthesise the target polymer with the radioactive labels in the appropriate locations, with representative molecular weights, or with representative morphological characteristics.

3.6   Laboratory-scale simulated accelerating environments

3.6.1   Principle

Biodegradation of a polymer material is usually associated with changes in the physical, chemical and mechanical properties of the material. It is indeed these changes, rather than the chemical reactions, which make the biodegradation process so interesting from an application point of view. These useful properties might be measured as a function of the duration of exposure to a biotic medium, to follow the consequences of the biodegradation process on material properties. The biotic media can be specifically designed in a laboratory scale as to mimic natural systems but with a maximum control of variables such as temperature, pH, microbial community, mechanical agitation and supply of oxygen. Regulating these variables improves the reproducibility and may accelerate the degradation process. Laboratory simulations can also be used for the assessment of long term effects due to continuous dosing on the activity and the environment of the disposal system[40].

3.6.2   Applications

The OECD Coupled Unit test[88] simulates an activated sludge sewage treatment system, but its application for polymers would be difficult as DOC is the parameter used to assess biodegradability. Krupp and Jewell[89] described well controlled anaerobic and aerobic aquatic bioreactors to study degradation of a range of commercially available polymer films. A relatively low loading rate of the semi-continuous reactors and a long retention time were maintained to maximise the efficiency of biodegradation. Experimental set-ups have also been designed to simulate marine environments[90], soil burial conditions[90-92], composting environments[93-96], and landfill conditions[97] at laboratory scale, with controlled parameters such as temperature and moisture level, and a synthetic waste, to provide a standardised basis for comparing the degradation kinetics of films.

A wide choice of material properties can be followed during the degradation process. However, it is important to select one which is relevant to the end-use of the polymer material or provides fundamental information about the degradation process. Weight loss is a parameter frequently followed because it clearly demonstrates the disintegration of a biodegradable product[98-100]. Tensile properties are also often monitored, due to the interest in the use of biodegradable plastics in packaging applications[44,101,102]. In those polymers where the biodegradation involves a random scission of the macromolecular chains, a decrease in the average molecular weight and a general broadening of the molecular weight distribution provide initial evidence of a breakdown process[71,103,104]. Visual examination of the surface with various microscopic techniques can also give information on the biodegradation process[105-108]. Likewise, chemical and/or physical changes in the polymer may be followed by (combinations of) specific techniques such as infrared[7,109] or UV spectroscopy[69,110], nuclear magnetic resonance measurements[104,111], x-ray diffractometry[112,113], and differential scanning calorimetry[114,115].

3.6.3   Drawbacks

An inherent drawback in the use of mechanical properties, weight loss, molecular weights, or any other property which relies on the macromolecular nature of the substrate is that in spite of their sensitivity, these can only address the early stages of the biodegradation process. Furthermore, these parameters can give no information on the extent of mineralisation. Especially in material blends or copolymers, the hydrolysis of one component can cause significant disintegration (and thus loss of weight and tensile properties) whereas other components may persist in the environment, even in disintegrated form[13]. Blends of starch, poly(3-hydroxy butyrate) or poly(ε-caprolactone) with polyolefins are examples of such systems[8,33,116].

3.7   Natural environments, field trials

Exposures in natural environments provide the best true measure of the environmental fate of a polymer, because these tests include a diversity of organisms and achieve a desirable natural closeness of fit between the substrate, microbial agent and the environment. However, the results of that exposure are only relevant to the specific environment studied, which is likely to differ substantially from many other environments. An additional problem is the time scale for this method, since the degradation process, depending on the environment, may be very slow (months to years)[14]. Moreover, little information on the degradation process can be gained other than the real time required for total disintegration.

Nevertheless, field trials in natural environments are still used to extrapolate results acquired in laboratory tests to biodegradation behaviour under realistic outdoor conditions[105,117]. Recent German regulations for the assessment of compostability of plastics even impose exposure of the product to a full scale industrial composting process to ensure that total disintegration will occur in real-life waste-processing[118].

4.   CONCLUSION

The overview presented above makes clear that there is no such thing as a single optimal method for determining biodegradation of polymeric materials. First of all, biodegradation of a material is not only determined by the chemical composition and corresponding physical properties; the degradation environment in which the material is exposed also affects the rate and degree of biodegradation. Furthermore, the method or test to be used depends on what information is requested.

Biodegradability is usually not of interest by itself. It is often just one aspect of health and environmental safety issues or integrated waste management concepts. It is fairly obvious but often neglected that one should always consider why a particular polymeric material should be (or not be) biodegradable when contemplating how to assess its biodegradability. After all, it is the intended application of the material that governs the most suitable testing environment and which parameters should be measured during exposure. For example, investigating whether biodegradation of a plastic material designed for food packaging could facilitate undesired growth of (pathogenic) micro-organisms requires a completely different approach from investigating whether its waste can be discarded via composting (i.e. whether it degrades sufficiently rapid to be compatible with existing biowaste composting facilities).

In most cases, it will not be sufficient to ascertain macroscopic changes, such as weight loss and disintegration, or growth of micro-organisms, because these observations may originate from biodegradation of just one of separate components. The ultimate fate of all individual components and degradation products must be included in the investigations. This implies that it is essential that both the polymeric materials and also intermediate degradation products have to be well characterised in order to understand the degradation process.

REFERENCES

1. Swift, G., 1992, Biodegradability of polymers in the environment: complexities and significance of definitions and measurements. *FEMS Microbiol. Rev.* **103**, 339-346.
2. Göpferich, A., 1996, Mechanisms of polymer degradation and erosion. *Biomaterials* **17**: 103-114.
3. Albertsson, A.-C. and Karlsson, S., 1990, Biodegradation and test methods for environmental and biomedical applications of polymers. In *Degradable Materials - Perspectives, Issues and Opportunities* (S.A. Barenberg, J.L. Brash, R. Narayan, and A.E. Redpath, eds), The first international scientific consensus workshop proceedings, CRC Press, Boston, pp. 263-286.
4. Palmisano, A.C., and Pettigrew, C.A., 1992, Biodegradability of plastics, consistent methods for testing claims of biodegradability need to be developed. *Bioscience* **42**: 680-685.
5. Albertsson, A.-C., and Rånby, B., 1979, Biodegradation of synthetic polymers. IV. The $^{14}CO_2$ method applied to linear polyethylene containing a biodegradable additive. *J. Appl. Polym. Sci.: Appl. Polym. Symp.* **35**: 423-30.
6. Austin, R.G., 1990, Degradation studies of polyolefins. In *Degradable Materials - Perspectives, Issues and Opportunities* (S.A. Barenberg, J.L. Brash, R. Narayan, and A.E. Redpath, eds), The first international scientific consensus workshop proceedings, CRC Press, Boston, pp. 209-229.
7. Goheen, S.M., and Wool, R.P., 1991, Degradation of polyethylene starch blends in soil. *J. Appl. Polym. Sci.* **42**: 2691-2701.
8. Breslin, V.T., 1993, Degradation of starch-plastic composites in a municipal solid waste landfill. *J. Environ. Polym. Degrad.* **1**: 127-141.
9. Anonymous, 1992, *Towards Common Ground*. The International Workshop on Biodegradability, Annapolis, Maryland, 20-21 October 1992, Meeting summary.
10. Van der Zee, M., Stoutjesdijk, J.H., Van der Heijden, P.A.A.W. and De Wit, D., 1995, Structure-biodegradation relationships of polymeric materials .1. Effect of degree of oxidation of carbohydrate polymers. *J. Environ. Polym. Degrad.* **3**, 235-242.
12. Eggink, G., Van der Zee, M., and Sijtsma, L., 1995, Biodegradable polymers from plant materials. *International edition of the IOP on Environmental Biotechnology*: 7-8.
13. Andrady, A.L., 1994, Assessment of environmental biodegradation of synthetic polymers. *J.M.S. - Rev. Macromol. Chem. Phys.* **C34**: 25-76.
14. Mayer, J.M., and Kaplan, D.L., 1993, Biodegradable materials and packaging - Environmental test methods and needs. In: Ching, C., Kaplan, D.L., and Thomas, E.L., eds., *Biodegradable Polymers and Packaging*. Chapter 16. Technomic Publishing Co. Inc., Lancaster-Basel, 233-245.

15. Tokiwa, Y., and Suzuki, T., 1981, Hydrolysis of copolyesters containing aromatic and aliphatic ester blocks by lipase. *J. Appl. Polym. Sci.* **26**: 441-448.

16. Tokiwa, Y., Suzuki, T., and Takeda, K., 1986, Hydrolysis of polyesters by *Rhizopus arrhizus* lipase. *Agric. Biol. Chem.* **50**: 1323-1325.

17. Arvanitoyannis, I., Nakayama, A., Kawasaki, N., and Yamamoto, N., 1995, Novel polylactides with aminopropanediol or aminohydroxymethylpropanediol using stannous octoate as catalyst - Synthesis, characterization and study of their biodegradability .2. *Polymer* **36**: 2271-2279.

18. Nakayama, A., Kawasaki, N., Arvanitoyannis, I., Iyoda, J., and Yamamoto, N., 1995, Synthesis and degradability of a novel aliphatic polyester - Poly(_-methyl-_-valerolactone-co-L-lactide, *Polymer* **36**: 1295-1301.

19. Walter, T., Augusta, J., Müller, R.-J., Widdecke, H., and Klein, J., 1995, Enzymatic degradation of a model polyester by lipase from *Rhizopus delemar. Enzym. Microb. Technol.* **17**: 218-224.

20. Nagata, M., Kiyotsukuri, T., Ibuki, H., Tsutsumi, N., and Sakai, W., 1996, Synthesis and enzymatic degradation of regular network aliphatic polyesters. *React. Funct. Polym.* **30**: 165-171.

21. Jun, H.S., Kim, B.O., Kim, Y.C., Chang, H.N., and Woo, S.I., 1994, Synthesis of copolyesters containing poly(ethylene terephthalate) and poly(_-caprolactone) units and their susceptibility to *Pseudomonas* sp. lipase. *J. Environ. Polym. Degrad.* **2**: 9-18.

22. Chiellini, E., Corti, A., Giovannini, A., Narducci, P., Paparella, A.M., and Solaro, R., 1996, Evaluation of biodegradability of poly(_-caprolactone)/poly(ethylene terephthalate) blends. *J. Environ. Polym. Degrad.* **4**: 37-50.

23. Nagata, M., Kiyotsukuri, T., Minami, S., Tsutsumi, N., and Sakai, W., 1996, Biodegradability of poly(ethylene terephthalate) copolymers with poly(ethylene glycol)s and poly(tetramethylene glycol). *Polym. Int.* **39**: 83-89.

24. Nagata, M., and Kiyotsukuri, T., 1994, Biodegradability of copolyesteramides from hexamethylene adipate and hexamethylene adipamide. *Eur. Polym. J.* **30**: 1277-1281.

25. Nagata, M., 1996, Enzymatic degradation of aliphatic polyesters copolymerized with various diamines. *Macromol. Rap. Commun.* **17**: 583-587.

26. Arvanitoyannis, I., Nikolaou, E., and Yamamoto, N., 1994, Novel biodegradable copolyamides based on adipic acid, bis(p-aminocyclohexyl) methane and several alpha-amino acids - Synthesis, characterization and study of their degradability for food packaging applications - 4. *Polymer* **35**: 4678-4689.

27. Arvanitoyannis, I., Nikolaou, E., and Yamamoto, N., 1995, New copolyamides based on adipic acid, aliphatic diamines and amino acids - Synthesis, characterization and biodegradability .5. *Macromol. Chem. Phys.* **196**: 1129-1151.

28. Matsumura, S., Shimura, Y., Toshima, K., Tsuji, M., and Hatanaka, T., 1995, Molecular design of biodegradable functional polymers .4. Poly(vinyl alcohol) block as biodegradable segment. *Macromol. Chem. Phys.* **196**: 3437-3445.

29. Glasser, W.G., McCartney, B.K., and Samaranayake, G., 1994, Cellulose derivatives with low degree of substitution .3. The biodegradability of cellulose esters using a simple enzyme assay. *Biotechn. Progr.* **10**: 214-219.

30. Rivard, C., Moens, L., Roberts, K., Brigham, J., and Kelley, S., 1995, Starch esters as biodegradable plastics - Effects of ester group chain length and degree of substitution on anaerobic biodegradation. *Enzym. Microb. Technol.* **17**: 848-852.

31. Strantz, A.A., and Zottola, E.A., 1992, Stability of cornstarch-containing polyethylene films to starch-degrading enzymes. *J. Food Protect.* **55**: 736-738.

32. Coma, V., Couturier, Y., Pascat, B., Bureau, G., Cuq, J.L. and Guilbert, S., 1995, Estimation of the biofragmentability of packaging materials by an enzymatic method. *Enzyme Microb. Technol.* **17**: 524-527.
33. Imam, S.H., Gordon, S.H., Burgess-Cassler, A., and Greene, R.V., 1995, Accessibility of starch to enzymatic degradation in injection-molded starch-plastic composites. *J. Environ. Polym. Degrad.* **3**: 107-113.
34. Imam, S.H., Gordon, S.H., Shogren, R.L., Greene, R.V., 1995, Biodegradation of starch-poly(_-hydroxybutyrate-co-valerate) composites in municipal activated sludge. *J. Environ. Polym. Degrad.* **3**: 205-213.
35. Vikman, M., Itävaara, M., and Poutanen, K., 1995, Biodegradation of starch-based materials. *J.M.S. - Pure Appl. Chem.* **A32**: 863-866.
36. Vikman, M., Itavaara, M., and Poutanen, K., 1995, Measurement of the biodegradation of starch-based materials by enzymatic methods and composting. *J. Environ. Polym. Degrad.* **3**: 23-29.
37. ASTM G21, 1996, Standard practice for determining resistance of synthetic polymeric materials to fungi. *ASTM standard G21-96.* American Society for Testing and Materials (ASTM), Philadelphia (PA), USA.
38. ASTM G22, 1996, Standard practice for determining resistance of plastics of bacteria. *ASTM standard G22-76(1996).* American Society for Testing and Materials (ASTM), Philadelphia (PA), USA.
39. ISO 846, 1978, Determination of behaviour under the action of fungi and bacteria - Evaluation of visual examination or measurement of change in mass or physical properties. *International Standard ISO 846:1987(E).* International Organization for Standardization (ISO), Genève, Switzerland.
40. Seal, K.J., 1994, Test methods and standards for biodegradable plastics. In: Griffin, G.J.L. (ed., *Chemistry and Technology of biodegradable polymers.* Blackie Academic and Professional, London, 116-134.
41. Potts, J.E., 1978, Biodegradation. In: Jellinek, H.H.G. (ed., *Aspects of degradation and stabilization of polymers.* Chapter 14. Elsevier Scientific Publishing Co., Amsterdam, 617-657.
42. Seal, K.J., and Pantke, M., 1986, An interlaboratory investigation into the biodeterioration testing of plastics, with special reference to polyurethanes; Part 1: Petri dish test. *Mater. Org.* **21**: 151-64.
43. Delafield, F.P., Doudoroff, M., Palleroni, N.J., Lusty, C.J., and Contopoulos, R., 1965, Decomposition of poly-beta-hydroxybutyrate by *Pseudomonads. J. Bacteriol.* **90**: 1455-1466.
44. Gould, J.M., Gordon, S.H., Dexter, L.B., and Swanson, C.L., 1990, Biodegradation of starch-containing plastics. In *Agricultural and synthetic polymers - Biodegradability and utilization* (J.E. Glass, and G. Swift, eds), ACS Symposium Series 433, American Chemical Society, Washington DC, pp. 65-75.
45. Augusta, J., Müller, R.-J., and Widdecke, H., 1993, A rapid evaluation plate-test for the biodegradability of plastics. *Appl. Microbiol. Biotechnol.* **39**: 673-678.
46. Nishida, H., and Tokiwa, Y., 1994a, Confirmation of poly(1,3-dioxolan-2-one) degrading microorganisms in environment. *Chem. Lett.* **3**: 421-422.
47. Nishida, H., and Tokiwa, Y., 1994, Confirmation of anaerobic poly(2-oxepanone) degrading microorganisms in environments. *Chem. Lett.* **7**: 1293-1296.
48. Crabbe, J.R., Campbell, J.R., Thompson, L., Walz, S.L., Schultz, W.W., 1994, Biodegradation of a colloidal ester-based polyurethane by soil fungi. *Int. Biodeter. Biodegrad.* **33**: 103-113.

49. ISO 9408, 1999, Water quality - Evaluation of ultimate aerobic biodegradability of organic compounds in an aqueous medium by determination of oxygen demand in a closed respirometer. *International Standard ISO 9408:1999(E)*, International Organization for Standardization (ISO), Genève, Switzerland.

50. ISO 10708, 1997, Water quality - Evaluation in an aqueous medium of the 'ultimate' aerobic biodegradability of organic compounds - Method by determining the biochemical oxygen demand in a two-phase closed bottle test. *International Standard ISO 10708:1997*, International Organization for Standardization (ISO), Genève, Switzerland.

51. OECD 301D, 1993, Ready biodegradability: Closed Bottle Test - 301 D. *Guidelines for testing of chemicals*. Organization for Economic Cooperation and Development (OECD), Paris, France.

52. OECD 302C, 1993, Inherent biodegradability: Modified MITI Test (II) - 302 C. *Guidelines for testing of chemicals*. Organization for Economic Cooperation and Development (OECD), Paris, France.

53. ISO 6060, 1989, Water Quality - Determination of the chemical oxygen demand. *International Standard ISO 6060:1989(E)*. International Organization for Standardization (ISO), Genève, Switzerland.

54. ISO 10707, 1994, Water quality - Evaluation in an aqueous medium of the 'ultimate' aerobic biodegradability of organic compounds - Method by analysis of biochemical oxygen demand (closed bottle test, *International Standard ISO 10707:1994(E)*, International Organization for Standardization (ISO), Genève, Switzerland.

55. ISO 14851, 1999, Determination of the ultimate aerobic biodegradability of plastic materials in an aqueous medium - Method by measuring the oxygen demand in a closed respirometer. *International Standard ISO 14851:1999(E)*, International Organization for Standardization (ISO), Genève, Switzerland.

56. prEN 14048, 2000, Packaging - Determination of the ultimate aerobic biodegradability of packaging materials in an aqueous medium - Method by measuring the oxygen demand in a closed respirometer. *Draft European Standard prEN 14048:2000*, European Committee for Standardization (CEN), Brussels, Belgium.

57. OECD 301F, 1993, Manometric respirometry test. *Guidelines for testing of chemicals*. Organization for Economic Cooperation and Development (OECD), Paris, France.

58. Tilstra, L., and Johnsonbaugh, D., 1993a, A test method to determine rapidly if polymers are biodegradable. *J. Environ. Polym. Degrad.* 1: 247-255.

59. ASTM D5209, 1992, Standard test method for determining the aerobic biodegradation of plastic materials in the presence of municipal sewage sludge. *ASTM standard D5209-92*. American Society for Testing and Materials (ASTM), Philadelphia (PA), USA.

60. ISO 9439, 1999, Water quality - Evaluation of ultimate aerobic biodegradability of organic compounds in an aqueous medium - Carbon dioxide evolution test. *International Standard ISO 9439:1999(E)*, International Organization for Standardization (ISO), Genève, Switzerland.

61. ISO 14852, 1999, Determination of the ultimate aerobic biodegradability of plastic materials in an aqueous medium - Method by analysis of evolved carbon dioxide. *International Standard ISO 14852:1999 (E)*, International Organization for Standardization (ISO), Genève, Switzerland.

62. prEN 14047, 2000, Packaging - Determination of the ultimate aerobic biodegradability of packaging materials in an aqueous medium - Method by analysis of evolved carbon dioxide. *Draft European Standard prEN 14047:2000*, European Committee for Standardization (CEN), Brussels, Belgium.

63. OECD 301B, 1993, Ready biodegradability: Modified Sturm test - 301 B. *Guidelines for testing of chemicals*. Organization for Economic Cooperation and Development (OECD), Paris, France.

64. ASTM D5338, 1998, Standard test method for determining aerobic biodegradation of plastic materials under controlled composting conditions. *ASTM Standard D5338-98-e1*. American Society for Testing and Materials (ASTM), Philadelphia (PA), USA.

65. ISO 14855, 1999, Determination of the ultimate aerobic biodegradability and disintegration of plastics under controlled composting conditions - Method by analysis of evolved carbon dioxide. *International Standard 14855:1999(E)*, International Organization for Standardization (ISO), Genève, Switzerland.

65. prEN 14046, 2000, Packaging - Determination of the ultimate aerobic biodegradability and disintegration of packaging materials under controlled composting conditions - Method by measuring analysis of evolved carbon dioxide. *Draft European Standard prEN 14046:2000*, European Committee for Standardization (CEN), Brussels, Belgium.

66. ASTM D5210, 1992, Standard test method for determining the anaerobic biodegradation of plastic materials in the presence of municipal sewage sludge. *ASTM standard D5210-92*. American Society for Testing and Materials (ASTM), Philadelphia (PA), USA.

67. ISO 11734, 1995, Water quality - Evaluation of the 'ultimate' anaerobic biodegradability of organic compounds in digested sludge - Method by measurement of the biogas production. *International Standard ISO 11734:1995(E)*. International Organization for Standardization (ISO), Genève, Switzerland.

68. ASTM D5511, 1994, Standard Test Method for Determining the Anaerobic Biodegradation of Plastic Materials under high-solids anaerobic-digestion conditions. *ASTM standard D5511-94*. American Society for Testing and Materials (ASTM), Philadelphia (PA), USA.

69. Day, M., Shaw, K., and Cooney, J.D., 1994, Biodegradability: an assessment of commercial polymers according to the Canadian method for anaerobic conditions. *J. Environ. Polym. Degrad.* **2**: 121-127.

70. Puechner, P., Mueller, W.-R., and Bardtke, D., 1995, Assessing the biodegradation potential of polymers in screening- and long-term test systems. *J. Environ. Polym. Degrad.* **3**, 133-143.

71. Bartha, R., and Yabannavar, A., 1995, Methods of assessment of the biodegradation of polymers in soil. Fourth international workshop on biodegradable plastics and polymers and fourth annual meeting of the Bio-Environmentally Degradable Polymer Society, October 11-14, Durham, New Hampshire, USA.

72. Andrady, A.L., Pegram, J.E., and Tropsha, Y., 1993, Changes in carbonyl index and average molecular weight on embrittlement of enhanced-photodegradable polyethylenes. *J. Environ. Polym. Degrad.* **1**: 171-180.

73. Yabannavar, A., and Bartha, R., 1993, Biodegradability of some food packaging materials in soil. *Soil Biol. Biochem.* **25**: 1469-1475.

74. Yabannavar, A.V., and Bartha, R., 1994, Methods for assessment of biodegradability of plastic films in soil. *Appl. Environ. Microbiol.* **60**: 3608-3614.11. Swift, G., 1995, Opportunities for environmentally degradable polymers. *J.M.S. - Pure Appl.Chem.* **A32**: 641-651.

75. Urstadt, S., Augusta, J., Müller, R.-J., and Deckwer, W.-D., 1995, Calculation of carbon balances for evaluation of the biodegradability of polymers. *J. Environ. Polym. Degrad.* **3**, 121-131.

76. Itävaara, M., and Vikman, M., 1995, A simple screening test for studying the biodegradability of insoluble polymers. *Chemosphere* **31**: 4359-4373.

77. Spitzer, B., Mende, C., Menner, M., and Luck, T., 1996, Determination of the carbon content of biomass - a prerequisite to estimate the complete biodegradation of polymers. *J. Environ. Polym. Degrad.* **4**: 157-171.

78. Allen, A.L., Mayer, J.M., Stote, R., and Kaplan, D.L., 1994, Simulated marine respirometry of biodegradable polymers. *J. Environ. Polym. Degrad.* **2**: 237-244.

79. Courtes, R., Bahlaoui, A., Rambaud, A., Deschamps, F., Sunde, E., and Dutriex, E., 1995, Ready biodegradability test in seawater - A new methodological approach. *Ecotox. Environ. Saf.* **31**: 142-148.

80. Barak, P., Coquet, Y., Halbach, T.R.. and Molina, J.A.E., 1991, Biodegradability of polyhydroxybutyrate (co-hydroxyvalerate) and starch-incorporated polyethylene plastic films in soils. *J. Environ. Qual.* **20**: 173-179.

81. Pagga, U., Beimborn, D.B., Boelens J., and De Wilde, B., 1995, Determination of the aerobic biodegradability of polymeric material in a laboratory controlled composting test. *Chemosphere* **31**: 4475-4487.

82. Pagga, U., Beimborn, D.B., and Yamamoto, M., 1996, Biodegradability and compostability of polymers - Test methods and criteria for evaluation. *J. Environ. Polym. Degrad.* **4**: 173-178.

83. Tosin, M., Degli-Innocenti, F. and Bastioli, C., 1998, Detection of a toxic product released by a polyurethane-containing film using a composting test method based on a mineral bed. *J. Environ. Polym. Degr.* **6**: 79-90.

84. Bellia, G., Tosin, M., Floridi, G. and Degli-Innocenti, F., 1999, Activated vermiculite, a solid bed for testing biodegradability under composting conditions. *Polym. Degrad. Stabil.* **66**: 65-79.

85. Albertsson, A.-C., Barenstedt, C., and Karlsson, S., 1993, Increased biodegradation of a low-density polyethylene (LDPE) matrix in starch-filled LDPE materials. *J. Environ. Polym. Degrad.* **1**: 241-245.

86. Komarek, R.J., Gardner, R.M., Buchanan, C.M., and Gedon, S., 1993, Biodegradation of radiolabeled cellulose acetate and cellulose propionate. *J. Appl. Polym. Sci.* **50**: 1739-1746.

87. Buchanan, C.M., Dorschel, D., Gardner, R.M., Komarek, R.J., Matosky, A.J., White, A.W. and Wood, M.D., 1996, The influence of degree of substitution on blend miscibility and biodegradation of cellulose acetate blends. *J. Environ. Polym. Degrad.* **4**, 179-195.

88. OECD 303A, 1993, Simulation test - Aerobic sewage treatment: Coupled Units Test - 303 A. *Guidelines for testing of chemicals*. Organization for Economic Cooperation and Development (OECD), Paris, France.

89. Krupp, L.R., and Jewell, W.J., 1992, Biodegradability of modified plastic films in controlled biological environments. *Environ. Sci. and Technol.* **26**: 193-198.

90. Kaplan, D.L., Mayer, J.M., Greenberger, M., Gross, R., and McCarthy, S., 1994, Degradation methods and degradation kinetics of polymer films. *Polym. Degrad. Stab.* **45**: 2 165-172.

91. Dale, R., and Squirrell, D.J., 1990, A rapid method for assessing the resistance of polyurethanes to biodeterioration. *Int. Biodeterior.* **26**: 355-67.

92. Seal, K.J., and Pantke, M., 1990, An interlaboratory investigation into the biodeterioration testing of plastics, with special reference to polyurethanes; Part 2: Soil burial tests.. *Mater. Org.* **25**: 87-98.

93. Gardner, R.M., Buchanan, C.M., Komarek, R., Dorschel, D., Boggs, C., and White, A.W., 1994, Compostability of cellulose acetate films. *J. Appl. Polym. Sci.* **52**, 1477-1488.

94. Buchanan, C.M., Dorschel, D.D., Gardner, R.M., Komarek, R.J., and White, A.W., 1995, Biodegradation of cellulose esters - composting of cellulose ester diluent mixtures. *J.M.S. - Pure Appl. Chem.* **A32**: 683-697.

95. Gross, R.A., Gu, J.-D., Eberiel, D., and McCarthy, S.P., 1995, Laboratory-scale composting test methods to determine polymer biodegradability - Model studies on cellulose acetate. *J.M.S. - Pure Appl. Chem.* **A32**, 613-628.

96. prEN 14045, 2000, Packaging - Evaluation of the disintegration of packaging materials in practical oriented tests under defined composting conditions. *Draft European Standard prEN 14045:2000*, European Committee for Standardization (CEN), Brussels, Belgium.

97. Smith, G.P., Press, B., Eberiel, D., McCarthy, S.P., Gross, R.A., Kaplan, D.L., 1990, An accelerated in-laboratory test to evaluate the degradation of plastics in landfill environments. *Polym. Mater. Sci. Eng.* **63**: 862-866.

98. Coma, V., Couturier, Y., Pascat, B., Bureau, G., Guilbert, S., and Cuq, J.L., 1994, Estimation of the biodegradability of packaging materials by a screening test and a weight-loss method. *Pack. Techn. Sci.*, **7**: 27-37.

99. Buchanan, C.M., Boggs, C.N., Dorschel, D., Gardner, R.M., Komarek, R.J., Watterson, T.L., and White, A.W., 1995, Composting of miscible cellulose acetate propionate-aliphatic polyester blends. *J. Environ. Polym. Degrad.* **3**: 1-11.

100. Goldberg, D., 1995, A review of the biodegradability and utility of poly(caprolactone, *J. Environ. Polym. Degrad.* **3**: 61-67.

101. Iannotti, G., Fair, N., Tempesta, M., Neibling, H., Hsieh, F.H., and Mueller, M., 1990, Studies on the environmental degradation of starch-based plastics. In *Degradable Materials - Perspectives, Issues and Opportunities* (S.A. Barenberg, J.L. Brash, R. Narayan, and A.E. Redpath, eds), The first international scientific consensus workshop proceedings, CRC Press, Boston, pp. 425-439.

102. Mergaert, J., Webb, A., Anderson, C., Wouters, A., and Swings, J., 1993, Microbial degradation of poly(3-hydroxybutyrate) and poly(3-hydroxybutyrate-co-3-hydroxyvalerate) in soils. *Appl. Environ. Microbiol.* **59**: 3233-3238.

103. Tilstra, L., and Johnsonbaugh, D., 1993, The biodegradation of blends of polycaprolactone and polyethylene exposed to a defined consortium of fungi. *J. Environ. Polym. Degrad.* **1**: 257-267.

104. Hu, D.S.G., and Liu, H.J., 1994, Structural analysis and degradation behavior in polyethylene glycol poly (l-lactide) copolymers. *J. Appl. Polym. Sci.* **51**: 473-482.

105. Greizerstein, H.B., Syracuse, J.A., and Kostyniak, P.J., 1993, Degradation of starch modified polyethylene bags in a compost field study. *Polym. Degrad. Stab.* **39**: 251-259.

106. Lopez-Llorca, L.V., and Colom Valiente, M.F., 1993, Study of biodegradation of starch-plastic films in soil using scanning electron microscopy. *Micron* **24**: 457-463.

107. Nishida, H., and Tokiwa, Y., 1993, Distribution of poly(3-hydroxybutyrate) and poly(_-caprolactone) aerobic degrading microorganisms in different environments. *J. Environ. Polym. Degrad.* **1**: 227-233.

108. Bastioli, C., Cerutti, A., Guanella, I., Romano, G.C., and Tosin, M., 1995, Physical state and biodegradation properties of starch-polycaprolactone systems. *J. Environ. Polym. Degrad.* **3**: 81-95.

109. Kay, M.J., McCabe, R.W., and Morton, L.H.G., 1993, Chemical and physical changes occurring in polyester polyurethane during biodegradation. *Int. Biodeter. Biodegrad.* **31**: 209-225.

110. Allen, N.S., Edge, M., Mohammadian, M., and Jones, K., 1994, Physicochemical aspects of the environmental degradation of poly(ethylene terephthalate, *Polym. Degrad. Stab.* **43**: 229-237.

111. Löfgren, A., and Albertsson, A.-C., 1994, Copolymers of 1,5-dioxepan-2-one and L-dilactide or D,L- dilactide - hydrolytic degradation behavior. *J. Appl. Polym. Sci.* **52**: 1327-1338.

112. Albertsson, A.-C., and Karlsson, S., 1995, New tools for analysing degradation. *Macromol. Symp.* **98**: 797-801.

113. Schurz, J., Zipper, P., and Lenz, J., 1993, Structural studies on polymers as prerequisites for degradation. *J.M.S. - Pure Appl.Chem.* **A30**: 603-619.

114. Albertsson, A.-C., Barenstedt, C., and Karlsson, S., 1994, Degradation of enhanced environmentally degradable polyethylene in biological aqueous media - Mechanisms during the 1st stages. *J. Appl. Polym. Sci.* **51**: 1097-1105.

115. Santerre, J.P., Labow, R.S., Duguay, D.G., Erfle, D., and Adams, G.A., 1994, Biodegradation evaluation of polyether and polyester-urethanes with oxidative and hydrolytic enzymes. *J. Biomed. Mat. Res.* **28**: 1187-1199.

116. Iwamoto, A., and Tokiwa, Y., 1994, Enzymatic degradation of plastics containing polycarprolactone. *Polym. Degrad. Stab.* **45**: 205-213.

117. Leonas, K.K., Cole, M.A., and Xiao, X.-Y., 1994, Enhanced degradable yard waste collection bag behaviour in a field scale composting environment. *J. Environ. Polym. Degrad.* **2**: 253-261.

118. DIN 54900, 1998, Testing of the compostability of polymeric materials. *Entwurf Deutsche Norm DIN 54900*. Deutsches Institut für Normung e.V. (DIN), Berlin, Germany.

119. Mochizuki, M., Hirano, M., Kanmuri, Y., Kudo, K., and Tokiwa, Y., 1995, Hydrolysis of polycaprolactone fibers by lipase - Effects of draw ratio on enzymatic degradation. *J. Appl. Polym. Sci.* **55**: 289-296.

Comparison of Test Systems for the Examination of the Fermentability of Biodegradable Materials

[1]JOERN HEERENKLAGE, [2]FRANCESCA COLOMBO and [1]RAINER STEGMANN
[1] Technical University of Hamburg-Harburg, Department of Waste Management, Harburger Schloßstraße 37, 21079 Hamburg, German, [2] Politecnico di Milano, Facolta di Ingeneria, Milano, Italy

Abstract: The examination of the biodegradability of biodegradable materials is of crucial importance for the product launch and for the disposal of these products. Biodegradable materials can be biodegraded under aerobic as well as under anaerobic environmental conditions. The examination of biodegradability under aerobic environmental conditions alone is not sufficient.There are some materials, which are biodegradable under aerobic environmental conditions, but an anaerobic biodegradation cannot be observed. Anaerobic biodegradation is to be investigated by use of standardised test methods for different materials. For the examination of the biodegradability under anaerobic conditions bench scale studies were carried out on two different test systems respecting the ISO DIS 14853. On the one hand gas production was measured with a volumetric test system using the principle of displacement volume, on the other hand in a manometric test system the determination of gas production was carried out via determination of gas pressure. Both test systems were compared. Materials tested were Polyhydroxybutyrate (PHB), Poly-e-caprolacton (PCL) and Ecoflex®. The assessment of the degree of their anaerobic biodegradability was carried out via determination of the gas production. The investigations have demonstrated, that with both test systems comparable results can be achieved. In comparison with the manual determination of the gas displacement volume when using the volumetric test method the rates of gas production can thus be registered automatically when using the manometric method. The biodegradability of PHB was confirmed by both test systems. PCL and Ecoflex® are not biodegradable under anaerobic conditions according to ISO DIS 14385.

Biorelated Polymers: Sustainable Polymer Science and Technology
Edited by Chiellini *et al.*, Kluwer Academic/Plenum Publishers, 2001

287

1.   INTRODUCTION

The application of plastic materials in the industrial society is constantly increasing. Plastic products are used to an increasing degree as food / fast food packaging material. After use, these products are to be recycled, reprocessed or disposed. Recycling of plastic materials is difficult, especially when it is polluted with other substances such as food residues for example. As an alternative to conventional plastic material, biodegradable plastic is being used to an increasing extent. These can be treated biologically (e.g. in a composting or anaerobic fermentation plant). Compared with composting, anaerobic fermentation plants have the advantage that the obtained biogas can supply energy.

However, the biodegradability of this plastic material has to be examined before launch with standardised test methods. Biodegradability should be tested under both aerobic and anaerobic conditions.

First anaerobic tests are described in this paper. Comparative investigations into biodegradability of plastic material were carried out according to ISO DIS 14853 at the Department of Waste Management of the Technical University Hamburg-Harburg. The biodegradability of three kinds of plastic material was examined by means of two different testing methods (volumetric and manometric method), to testify the applicability of both test systems.

2.   MATERIAL AND METHODS

Within the scope of the comparative investigations between manometric and volumetric test system, three different kinds of material were investigated with regard to their biodegradability. For the examination of test conditions, Poly-β-hydroxybutyrate (PHB) produced by SENECA (ICI) with specification BX GO8 (powder) was used as positive reference substance. Poly-ϵ-caprolactone (PCL tone polymer 767) produced by UNION CARBIDE and Ecoflex® produced by BASF were used as test material. Both test materials are biodegradable under aerobic conditions[6,7]. The test materials were brought into an inoculated test medium. Test medium and inoculum have been prepared according to ISO DIS 14853 in a glove box under anaerobic conditions. Stabilised sewage sludge obtained from the wastewater treatment plant at Hamburg-Köhlbrandhöft with an ignition loss of 1.9 (w/v) and a pH value of 7.6 was used as inoculum. The sludge was washed with test medium (solution A) to attain a dissolved organic carbon content (DOC) of $C_{DOC}<10$ mg/l. The composition of the test medium (solution A) is specified in Table 1. The total solid content of washed sludge in the medium (1:10 v/v) after having been inserted amounted to DS=2.96 g/L.

Table 1. Composition of the test medium, solution A (ISO/DIS 14853)

Substance	Weight
Anhydrous potassium dihydrogen phosphate (KH_2PO_4)	0.27 g
Disodium hydrogen phosphate ($Na_2HPO_4 * 12 H_2O$)	1.12 g
Ammoniumchloriden (NH_4Cl)	0.53 g
dissolved in 950 ml deionised water	

For the determination of gas production resulting from the biodegradation of the test materials, two different test systems each containing 12 reaction vessels were set up. The volumetric test system of TUHH (Fig 1) and a manometric test system (Sensomat, Fig 2) of the company Aqualytic, 63201 Langen, Germany, were applied as test systems. The test vessels used were modified 500 ml glass flasks (Schott).

The test vessel of the volumetric test system is connected to a gas collection tube with a capacity of 200 ml by a Viton tube (Fig 1). The pH value of the barrier liquid (including an pH-indicator) is pH 2.0. The reading precision of gas volume is 1.3 ml per scale unit. For the determination of produced gas volume, the reservoir tank is balanced by a flexible connecting tube until liquid columns in the gas collection tube resp. in the reservoir tank are equal. Gas sampling in order to determine the concentration of the various gas contents is carried out discontinuously with a gas sampling syringe which is inserted into the septum of the gas collection tube. Sampling of liquids is carried out via the septum at the reaction vessel itself. Samples are mixed discontinuously using magnetic stirrers.

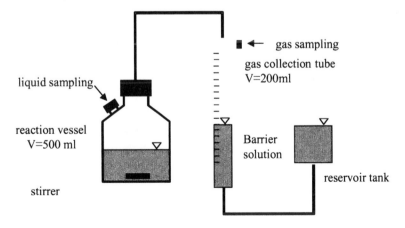

Figure 1. Principle outline of the volumetric test system[3]

The gas production in the manometric test system were measured automatically and semi-continuosly every 120 minutes by measuring the pressure in the headspace of the reaction vessel. The limit of the measured pressure is 350 mbar. By using an IR-sensor the data can be registered by a data logger and transferred to a computer. Sampling of liquids is carried out via the septum at the reaction vessel itself. Samples are mixed discontinuously using magnetic stirrers.

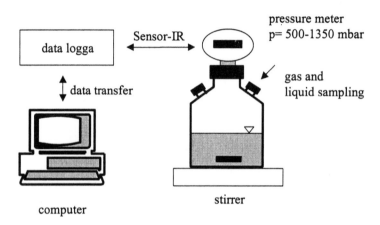

Figure 2. Principle outline of the manometric test system: Sensomat, Company Aqualytic, Germany[3]

All tests were carried out in a climatic chamber with a temperature of T=35°C±1°C during a testing period of d=30 days resp. d=60 days. The tested variants for both test system are shown in Table 2. The three kinds of material and the inoculum were examined each as a triple test. The material to be tested is introduced in a pulverised form at a concentration of 100 mg TOC/l into a medium/inoculum mixture. Test material Ecoflex® was taken from biowaste bags and ground to particle sizes d<5 mm for the investigations. 500 ml of the medium and 50 ml of the inoculum were inserted into the volumetric test system. 50 ml of suspension were sampled for analysis. The manometric test system was filled with 300 ml of the tested suspension. All reaction vessels were purged with nitrogen before being incubated.

Table 2. Description of variants tested with the volumetric and manometric test system

Flasks:	Test material	Reference material	Inoculum
F_{T1-T3} Test	Poly-ε-caprolacton	-	+
F_{T4-T6} Test	Ecoflex®	-	+
F_{B1-B3} Blank	-	-	+
F_{c1-c3} Inoculum check	-	Poly-β-hydroxybutyrate	+

Determination of pH value, total organic carbon content (TOC), dissolved organic carbon content (DOC), dissolved inorganic carbon content (DIC) and of solid content (DS) in the suspension was carried out both at the outset and at the end of tests. Analysis of the gas volume was carried out on each working day for the volumetric test system and discontinuously (every 120 minutes each day) for the manometric test system. The gas concentrations of carbon dioxid (CO_2), methane (CH_4), nitrogen (N_2) and oxygen (O_2) were measured discontinuously (twice per week) by gaschromatographie (GC-WLD: HP 5890 II, Hayesep N-column, molar sieve column, WLD) during the whole testing period.

Investigation conditions and criteria according to ISO/DIS 14853 are shown in Table 3.

Table 3. Overview of investigation criteria according to ISO/DIS 14853

Norm	ISO/DIS 14853
Test method	mineralisation of plastic material in aqueous medium under anaerobic conditions
Matrix	aqueous inorganic medium with vitamins and trace elements
Sludge	from a wastewater treatment plant (anaerobic digestor, age: 21 days), volatile organic solids: 1-2 %
Inoculum	washed sludge, DOC < 10 mg/l
Test material	plastic
Concentration of test substance	100 mg/l TOC
pH	7.0
Temperature	35°C ± 2°C
Duration time	30 or 60 days
Measuring method for gas production	manometric or volumetric
C-balancing	optional

The maximum theoretic amount of carbon (*ThC*) contained in the test material and may to be transformed into the gaseous phase during anaerobic degradation is given by equation 1.

$$ThC = \frac{m * X_c}{M_c} \qquad [1]$$

ThC	theoretic conversion org C into gas phase, [mol]
m	mass of test material, [g]
X_c	carbon content of test material, [-]
M_c	relative atomic mass of carbon, [12 g*mol^{-1}]

The maximum theoretic amount of biogas produced (*Thbiogas*) (expressed in cm^3 at standard conditions) is given by equation 2.

$$Thbiogas = ThC * V_m \qquad\qquad [2]$$

Thbiogas   Theoretic amount of biogas under standard conditions), [cm^3]
ThC   Theoretic gaseous carbon, [mol]
V_m Molar volume of an ideal gas under standard conditions, [22,414 cm^3/mol]

The maximum theoretic amount of CO_2 and CH_4 produced is given by equations 3, 4 and 5 according to *Buswell*[1].

$$C_cH_hO_o + \left(c - \frac{h}{4} - \frac{o}{2}\right)H_2O \rightarrow \left(\frac{c}{2} + \frac{h}{8} - \frac{o}{4}\right)CH_4 + \left(\frac{c}{2} - \frac{h}{8} + \frac{o}{4}\right)CO_2$$

$$[3]$$

$$ThCH_4 = \left(\frac{nCH_4}{c * Mc + h * Mh + o * Mo}\right) * m * Vm$$

$$[4]$$

ThCH$_4$ theoretical methane production at STP, [cm^3]
n   number of mole, [mol]
M_c relative atomic mass of carbon [12 g/mol]
M_h relative atomic mass of hydrogen [1 g/mol]
M_o relative atomic mass of oxygen [16 g/mol]
m   mass of the test material, [g]
V_m molar volume of an ideal gas at standard conditions [22,414 cm^3/mol]

$$ThCH_2* = \left(\frac{nCO_2}{c * Mc + h * Mh + o * Mo}\right) * m * Vm$$

$$[5]$$

ThCO$_2$* theoretical carbon dioxide production at standard conditions, [cm^3]
n   number of mole, [mol]
M_c relative atomic mass of carbon [12 g/mol]
M_h relative atomic mass of hydrogen [1 g/mol]
M_o relative atomic mass of oxygen [16 g/mol]
m   mass of the test material, [g]
V_m molar volume of an ideal gas at standard conditions [22414 cm^3/mol]

The amount of produced biogas under standard conditions in cm^3 is given by equations 6 and 7

$$V_N = \frac{\left(p_t - p_w\right)\left[mbar\right] \cdot V_t\left[cm^3\right] \cdot T_N\left[K\right]}{T_t\left[K\right] \cdot p_N\left[mbar\right]} \tag{6}$$

$$V_{NBG} = V_{Nt} - V_{Nt0} \tag{7}$$

V_N Amount of biogas at standard conditions, corrected for water vapour pressure expressed in cm^3 at time t.

p_t Atmospheric pressure (volumetric method) or pressure in the headspace of the test flask (manometric method) at time of reading t of produced biogas, expressed in mbar

p_w Water vapour pressure at given incubation temperature T_t expressed in mbar

V_t Volume of gas phase at time of reading t expressed in cm^3

T_N Temperature at standard conditions, expressed in Kelvin (=273.15 K)

p_N Pressure under standard conditions, expressed in mbar (=1013.25 mbar)

V_{NBGt} Volume of biogas produced at time t, expressed in cm^3 under standard conditions

V_{Nt} Volume of sample at time t, expressed in cm^3 under standard conditions

V_{Nt0} Volume of sample at time t_0 (beginning of the test), expressed in cm^3 under standard conditions

Calculation of the percentage of biodegradation D_t (%) for both test systems based on amounts of biogas produced is given in equation 8.

$$D_t = \frac{\left(\Sigma\left(V_{NBGt}\right)_{T1...Tn} - \Sigma\left(V_{NBGt}\right)_{Bl...Bn}\right)}{\Sigma\left(Thbiogas\right)_{1...n}} \tag{8}$$

n Number of replicates

$(V_{NBGt})_{T1,2..n}$ Amount of biogas produced at time t at standard conditions in test flasks F_{T1}, F_{T2},...F_{Tn} expressed in cm^3

$(V_{NBGt})_{B1,2,..n}$ Amount of biogas produced at time t at standard conditions in blank flasks F_{B1}, F_{B2},...F_{Bn} expressed in cm^3

$Thbiogas_{1,2,..n}$ Amount of theoretic biogas calculated by equation 2 in respective test flasks F_{T1}, F_{T2},...F_{Tn}

The carbon content of the different kinds of test material is shown in Table 4. In addition, theoretic rates of biogas production *Thbiogas* and theoretic rates of CH_4 and CO_2 production *ThCH₄* and *ThCO₂* respectively of the different kinds of material are compared. Degradability of a material is proved when 70% of theoretically possible gas production are ascertained.

Table 4. Characteristics of the tested materials with respect to the TOC and the theoretic biogas production

Test material		PHB $C_4H_6O_2$	PCL $C_6H_{10}O_2$	Ecoflex®
Carbon content	[%]	55.81	63.16	62
Theoretic biogas production	[cm^3*mg^{-1}]	1.042	1.179	1.158
Theoretic methane production	[cm^3*mg^{-1}]	0.585	0.738	-
Theoretic carbon dioxide production	[cm^3*mg^{-1}]	0.455	0.442	-

3.   RESULTS AND DISCUSSION

3.1   Preliminary tests

3.1.1   Preparation of inoculum

It is important that gas volume as well as concentration of soluble carbon components originating from the inoculum are low – in comparison to the setup containing the test substance. To reduce DOC and DIC content in the inoculum, it is washed several times with medium A (see Chapter 2 and Table 1) in a pretreatment step. Instructions for washing the inoculum were given in the ISO DIS14385. By washing it six times, a reduction of 78% (see Figure 3) of both parameters could be attained in preliminary tests. After the washing procedure, DOC concentration in the inoculum was at 48 mg/l and DIC concentration was at 273 mg/l. For DOC in the solution (medium with inoculum) of test setup, a DOC concentration of 4.4 mg/l results from a mixing ratio of 10:1 (medium/inoculum). The Maximum should not exceed 10 mg/l DOC according ISO DIS 14385. For DIC, concentration should not remain under 10 mg/l because of possible inhibitory effects[5].

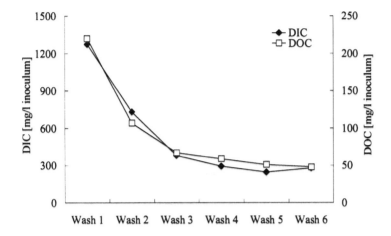

Figure 3. Dependency of DOC and DIC in the inoculum of the pre-treatment (number of washing steps)

3.1.2 Determination of carbon dioxide and methane in the aqueous and gasphase

Determination of the dissolved part of CO_2 in the aqueous phase is necessary for a complete determination of the biogas production. Compared to methane (solubility=18.35 mg/l at p=1,013 mbar and T=35°C) the maximum solubility of carbon dioxide in aqueous media is 1,141 mg/l. In preliminary tests, dissolved CO_2 was determined at the end of the test in the test setup by means of different detection methods. In the one case, the dissolved CO_2 quantity was determined by the acidification with HCl, in the other case, CO_2 determination was carried out by means of a DIC concentration analysis in the test setup. A comparison of both determination methods via HCl addition and DIC determination leads to similar results. *Püchner*[6] also ascertained similar results via DIC determination compared with HCl addition in her investigations. In this tests the determination of DIC was carried out by the HCl addition method.

The results of preliminary tests concerning the degradation of PHB are shown in Fig 4. After an investigation period of 22 days at a temperature of T=30°C a reduction of 85.4% according to the theoretical biogas production was attained. A portion of 30.5% of CO_2 obtained by the degradation of PHB was analysed in the aqueous phase, a portion of 8.1% was analysed in the gas phase, referred to the theoretic maximum biogas production (Thbiogas). Up to the end of the investigation period 46.8% of produced methane could be found in the gas phase, referred to the theoretic possible maximum biogas production (Thbiogas).

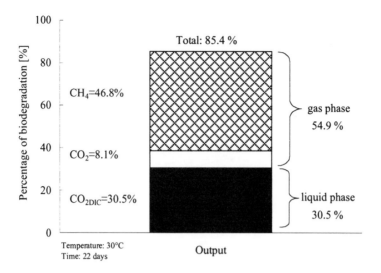

Figure 4. Conversion of PHB to CO2 and CH4 during anaerobic biodegradation

3.2   Comparison of test systems

3.2.1   Anaerobic degradation of PHB

To balance degradation of the reference material PHB (c=100mg TOC/l), the inoculum itself has to show only low biogas production. Biogas production curves of the test setups with PHB (F_{c1}-F_{c3}) and blanks (F_{b1} – F_{b3}) are shown in Fig 5 for the whole test period. After an increase of biogas production in blanks (F_{b1}-F_{b3}) at the beginning of the test, curve course is almost linear over the whole test period. At the end of the test, the mean cumulated gas production was at about 15 Nml. The course of gas production of tests with PHB is exponential and shows a mean gas production of approx. 78 Nml until the end of the test. The standard deviation of the triple is 9.9%. The inoculum shows a percentage of 19% of total gas production at the end of the examination. HCl was added to some test setups at the end of the investigation in order to collect the soluble part of mainly CO_2.

Fig 6 shows a comparison of the manometric and volumetric test method by means of cumulated biogas production during investigations of the biodegradability of PHB referred to the theoretic maximum biogas production. It can be said for both test methods that significant degradation of PHB starts simultaneously after 3 days. In the manometric test system, the exponential degradation phase ends after approx. 8 days – while in the volumetric test system biogas production takes another few days until a maximum is attained.

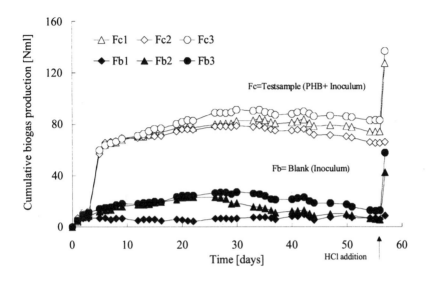

Figure 5. Biogas production for anaerobic degradation of PHB, acidification of samples with HCl at the end of the test (volumetric method)

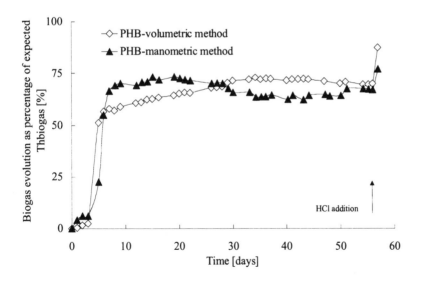

Figure 6. Comparison of the biogasformation determined with the manometric- and volumetric test method

Comparing both test systems different biogas quantities were analysed by addition of HCl at the end of the test. This can be attributed to the fact that test setups of both systems show different ratios between headspace and volume of inserted medium. In the manometric setup, ratio is approx. 1:1, in the volumetric setup average is 1:5. The CO_2 concentrations are lower in the gas phase of the manometric setup also leading to a smaller part of dissolved CO_2 in the aqueous phase. However, stripped gas quantities could clearly be recorded and quantified with both test systems.

Over the test period of 57 days, approx. 135 Nml biogas were recorded on average during PHB degradation by means of the manometric test system after addition of HCl. Thus, the reading precision of produced biogas in gas collection tube is 1%. A gas withdrawal for a gas collection volume of 200 ml is not necessary during test period. In the manometric test system, gas pressure in test setup would exceed measuring range of 350 mbar excess pressure at the end of the test. Therefore, it is necessary to perform a gas withdrawal during investigations. In the investigations in hand, a gas withdrawal took place when PHB was degraded within the area of exponential rise in pressure resp. gas production. During the whole test period of 60 days, leaks in single test setups could be ascertained for volumetric and also for manometric test system – especially in the second phase of investigation. In spite of gas tightness examinations before starting real degradation tests, gas leaks can occur. Therefore, at least a triple determination should be provided for each test setup.

Until the end of the test, biogasproduction is 67.2% (manometric test system) - resp. 70.2% (volumetric test system) of the theoretical maximum amount of biogas production. Biogas produced at the end of the test by HCl addition shows values of 10.0% (manometric test system) resp. 16.7% (volumetric test system) of Thbiogas. In total, 86.9% of Thbiogas were analysed during degradation of PHB in the volumetric test system and 77.2% of Thbiogas in the manometric test system (Fig 7). The measured and calculated biogas production in both test systems is greater than 70% of Thbiogas, therefor PHB is biodegradable.

Comparable degradation degrees were achieved within the scope of tests by different laboratories for ISO DIS 14853. Degradation degrees ranged from 68 to 96% of Thbiogas[4].

By means of DOC analyses in test setups, no dissolved components of inserted PHB could be detected at the end of the test. The balance gap can possibly be attributed to the fact that parts of inserted PHB were converted into microbial biomass. According to *Püchner*[6], 2.4 gDS biomass are produced on average by methane producers per mole of methane (assumption: acetate base product). This corresponds to 10% of substrate carbon.

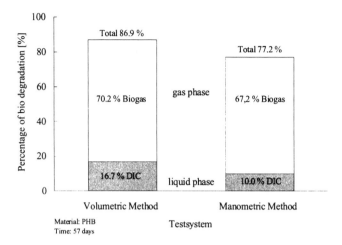

Figure 7. Comparison of measured biogas evolution expressed as a percentage of expected Thbiogas between manometric- and volumetric test method

3.2.2 Anaerobic degradation of PCL and Ecoflex®

Degradation tests with PCL and Ecoflex® showed that an anaerobic degradation according to ISO DIS 14853 does not take place (Fig 8). Biogas production rates of blanks and of attempts with material to be examined only show slight differences.

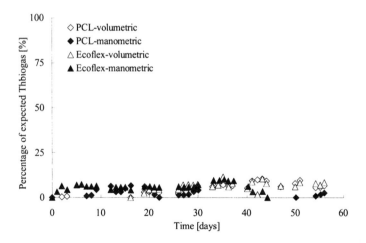

Figure 8. Percentage of expected Thbiogas of test material PCL and Ecoflex® (comparison of manometric- and volumetric method)

On the whole, biogas production values referred to the gas quantity theoretically produced, range between 5 and 15% of Thbiogas. Setups of blanks partly show higher gas formation rates probably due to inhomogeneities of the inoculum. However, in order to be in the position to make incontestable statements concerning degradation behavior under anaerobic conditions according to ISO DIS 14853, further investigations have to be carried out.

4. CONCLUSION

The biological degradation behaviour of Poly-ß-hydroxybutyrate (PHB), Poly-ε-caprolacton (PCL) and Ecoflex® were examined under anaerobic conditions according to ISO DIS 14853 by means of two different test methods. The test systems – a volumetric and a manometric determination method – were compared for determining biodegradability, gas production as well as gas composition with respect to CH_4 and CO_2. The volumetric measuring method of TUHH and the manometric measuring method (test system of the company Aqualytic) are both suitable for examination of biodegradability of plastic material according to ISO DIS 14853. Both test systems allow a recording/collection of low gas volumes (blank) and of higher gas volumes (reference material). In comparison with the manual determination of the displacement of gas volume in the volumetric test system, the determination of gas production through automatic analysis of the gas pressure shows some advantages. For instance the handling of the manometric test system is comfortable since there are no connecting pipes. The advantage of the volumetric test system in comparison to the manometric test system is the simpleness of configuration by yourself. Improvements of the manometric test system have to be carried out to enable in future an automatic let off the gas at too high gas pressures. This was done in that experiments manually.

The biodegradability under anaerobic conditions of PHB was confirmed by both test systems. No significant biodegradation of the materials PCL and Ecoflex® under anaerobic conditions could be detected.

The assessment of the biodegradability of so called biodegradable plastic materials under aerobic and anaerobic conditions is of crucial importance. With the expected increasing amounts of those materials they will be biologically treated in anaerobic treatment plants as well as in after-composting plants in the future.

REFERENCES

1. Tarwin, D., and A.M. Buswell, 1934, The methane fermentation of organic acids and carbohydrates. *J. Am. Chem. Coc.* **56**: 225-261
2. ISO CD 14853 (version 01.12.1997) Draft International Standard
3. Heerenklage, J., Colombo, and F., Stegmann, R., 2000, Comparison of test tsystems for the determination of biodegradability of organic materials under anaerobic conditions, *Proceedings of the international Conference on "Biodegradable Polymers"*, September 4-5, 2000, Wolfsburg, (W. Bidlingmaier and E. K. Papadimitriou eds), Bauhaus-Universität Weimar. Universitätsverlag, ISBN 3-86068-143-5
4. Mueller, W.-R. and Joerg, J., 1999, *Report of a ring test about aquatic biodegradability for plastics under strict anaerobic conditions based on ISO 11734 and ISO DIS 14853*, Draft Version, University of Stuttgart, Germany
5. Joerg, J., 1999, personality notes, University of Stuttgart, Germany
6. Püchner, P., 1995, *Srening-testmethoden zur Abbaubarkeit von Kunststoffen unter aeroben und aneroben Bedingungen, Stuttgarter Berichte zur Abfallwirtschaft*, Bd. 59, ISBN 3-503-03528-1, Erich Schmidt Verlag, Bielefeld, Germany
7. Seeliger, U., 1998, Ecoflex - Biologisch abbaubarer Kunststoff von BASF, Pressegespräch zur "Nowea-Journalistenreise K'98" am 17. März 1998 in Ludwigshafen

Structure-Biodegradability Relationship of Polyesters

ROLF-JOACHIM MÜLLER, ELKE MARTEN, and WOLF-DIETER
DECKWER
*Gesellschaft für Biotechnologische Forschung mbH, Mascheroder Weg 1, D-38124
Braunschweig, Germany*

Abstract: The biodegradability of polymers and also polyesters is solely determined by
              the structure and the morphology of the plastics. To ensure environmental
              safety of products and to be able to design new tailor made biodegradable
              plastics, it is important to know the correlation of structure and
              biodegradability. Based on especially synthesised aliphatic polyesters,
              aromatic polyesters, aliphatic-aromatic copolyesters and low molecular weight
              oligo-esters we studied the degradation behaviour with a lipase from
              Pseudomonas sp. For aliphatic polyesters the difference between melting
              temperature of the polymer and the degradation temperature turned out to be
              predominantly determining the degradation. The missing degradability of
              aromatic polyesters is obviously not caused by a steric hindrance of the
              polymer – enzyme complex but must be correlated with the high melting point
              and also the low flexibility of the polymer chains. The degradability of
              aliphatic-aromatic copolyesters is not determined by the number of aliphatic
              ester bonds and the length of aliphatic sequences, but the length of aromatic
              domains. The length of these domains correlates with the melting temperature
              of the materials.

1.   INTRODUCTION

Biodegradation of polymers is a complex process. Usually plastics are
not water soluble and microorganisms that want to use the polymeric
substrate as energy or carbon source have to excrete extracellular enzymes
being able to attach somehow to the plastics surface and cleave the
macromolecular chains until the polymer fragments become water soluble
and so can be transported into the cells. There, the intermediates can be

incorporated into the cellular metabolisms and finally transferred to water, carbon dioxide and other natural metabolic products.

Whether the extracellular enzymes can attack the polymer, solely depends on the complex structure of a plastics and not on the origin of the raw materials. Poly(propylene terephthalate), for instance, contains a di-alcohol component (1,3-propanediol) which can be obtained by anaerobic fermentation from glycerol. The aromatic polyesters, however, is inert to biodegradation.

To know the correlation between the structure of a polymer and its biodegradability is important, both for environmental safety reasons, and to be able to design new tailor made biodegradable plastics.

To do such investigations encounters basic difficulties. Parameters of a polymer, which are possibly relevant for biodegradability are often not independent of each other. If you change, for instance, the structure of a polymer chain, the melting point or the crystallinity of the material may change, too. Thus it is problematic to decide which change of a polymer property is responsible for a change in biodegradability.

We synthesised different model polyesters and also oligomeric esters with specific structures to study the correlation between biodegradability and the polymer-parameters responsible for the biological susceptibility. Polyesters are a suitable substance class for such investigations, because they can easily be synthesised in the lab, are potentially biodegradable due to their hydrolyzable ester bonds and they are of practical relevance, because most of the biodegradable plastics currently on the market are based on polyesters. A substantial part of our investigations is the detailed analysis of the structures of complex polyesters (e.g. copolyesters).

As biodegradation test we predominantly used the polyester hydrolysis with lipases. The free acids built during the ester cleavage were monitored via an automatic and improved titration system. Beside the enzyme test we also performed soil burial tests, compost simulations and tests with isolated microorganisms to check whether the results obtained with the lipase can be transferred to natural conditions.

In principle three simple questions should be answered by our work:

- Why are many aliphatic polyesters biodegradable and which parameters control the differences in the degradation rate?
- Why are aromatic polyesters usually not attacked by microorganisms?

- What is the degradation mechanism for aliphatic-aromatic copolyesters, where aliphatic and aromatic structures are combined in one polymer chain?

Results presented here where obtained during a PHD-work at the GBF[1]; details of the experiments are described there.

2. ALIPHATIC POLYESTERS

It is known for a long time that aliphatic polyesters can be hydrolysed by lipases and other hydrolases[2]. As parameters controlling biodegradability different parameters were discussed. The specific structure of the ester bond was supposed to be of importance, but also the hydrophobic-hydrophilic balance in neighbourhood to the ester and thermal properties like the melting point were regarded[2-4].

In literature it was often mentioned, that especially polyesters with adipic acid exhibit a good biodegradability. From the degradation of different aliphatic polyesters based on the di-alcohol 1,4-butanediol and di-acids with varying length (C4 – C12) with a lipase from *Pseudomonas sp.* showed, that the specificity of the lipase to a certain structure of the ester does not play the predominant role. When increasing the test temperature from 25°C to 50°C the maximum in degradation rate shifted from the C6-acid to the C8-acid. The degradation rates generally increased significantly, although the enzyme exhibited a rapid loss in activity at 50 °C. This indicated that the most important parameter is polymer related.

Tokiwa and Suzuki[2] already proposed in 1977 a correlation between the melting temperature of a polyester and its biodegradability. Based on this attempt, we plotted the degradation rate of different aliphatic polyesters against the difference of melting temperature (maximum of the melt peak in the DSC) (Fig 1).

We interpret this temperature difference as a measure for the "mobility" of the polyester chains. To fit into the active site of the lipase the chains have to build for an interim period of time a loop outside the polymer material.

The greater the difference between the melting temperature and the degradation temperature is, the higher are the forces binding the polymer chains in the material.

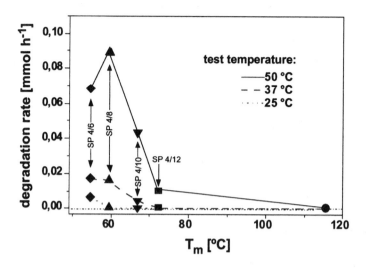

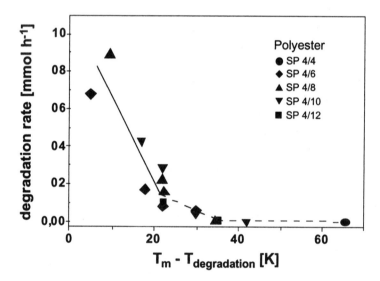

Figure 1. Rate of the hydrolysis of different aliphatic polyesters with a lipase from *Pseudomonas sp.* at different degradation temperatures. Left side: correlation with the melting point of the polyesters; right side: correlation with the difference between melting temperature and degradation temperature

3.   AROMATIC POLYESTERS

Aromatic polyesters such as PET or PBT provide excellent material properties and, thus, are widely used for many applications, but proved to be totally microbial resistant[2, 5].

One could estimate that the reason for that is the quite large aromatic rings near the ester bonds, hindering the proper alignment between active site of the enzyme and ester bond.

We could show with special synthesised aromatic and cycloaliphatic di-esters, that the lipase of *Pseudomonas sp.* in principle is able to cleave ester bonds located beside bulky chemical groups (Fig 2). However, the rate of hydrolysis is quite low, when compared to the cleavage of corresponding pure aliphatic di-esters.

One could further assume that the high melting temperatures of more than 200 °C of aromatic polyesters are the reason for their failing microbial susceptibility.

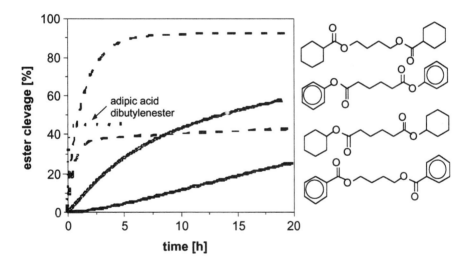

Figure 2. Degradation of aromatic and cyclo-aliphatic model esters with the lipase from *Pseudomonas sp.*

We synthesised an aromatic polyester from 1,4-butanediol and phenyl diacetic acid, which is comparable with poly(butylene terephthalate) where one methylene group is inserted between the acids and the aromatic ring of terephthalic acid. This aromatic polyester exhibits a melting point of only 70°C, which is comparable to the melting point of many aliphatic polyesters, but was not hydrolysed by the lipase. We assume, that the polymer chains of this aromatic polyester are "mobile" enough to be degraded in principle, but are not flexible enough the allow the aromatic ester bond properly fit into the active site of the enzyme (Fig 3).

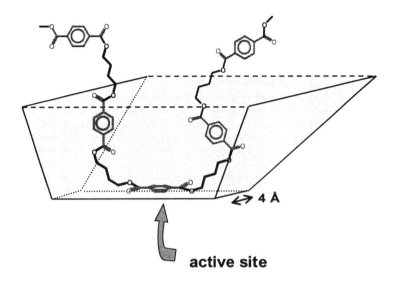

Figure 3. Polymer chain of an aromatic copolyester in the active site of a lipase from *Pseudomonas cepacia*

4.   ALIPHATIC-AROMATIC COPOLYESTERS

With the attempt to combine good material properties of aromatic polyesters and biodegradability of aliphatic polyesters, aliphatic aromatic copolyesters have been developed during the last years to be used as technical biodegradable plastics [6, 7]. The BASF AG / Germany is now producing a biodegradable material based on a copolyester of 1,4-butanediol, adipic acid and terephthalic acid (BTA-copolyester) under the trade name Ecoflex® in a several thousand tons per year scale.

In copolyesters like BTA the situation is quite complex. In principle four different kinds of ester exist (Fig 4) and there are aliphatic and aromatic domains of different length in the polymer chains.

Generally, the overall degradation rate of such copolyesters decreases, when the fraction of aromatic component (e.g. terephthalic acid) is increased.

One could assume that ester bonds near to the aliphatic diacid are rapidly degraded and determine the overall degradation rate and, thus, the degradation rate should correspond with the number of these kind of ester bonds (bonds 3 and/or 4 in Fig 4). However, the biodegradability drops much faster than it could be expected from the decreasing number of aliphatic ester bonds.

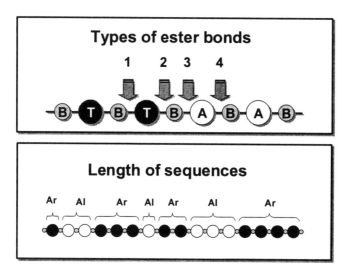

Figure 4. Types of ester bonds and distribution of sequences in the aliphatic-aromatic BTA-copolyesters (A: adipic acid, B:1,4-butanediol, T: terephthalic acid) (Ar: aromatic, Al: aliphatic)

To check the influence of the length of the aliphatic and aromatic sequences in the chains, we synthesised copolyesters of different structure, such as (blends), block copolymers, random copolymers and alternating BTA copolyester, all with the same fraction of terephthalic acid (50 mol % of the total acid components) (Fig 5).

Degradation tests with the lipase from *Pseudomonas sp.* showed, that the degradability is not increased by the length of the aliphatic sequences. While for block copolyesters (long aliphatic sequences) only very little degradation could be observed, the alternating copolyester with an aliphatic sequence length of one aliphatic acid, was rapidly degraded at 50°C [1]. Again here, the degradation is depending on the temperature difference between melting point and degradation temperature, and the melting point is correlated with the length of the aromatic sequences. That means, that long aromatic sequences hinder the aliphatic domains in the chain to be "mobile" enough for the enzymatic catalysed cleavage.

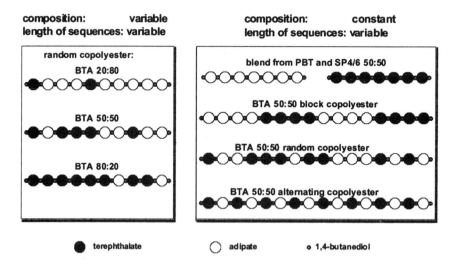

Figure 5. Aliphatic and aromatic sequences in copolyesters with different structure

5.   CONCLUSION

From our results a kind of flow chart of successive questions can be formulated by which the biodegradability or non-biodegradability of a polyester can be explained:

1) are the polyester chains mobile enough?

The mobility of the chains correlates with the difference between test temperature and melting temperature of the polyester.

2) are the polymer chains flexible enough?

The chains have to fold into the active site of the enzyme; for stiff chains the proper fitting of the ester bond into the catalytic centre may not be possible.

3) are there bonds in the polymer chain, which can be cleaved by enzymes, present in nature?

Ester or amide bonds, for instance, can by hydrolysed by hydrolytic enzymes, while for polymers with solely carbon bond in the main chain, obviously no extracellular enzymes exist (up to now) to cleave such bonds.

In some special cases also the solubility of the degradation intermediates can have an influence on the biodegradation.

Comparison of the results obtained with the enzymatic tests with biodegradation in natural environments proved that the conclusions drawn from the enzyme tests have also relevance for biodegradation under natural conditions [1].

REFERENCES

1. Marten, E., 2000. Korrelation zwischen der Struktur und der enzymatischen Hydrolyse von Polyestern, PhD Thesis, Technical University Braunschweig, Internet: http://opus.tu-bs.de/opus/volltexte/2000/136.
2. Tokiwa, Y., Suzuki, T., 1977. Hydrolysis of polyesters by lipases. *Nature* **270**: 76-78.
3. Bitritto, M.M., Bell, J.P., Brenckle, G.M., Huang, S.L., Kox, J.R. 1979. Synthesis and biodegradation of polymers derived from ✕-hydroxy acids, *J. Appl. Polym. Sci. Appl. Polym. Symp.* **35**: 405-414
4. Fields, R.D., Rodriguez F. 1976. *Microbial degradation of aliphatic polyesters.* Proceedings of the 3. Int. Biodegradation Symposium, I.M. Sharpley, A.M. Kaplan (eds), Appl. Sci., Barking, UK
5. Levefre, C., Mathieu, C., Tidjani, A., Dupret, A., Vander Wauven, C., De Winter, W., David, C., 1999. Comparative degradation by microorganisms of terephthalic acid, 2,6-naphthalene dicarboxylic acid, their esters and polyesters. *Polym. Degrad. Stab.* **64**: 9-16.
6. Witt, U., Müller, R.-J., Deckwer, W.-D., .1995. Biodegradation of polyester copolymers containing aromatic compounds. *J. Macromol. Sci. - Pure Appl. Chem.* **A32(4)**: 851-856.
7. Witt, U., Müller, R.-J., Deckwer, W.-D., 1995. New biodegradable polyester-copolymers from commodity chemicals with favorable use properties. *J. Environ. Polym. Degrad.* **3**: 215-223.

Biodegradation of the Blends of Atactic Poly[(R,S)-3-hydroxybutanoic Acid] in Natural Environments

[1]MARIA RUTKOWSKA, [1]KATARZYNA KRASOWSKA, [1]ALEKSANDRA HEIMOWSKA, and [2]MAREK KOWALCZUK
[1]Gdynia Maritime Academy, 81-225 Gdynia, Poland, [2]Polish Academy of Sciences, Centre of Polymer Chemistry, ul. Marii Curie Sklodowskiej 34, 41-819 Zabrze, Poland

Abstract: The paper presents a part of the results of the biodegradation of the blends of natural poly(3-hydroxybutyrate-co-3-hydroxyvalerate) and atactic poly[(R,S)-3-hydroxybutanoic] acid or natural poly [(R)-3-hydroxybutanoic acid] in natural environments such as a compost containing active sludge and under marine exposure conditions in sea water.

1.   INTRODUCTION

Environmental concerns are stimulating research in the development of new materials for packaging. Demand of polymeric packaging materials for the 21st century is directed toward environmentally degradable plastics. Selective use of biodegradable polymers in certain applications might help to reduce the environmental impact of plastic wastes. Under specific conditions encountered in compost, degradation of these materials may occur with the aid of specific enzymes formed by microorganisms such as bacteria or fungi.

Aliphatic polyesters are the most representative examples of biodegradable polymeric materials. Poly(3-hydroxyalkanoate)s, PHA, are well known biocompatible and biodegradable polyesters that are produced by various microorganisms as carbon and energy reserves. The physical properties of PHAs vary from crystalline–brittle to soft-sticky materials depending on the length of the side aliphatic chain on β carbon[1].

Recently, special emphasis is given to the atactic poly [(R,S)-3-hydroxybutanoic acid], a-PHB, synthesised *via* anionic ring opening polymerization of (R,S) β-butyrolactone mediated by activated anionic initiators, and its blends with natural polyesters. It was found that mechanical properties of the brittle poly(3-hydroxybutyrate-*co*-3-hydroxyvalerate), PHBV, biopolymer might be remarkably improved by blending with synthetic amorphous a-PHB. In the range of composition explored (10-50% a-PHB), blends of bacterial PHBV and a-PHB were miscible in the melt and solidified with spherulite morphology[2]. It was demonstrated earlier that this amorphous high molecular weight a-PHB (Mn=40000) does not biodegrade in the pure state but its binary blends undergo heterogeneous enzymatic attack (by PHB depolymerase A from *Pseudomonas lemoignei*) in the presence of a second crystalline component such as e.g. natural PHBV [2,3,4].

Several hundred thousand tonnes of plastics have been reported to be discarded into marine environments every year [5]. It has been estimated that over one million marine animals are killed every year either by choking on floating plastic items or becoming entangled in plastic debris[5,6]. The development of polymers biodegradable in sea water would be the key to solving problems caused by marine plastic debris. Degradation of microbial polyester in sea water may involve a simple hydrolytic degradation process in addition to a microbial (enzymatic) degradation. But unexpectedly no weight loss of films was observed in the pretreated sea water (15 min. at 120^0C), indicating that a simple hydrolytic degradation process does not contribute to the degradation of microbial polyesters in the marine environment[7].

The aim of the present study is an examination of the biodegradation of the blends of atactic poly[(R,S)-3-hydroxybutanoic] acid with selected biopolyesters i.e. PHBV or/and poly[(R)-3-hydroxybutanoic acid], PHB, in natural environments such as compost containing active sludge and under marine exposure conditions in sea water, respectively. The changes in macroscopic observations of surface and changes of weight of the samples were monitored during experiments performed. The characteristic parameters of compost and sea water are also investigated and their influence on the rate of biodegradation is discussed.

2.   EXPERIMENTAL

2.1   Materials

Bacterial PHBV (3HV content 12%, Mn = 336 000, Mw/Mn = 2,8) was supplied by Aldrich. Atactic PHB (a-PHB; Mn = 13 000, Mw/Mn = 1,3) was synthesized via anionic ring opening polymerization of racemic β-butyrolactone initiated with the 18-crown-6-supramolecular complexes of potassium hydroxide[8]. Natural PHB was supplied by Biomer (Germany) and the fraction soluble in chloroform (Mn = 103 000, Mw/Mn = 5,2) was used for experiments.. The binary PHBV/a-PHB blends (contained 30 and 40 wt% of a-PHB, respectively), PHBV/PHB blend (70/30 wt%) as well as PHBV/PHB/a-PHB blend (50/25/25 wt%) were prepared by casting from chloroform solution on glass plates at room temperature. Films with thickness of 0,2-0,3 mm were obtained. The films of PHB and PHBV were prepared similarly.

2.2   Environments

The incubation of polymer samples took place in sea water and in the compost with active sludge under natural conditions. The blend samples were located in the special basket on 2 meters depth under the surface of Baltic Sea, near the ship of Polish Ship Salvage Company, in Gdynia Harbour. The compost pile consisted of active sludge, burnt lime (CaO, in order to ravage pathogenic bacterium and eggs of parasites) and straw. It was prepared in natural conditions at sewage farm. The blend samples were put into the special basket and buried in the compost pile.

The incubation lasted 6 weeks and for some samples has been still continuing. The characteristic parameters of sea water environment according to Gdynia Water Management and Meteorology Institute are presented in Tab. 1.

Table 1. The characteristic parameters of sea water*

Parameter	Month		
	June	July	August
Temperature [°C]	17.6	20.3	19.3
pH	8.53	8.17	8.78
Cl content [g/kg]	2.92	3.28	3.21
Oxygen content [cm³/dm³]	7.55	7.57	7.42
Salt content [ppt]	5.36	5.58	6.01
Activity of Dehydrogenases [mol/mg d.m.]			6,2850

*Parameters from 1999

2.3 Methods

2.3.1 Investigation of compost

Dry mass in compost was done according to Polish Standard [9]. The samples of 10 g of compost were dried at the temperature 105°C for 8 hours and then weighed. The drying was continued until a difference of weight < 0,1% was achieved. When percent of dry mass in compost was known, moisture content was calculated.

Activity of dehydrogenases (the redox enzyme) was done according to Polish Standard [10].

This is the method for an estimation of biochemical activity of microorganisms in active sludge by oxidation processes of organic compounds. It is based on the dehydration of glucose added to the compost.

pH of compost was determined using a pH-meter. The sample of 50 g of compost and 100 cm^3 of distilled water were homogenised for 30 min. and put aside for an 1 hour. Then the pH of the solution was determined using N 5172 f. Teleko pH-meter.

The characteristic parameters of compost with active sludge are presented in Tab. 2.

Table 2. The characteristic parameters of compost with active sludge

Parameter	Month		
	June	July	August
Temperature [°C]	17.6	18.2	19.2
pH	7.84	5.78	6.30
Moisture content [%]	42.4	45.8	43.5
Activity of dehydrogenases [mol/mg d.m.]	0.0197	0.0568	0.0485

2.3.2 Investigation of blends

The macroscopic observations of blend samples and changes of their weight were tested during experiment.

The weight changes (%) were estimated using an electronic balance Gibertini E 42s.

The results obtained for clean and dried samples of blends after biodegradation experiments were compared with those of the respective samples before biodegradation.

3.   RESULTS AND DISCUSSION

The comparison of characteristic parameters of compost and sea water, presented in Table 1 and 2, indicates that the temperature in both environments was almost on the same level and was lower as that preferred for enzymatic degradation [11]. The pH values were however different and the average pH in see water was alkaline (8.49) while that of active sludge was slightly acidic (6.64).

The activity of dehydrogenases depends on the degree of growth microorganisms of populations, which are producing enzymes involved in biodegradation process. During the experiment the activity of dehydrogenases in compost had been changing a lot. With decreasing of pH the higher absolute value of activity of dehydrogenases was observed. It means that in this case (low temperature and oxygen presence in biodegradation process) psychrotrophic aerobic acidofilic microorganisms (mould) play the main role.

The results of the weight changes of the polymer samples studied during biodegradation in compost and in sea water are presented in Table 3 and Table 4, respectively.

Table 3. The weight losses of polymer samples after biodegradation in compost with active sludge

Entry No.	Polymer Sample	Time [weeks]/weight loss [%]				
		2	3	4	5	6
1.	Natural PHB	-12,29	-58,96	-73,71	complete disintegration	
2.	Natural PHBV	-1,78	-38,94	complete disintegration		
3.	Blend of natural PHBV/PHB (70/30 wt%)	-1,87	-11,99	-55,73	-66,09	-67,38
4.	Blend of PHBV/a-PHB (70/30 wt%)	-3,74	-55,43	-68,01	complete disintegration	
5.	Blend of PHBV/a-PHB (60/40 wt%)	-0,69	-9,35	-24,16	-54,89	-81,58
6.	Blend of PHBV/PHB/a-PHB (50/25/25 wt%)	-0,72	-5,3	-12,81	-23,13	-34,84

The small changes of weight for almost all polymer samples at the beginning of experiment were caused by the low activity of dehydrogenases in compost in June. The obtained results indicate that natural PHB and PHBV degrade in compost during 4-5 weeks. The natural PHB was completely disintegrated in compost after 5 weeks, even though the weight changes at the beginning of incubation were higher than that of PHBV. It is interesting to notice that the binary blend consisted of PHBV and PHB biopolymers degrades slower than both natural components (comp. Table 3, entries 1-3). However, when the natural PHB was replaced in the binary blend by synthetic a-PHB the blend of the same weight ratio degrades much faster and it was destroyed after 5 weeks. The increase of the a-PHB content in the blend from 30 to 40 wt% caused decrease of the degradation rate (comp. Table 3, entries 3-5). The lowest degradation rate was observed for the blend consisted of both natural PHA and a-PHB (Table 3, entry 6). The process of biodegradation is less intensive for blends containing natural crystalline PHB and after 6 weeks of incubation the process has been still continuing.

Table 4. The weight losses of polymer samples after biodegradation in sea water

Entry No.	Polymer Sample	Time [weeks]/weight loss [%]				
		2	3	4	5	6
1	Natural PHB	-22,7	-77,6	-89.9	complete disintegration	
2.	Natural PHBV	-15,5	-25,2	-35,1	-49,1	-67,1
3.	Blend of natural PHBV/PHB (70/30 wt%)	-4,7	-8,1	-14,8	-20,5	-27,8
4.	Blend of PHBV/a-PHB (70/30 wt%)	-36,9	-56,7	-69,6	-85,1	complete disintegration
5.	Blend of PHBV/a-PHB (60/40 wt%)	-12,8	-14,0	-23,1	-34,6	-49,8
6.	Blend of PHBV/PHB/a-PHB (50/25/25 wt%)	-5,3	-11,2	-	-24,6	-32,6

Similar trends were observed during the degradation of polymer samples in sea water as compared with degradation in compost (see Tables 3 and 4). The natural PHB was degraded in compost and in sea water during 4 weeks. However, in sea water PHBV degrades slower than in compost. After 6 weeks the sample of PHBV lost 67% of its weight. Because the activity of the dehydrogenases was higher in this period, the slower degradation may be due to different kind of mocro-organisms which are not involved in enzymatic degradation of this poyester. Again, the binary blend consisted of PHBV and PHB biopolymers degrades slower than both natural components (comp. Table 4, entries 1-3). The replacement of natural PHB with amorphous a-PHB accelerate the degradation process and the blend of the same weight ratio was destroyed after 5 weeks. The similar effect of increase of the a-PHB content in the blend on the degradation rate was observed and again the lowest degradation was observed for the blend consisted of both natural PHA and a-PHB (comp. Table 4, entries 3-6). The blends with natural crystalline PHB are the most resistant and after 6 weeks of incubation the weight loss is ~ 30% - unexpectedly.

The observed differences in the degradation rate of the polymer samples investigated may be caused by their surface morphology. Therefore, the further studies on the influence of polymer films preparation (solvent casting and blowing) as well as the influence of degradation environment on the changes of surface morphology, sample molecular weight, composition and mechanical properties are under way in our laboratories. The respective results will be published soon.

4.   CONCLUSIONS

The results of the present study revealed that natural poly(3-hydroxyalkanoate)s and their blends with synthetic atactic poly [(R,S)-3-hydroxybutanoic acid] degrade in natural environment, and the rate of biodegradation process depends on their composition, the nature of environment and the specific microorganisms.

ACKNOWLEDGEMENT

This work was partially supported by Polish State Committee for Scientific Research (Grant EUREKA E! 2004 "MICROPOL").

REFERENCES

1. Doi, Y. 1990, *Microbial Polyesters*, VCH Publishers New York
2. Scandola, M., Focarete, M. L., Adamus, G., Sikorska, W., Baranowska, I., Swierczek, S., Gnatowski., Kowalczuk, M., Jedlinski, Z., 1997, Polymer Blends of Natural Poly(3-Hydroxybutyrate-co-3-Hydroxyvalerate) and Synthetic Atactic Poly(3-Hydroxybutyrate). Characteryzation and Biodegradation studies, *Macromolecules* 30: 2568
3. Focarte, M. L., Ceccorulli, G., Scandola, M., Kowalczuk, M. 1998, Futher evidence of crystallinity-induced biodegradation of synthetic atactic poly(3-hydroxybutyrate) by PHB-depolymerase A from Pseudomonas lemoignei. Blends of atactic poly[(R,S)-3 hydroxybutyrate] with crystalline polyesters, *Macromolecules* 31: 8485
4. Scandola, M., Focarete, M. L., Gazzano, M., Matuszowicz, A., Sikorska, W., Adamus, G., Kurccok, P., Kowalczuk, M., Jedlinski, Z., 1997, Crystallinity-Induced Biodegradation of Novel [(R, S)-Butyrolaktone]-β-Pivalolactone Copolymers, *Macromolecules* 30: 7743
5. Pruter, A., 1987, Sources, Quantities and Distribution of Persistent Plastics in the Marine Environment, *Marine Pollution Bulletin* 18(6B): 305-310
6. Laist, D., 1987, Overview of the Biological Effects of Lost and Discarded Plastic Derbis in the Marine Environment, *Marine Pollution Bulletin* 18(6B): 319-326
7. Doi, Y., Kanesawa, Y., Tanahashi, N., Kumagai, Y., 1992, Biodegradation of microbial polyesters in the marine environment, *Polymer Degradation and Stability* 36: 173-177

8.   Jedlinski, Z.; Kurcok, P.; Lenz, R. W., 1995, Synthesis of potentially biodegradable polymers, *J. Macromol. Sci. Pure Appl. Chem.* **A32**: 797
9.   Polish Standard: BN-88/91103-07
10. Polish Standard: PN-82/C-04616.08
11. Lenz, W.R., 1993, Biodegradable polymers, In *Advances in Polymer Science* Vol. 107, Springer-Velay Berlin Heidelberg, pp. 1-40

Biodegradable Mater Bi Attacked by Trypsin

[1]MARIA RUTKOWSKA, [2]HELENA JANIK, [1]JOWITA TWARDOWSKA, and [1]HANNA MILLER
[1] Gdynia Maritime Academy, Morska 83, 81-225 Gdynia, Poland; [2] Technical University of Gdańsk, Chemical Faculty, Polymer Technology Department, Narutowicza 11/13, 80-952 Gdańsk, Poland

Abstract: The Mater Bi samples made of starch and poly-ε-caprolactone or starch and cellulose derivatives were studied. Enzymatic degradation of Mater Bi samples was investigated by incubating the biomaterial in concentrated trypsin solution. The weight changes, tensile strength and microscopic observations were registered before and after trypsin treatment. Mater Bi composed of thermoplastic starch and poly-ε-caprolactone, having high hydrolytic stability, is also more resistant against attack of trypsin than Mater Bi, composed of starch and cellulose.

1.   INTRODUCTION

Biodegradable polymers require intensive research and development before they become practical.

It has been known that Mater Bi is the family of biodegradable thermoplastic materials developed and marketed by Novamont S.p.A. Italy. They are a new generation of bioplastics from natural sources which retain their properties while in use and, when disposed properly, completely biodegrade[1].

Their physical properties are similar to those of conventional plastics, but they are biodegradable in different environments. Living micro-organisms transform Mater Bi products into water, carbon dioxide or methane. Mater Bi can be used in a wide range of applications such as packaging, disposable items, personal care and hygiene, agriculture and floriculture.

Biorelated Polymers: Sustainable Polymer Science and Technology
Edited by Chiellini et al., Kluwer Academic/Plenum Publishers, 2001

322   *MARIA RUTKOWSKA et al.*

It seems to be interesting to develop their applications into the biomedical field. Trypsin is a proteolytic enzyme, which is present in living organisms. Numerous proteases may be present at the wound site[2]. One of the application of Mater Bi in biomedical field could be plastic implants. That is why the study of the resistance of Mater Bi against the attack of trypsin was undertaken.

The Mater Bi samples made of starch and poly-ε-caprolactone or cellulose derivatives were studied. Enzymatic degradation of Mater Bi samples was investigated by incubating the biomaterial in concentrated trypsin solution to simulate the in vitro interaction between the polymer and the degradative enzymes during the inflammatory response.

2.   EXPERIMENTAL

2.1   Materials

The following, two types of Mater Bi samples were studied:
- Mater Bi Grade YIO1U, biodegradable, compostable, containing starch and cellulose derivatives, for rigid and dimensionally stable injection moulded items, from Novamont S.p.A.
- Mater Bi ZIO1U/C, thermoplastic starch - poly-ε-caprolactone, biodegradable and compostable, mainly for films and sheets; no sensitivity to humidity, no ageing, high UV stability, high hydrolytic stability, from Novamont S.p.A.

2.2   In vitro treatment

Samples of Mater Bi ZIO1U/C film, in a size of 15x2cm, and Mater Bi YIOU plate, were placed in a test tube containing trypsin (from bovine pancreas, Wytwórnia Surowic i Szczepionek, Warsaw) solution and incubated in the dark in incubator at 37 ^{0}C for 3 months.

The stock of 5% by weight trypsin solution consisted of the enzyme dissolved in 0.04M Tris-HCl buffer (pH=8.1) containing 0.01M $CaCl_2$ (1.1g/1 l of solution)[3] . Sodium azide (0.02% w/v) was added to incubation media to inhibit bacterial growth. The trypsin solution was prepared once before the experiment and the activity measured by caseinolytic method[5] was 2.844 U/ccm. After six days of the experiment it decreased to 1.558 U/ccm The samples after treatment in trypsin were rinsed with deionized water.

2.3   Investigation of Mater Bi samples

The weight changes (%) were determined using an electronic balance Gibertini E 42s. Clear and dried samples of Mater Bi after degradation were compared with samples before degradation.

Tensile strength of Mater Bi was measured on an Instron 1122 according to Polish Standard which are comparable to ISO standarts[4].

Microscopic observation of polymer surface changes was performed at a magnification of 1:400 with a metallographic microscope equipped with polarizers.

3.   RESULTS AND DISCUSSION

Results of weight loss of Mater Bi samples during *in vitro* treatment are presented in Fig 1.

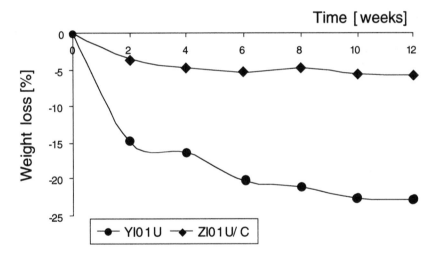

Figure 1. Weight changes of Mater Bi after incubation in trypsin solution v/s incubation time.

A decrease of weight was observed for both Mater Bi samples after treatment in trypsin solution. For both polymers the highest decrease of weight was observed at the beginning of the experiment, when the trypsin activity was at the highest level. The activity of trypsin, was found to have decreased by 50% after the 6-day treatment (look into 2.2 point) [6].

The weight changes of the sample Mater Bi YIO1U was higher than for the sample Mater Bi ZIO1U/C. It confirms the Novamont statement, that Mater Bi ZIO1U/C, which is thermoplastic starch - poly-ε-caprolactone, has high hydrolytic stability [1]. The weight of ZIO1U/C (polycaprolactone based) after very small decrease at the beginning, has not changed for 4 weeks of treatment. Probably the activity of trypsin was too low (less than 1.558 U/ccm) [6] and polymer could only be depredated from acid- or base-catalysed hydrolysis. Our studies on pure polycaprolactone have shown its good resistance against chemical hydrolysis [7]. Nevertheless the results have shown that trypsin in buffer solution has ability to induce degradation of Mater Bi ZIO1U/C. But in the case of Mater Bi YIO1U, containing starch and cellulose derivatives, trypsin shows better ability to catalyse enzymatic degradation, which is observed in Fig 1.

The surfaces of Mater Bi YIO1U samples were observed under metallographic microscope (Fig 2).

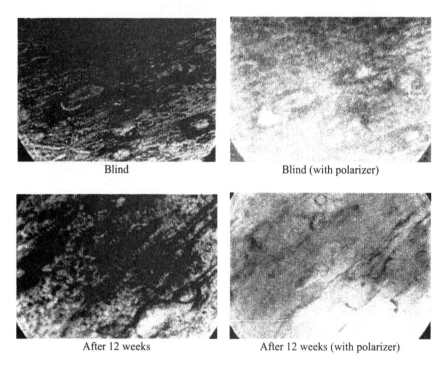

| Blind | Blind (with polarizer) |
| After 12 weeks | After 12 weeks (with polarizer) |

Figure 2. Micrographs of Mater Bi Y1101U before and after incubation in trypsin solution.

The primary sample structure was oriented with some globular dispersed phase oriented as well in the same direction as the matrix. The dispersed phase was in some places highly birefringent. After incubation of the samples in trypsin solution the oriented structure of the matrix was again observed, but had weaker birefringence. There was observed a slight decay of dispersed phase, probably due to cellulose degradation.

The primary samples of Mater Bi ZIO1U/C, observed under metallographic microscope, were slightly birefringent with co-continous structure (Fig 3).

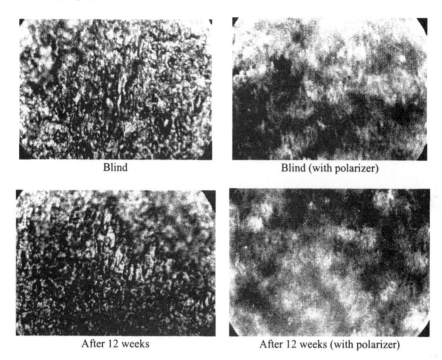

Blind   Blind (with polarizer)

After 12 weeks   After 12 weeks (with polarizer)

Figure 3. Micrographs of Mater Bi Z1101U/C before and after incubation in trypsin solution.

After incubation in trypsin there was not observed any well visible changes in the structure both with the use of polarizers and without.

The results of tensile strength of Mater Bi ZIO1U/C are shown in Fig 4.

The decrease in tensile strength of Mater Bi ZIO1U/C samples after enzyme degradation can be observed explicitly, although the loss weight of this sample was not very distinguished. After 3 months of trypsin treatment Mater Bi ZIO1U/C lost almost its tensile strength, what can be the effect of breaking some bonds in macrochains of poly-ε-caprolactone, but it seems, that they were still long enough not to cause any serious changes in weight loss. This behaviour is confirmed by microscopic observation were there was no any visible changes in the structure observed. Thus we can tell that some biodetorioration is present after action of trypsin but not very much.

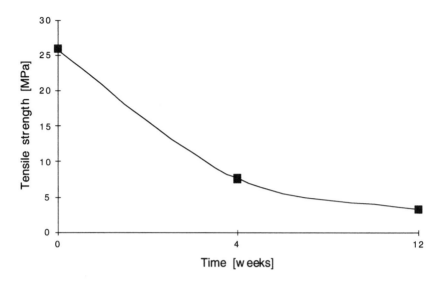

Figure 4. Changes of tensile strength [MPa] of Mater Bi ZIO1U/C after incubation in trypsin solution.

4.   CONCLUSIONS

The results of our investigation on biodegradation studies presented in the paper have shown that trypsin has the ability to induce degradation in some Mater Bi samples. The effect of trypsin on the physical and mechanical properties of the biodegradable thermoplastic materials Mater Bi is obvious, but to have the conditions of treatment by trypsin more biomimic, the fresh solution should be prepared and used for biodegradation each few days in the future.

Mater Bi composed of thermoplastic starch and poly-ε-caprolactone, having high hydrolytic stability, is also more resistant against attack of trypsin than Mater Bi composed of starch and cellulose.

REFERENCES

1. Bastioli, C., and Floridi, G., 1998, Starch-based Biodegradable Materials: Present Situation and Future Perspectives, Novamont S.p.A. Novara, Italy, *Proceedings of ICS-UNIDO Workshop, Environmentally Degradable Polymers: Environmental & Biomedical Aspects*, September 1998, Antalya, Turkey.
2. Dickinson, H., and Hiltner, A., 1981, Biodegradation of poly-(α-amino acid) Hydrogel, *J. of Biomed. Mat. Res.* **15**: 591-603.
3. Bouvier, M., Chawla, A.S., and Hinberg, I., 1991, In vitro Degradation of a poly(ether urethane) by Trypsin, *J. Biomed. Mat. Res.* **25**: 773-789.

4. Polish Standard PN-81/C89092.
5. Kunitz, M., 1947, Crystalline Soybean Trypsin Inhibitor 2. General properties, *J. Gen. Physiol.*, **31**: 291-298.
6. Twardowska, J., 1999, Master Thesis, Gdynia Maritime Academy
7. Heimowska, A., 1998, Master Thesis, Gdynia Maritime Academy

Biodegradation of Poly(Vinyl Alcohol)/ Poly(β-Hydroxybutyrate) Graft Copolymers and Relevant Blends

[1]EMO CHIELLINI, [1]ANDREA CORTI, [2]MAREK KOWALCZUK, and [1]ROBERTO SOLARO
[1]*Department of Chemistry and Industrial Chemistry, University of Pisa, via Risorgimento 35, Pisa, Italy:* [2]*Centre of Polymer Chemistry, Polish Academy of Sciences, ul. Marii Curie Sklodowskiej 34, Zabrze, Poland*

Abstract   In the present paper, the results are reported of an investigation on the biodegradability of poly(vinyl alcohol) (PVA) based systems containing poly(β-hydroxybutyrate) (PHB) of both microbial and synthetic source, as attained by polymerization of racemic β-butyrolactone. These materials, in the form of graft copolymers and blends, were submitted to the biological attack of natural soil and specific PVA-degrading microorganisms in respirometric tests. Significant levels of mineralization were recorded under soil burial conditions, though still lower than the biodegradation extent recorded by a filter paper sample utilized as biodegradable reference material. High levels of mineralization of PVA based blend and graft copolymer with atactic PHB occurred in a aqueous medium inoculated with PVA-degrading microorganisms. These results confirmed that amorphous PHB oligomers can be bioassimilated by non-PHB-degrading bacteria, also evidencing that high molecular weight atactic PHB can be degraded by the same selected microbial strains when blended with PVA.

1.   INTRODUCTION

The wide spread use of biosynthetic poly(β-hydroxybutyrate) (PHB) appears yet limited by the relatively high price when compared with synthetic polyolefins displaying comparable physical properties. Whereas the cost-bound features could be very much mitigated by economy of scale, the poor mechanical properties of PHB such as stiffness and brittleness, due to the relatively high melting temperature and crystallinity, appear to be more difficult to overcome, even though the versatility of the microbial mediated synthesis of poly(hydroxy alkanoate)s (PHA) allows to envisage a structural tailoring that should lead to improvement in processing and mechanical-dynamic properties of PHA based items. Accordingly the possibility to blend PHB with other less expensive materials, also capable of improving the mechanical performances, represents an interesting issue for the production of environmentally degradable plastics. The increase of mechanical properties of PHB-based materials was attempted by blending the bacterial polyester with several synthetic polymers such as poly(vinylidene fluoride)[1], poly(epichlorohydrin)[2], and poly(vinyl acetate) (PVAc)[3]. Miscible blends, such as PHB/PVAc combinations[3], were obtained, in this case however the positive improvement of the mechanical performance was partially discontented by the limited biodegradation[4]. Blends of PHB with biodegradable synthetic polymers, such as poly(β-propiolactone) or poly(ethylene adipate)[5] and poly(ε-caprolactone)[6], even though not miscible, are characterized by good biodegradability both in the presence of extracellular PHB depolymerase[5,6] and in soil environment[7]. PHB blends with water soluble polymers such as poly(ethylene oxide) (PEO)[8,9] and more recently with poly(vinyl alcohol) (PVA)[10] were also made by solution casting technique. In the former case miscibility and compatibility of the homopolymers in the blend were demonstrated on the basis of thermal analysis[8]. PHB/PEO blends resulted completely degradable in the presence of PHB-depolymerase[11] most likely because the PEO dissolution in the aqueous phase allows for an easier access of the degrading enzyme[6]. The influence of PVA tacticity on the compatibility and crystallization of cast films of PHB/PVA blend was also investigated[10,12], evidencing that syndiotactic PVA yields compatible blends with PHB over a wide range of composition[13]. Their biodegradation behavior was affected by structure, crystallinity, and surface composition[14].

In the present contribution an investigation on the biodegradation of blends and graft copolymers of PHB of both natural and synthetic origin[15,16] and PVA/PVAc blends is reported. The evaluation of the effects of each component in blends and copolymers on both the overall extent of biodegradation and the biodegradation of counterparts, was investigated in aqueous media and in soil by means of respirometric tests.

2.   MATERIAL AND METHODS

2.1   Polymer Samples

Poly(vinyl alcohol) samples containing 12% of acetyl groups were either kindly supplied by Idroplast S.p.A. (Altopascio, Lucca-Italy) (PVA88M, Mw 34) or purchased from Sigma (PVA88S, Mw 52). A sample of poly(vinyl acetate) (PVAc, Mw 310 kD) was purchased from Aldrich. Bacterial poly(3-hydroxybutyric acid) (PHB, Mw 260 kD) was kindly supplied by Copersucar (São Paulo-Brazil).

Atactic poly(3-hydroxybutyrate) (PHB) samples having Mn 0.6 (*a*PHB1) and 10 kD (*a*PHB2) were obtained via anionic oligomerization and polymerization of (R,S)-β-butyrolactone initiated by the 18-crown-6 complex of the sodium salt of (R,S)-3-hydroxybutyric acid[15,16]. The PVA88S-*g*-*a*PHB1 graft copolymer was synthesized by oligotrans-esterification reaction of (R,S)-β-butyrolactone with PVA88S at 35 °C, in the presence of porcine pancreatic lipase as a catalyst[17,18], whereas the PVA88S/*a*PHB2 blend (60/40 wt %) was obtained by casting from a mixture of ethanol and water solutions containing 1g of *a*PHB2 and 1.5g of PVA88S, respectively.

2.2   Film Preparation

Thick films (0.5 mm) were obtained by hot pressing at 180 °C and 1.5 tons for 1 min of powdered mixtures of PVA based blends with microbial PHB and PVAc containing low molecular weight polyols.

A film of PVA88S/*a*PHB2 polymer blend (PVA88S/*a*PHB2) cast from hexafluoro*iso*propanol solution was also utilized in the present investigation.

2.3   Respirometric Biodegradation Tests

2.3.1   Simulated Soil Burial[19]

The test was carried out in 1 l Erlenmeyer flasks containing the following multilayer substrate: the bottom of each flask was covered with 20 g of a mineral hygroscopic bed represented by heat-expanded aluminosilicate perlite, the middle layer was filled with a mixture of 20 g milled perlite and 10 g garden soil, and the top was covered with further 20 g perlite. Polymeric film samples and filter paper (positive control) were settled in the middle layer. Test flasks were incubated in the dark at room temperature for 6 months.

2.3.2   Aqueous Medium

The test was carried out in 500 ml cylindrical vessels containing 100 ml of a mineral based culture medium, inoculated with a selected PVA-degrading mixed bacterial culture[20]. Incubation was carried out under static conditions at 20 °C.

Test flasks and vessels were purged periodically with carbon dioxide free air. CO_2 evolved from the test cultures was monitored within the incubation time by adsorption in $Ba(OH)_2$ 0.025 N solutions that were titrated with 0.05 N HCl.

The extent of mineralization of polymeric samples was evaluated as percent of the theoretical amount of CO_2 from the test material, determined from its carbon content, corrected for the quantity of CO_2 produced in the blank during the incubation time.

3.   RESULTS AND DISCUSSION

3.1   Simulated Soil Burial

Three different films obtained by a hot pressing procedure of powder mixtures containing plasticized PVA, PVAc with and without bacterial PHB were submitted to the biological degradation of soil microorganisms in a respirometric test aimed at simulating soil burial conditions. The compositions of the blend films are summarized in Table 1.

About 550 mg of each blend film was utilized in the soil burial experiment, along with about 480 mg of filter paper utilized as truly biodegradable reference material.

Table 1. Characteristics of blends tested in simulated soil burial

Blend Type	PVA88M (g)	PVAc (g)	PHB (g)	Polyols (g)
C	2.0	0.0	0.0	0.78
D	1.5	0.5	0.0	0.59
E	1.5	0.1	0.4	0.59

The mineralization rate and extent of all the PVA based blend films was strictly similar, reaching about 30% of biodegradation after 6 months of incubation. The mineralization profile of the filter paper sample approached instead a plateau corresponding to 42% of the net theoretical CO_2, already at the end of the third month of incubation (Fig 1).

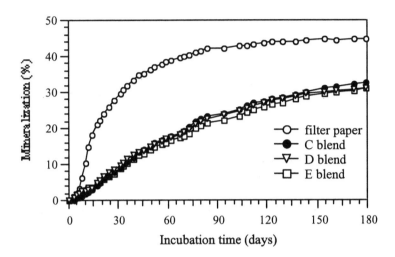

Figure 1. Biodegradation profiles of hot-pressed PVA based films in simulated soil burial test.

At the end of the test, significant amounts of degraded test films were recovered only in the case of the PVAc containing samples (Blends D and E), whereas the plasticized PVA-based C sample had completely disappeared. Film fragments derived from D blend were thoroughly washed in distilled water, dried, and then extracted with chloroform. Both the aqueous and chloroform solutions were then analyzed by GPC in order to evaluate the content of PVA and PVAc, as well as their relevant molecular weight distributions. For comparison, an analogous extraction procedure was carried out on the original blend D sample, and the retrieved polymer fractions were analyzed by GPC.

In the case of PVA, only small traces of low molecular weight fractions were recovered from the film fragments (Fig 2).

On the contrary, an almost quantitative recovery of the PVAc fraction was obtained from the D film submitted to degradation in soil. Moreover, the molecular weight distribution of PVAc retrieved from the soil burial respirometric experiment was practically unchanged (Fig 3).

These results therefore indicate that the carbon dioxide detected in the respirometric experiment in the soil containing flask supplemented with PVAc-containing D film, can be attributed only to the microbial assimilation of low molecular weight polyols and PVA.

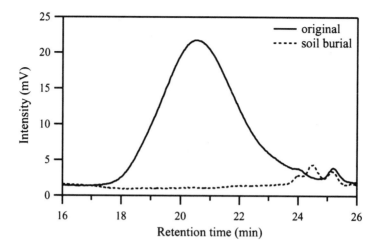

Figure 2. GPC chromatograms of the PVA fraction of D blend before and after biodegradation under simulated soil burial conditions.

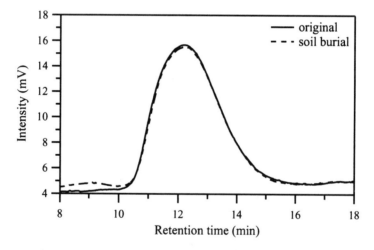

Figure 3. GPC chromatograms of the PVAc fraction of D blend before and after biodegradation under simulated soil burial conditions.

Accordingly, the net carbon dioxide emissions of the PVAc containing blends can be recalculated leaving out the contribution of PVAc carbon content. On this basis, a 6% higher mineralization of D blend can be outlined in comparison with the overall extent of biodegradation recorded for the C blend containing only plasticized PVA (Fig 4). Consequently, it seems that the presence of PVAc can promote the degradation of the counter part PVA in soil, whose limited biodegradability in this environment had been repeatedly ascertained[20-22].

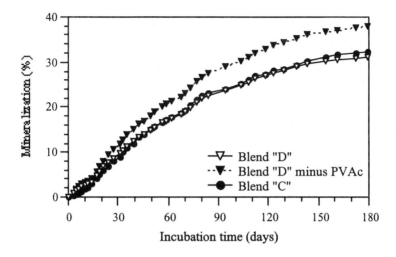

Figure 4. Biodegradation curves of C and D hot pressed PVA based films under simulated soil burial conditions, compared with the mineralization profile of D blend diminished of the PVAc-derived organic carbon.

3.2   Aqueous Medium

A preliminary investigation on the biodegradation of graft copolymer and blends of PVA with chemically synthesized atactic PHB was carried out in an aqueous medium by using a PVA-degrading selected bacterial culture as inoculum.

In this test, the biodegradation behavior of PVA88S-*g*-*a*PHB1 graft copolymer and PVA88S/*a*PHB2 blend cast film was compared with the corresponding *a*PHB1, *a*PHB2, and PVA88S homopolymers (Table 2).

Table 2. Characteristics of homopolymer, graft copolymer, and blend tested in an aqueous medium in the presence of PVA-degrading microorganisms.

Sample	Characteristic	Mn (kD)
PVA88S	PVA (88% degree of hydrolysis)	30-70
*a*PHB1	atactic PHB	0.6
*a*PHB2	atactic PHB	10
PVA88S-*g*-*a*PHB1	graft copolymer PVA88S and *a*PHB1	-
PVA88S/*a*PHB2	cast film of PVA88S and *a*PHB2 blend	-

In the case of *a*PHB2 sample, which resulted almost completely insoluble in the aqueous culture medium, molecular weight analyses were also carried out by GPC before and after the degradation test. All the other polymer samples quickly swelled and disintegrated after immersion in the culture medium, and were not recoverable at the end of the test.

An appreciable microbial assimilation of atactic PHB oligomers (aPHB1 sample), as well as of both graft copolymer (PVA88S-g-aPHB1 sample) and blend (PVA88S/aPHB2 sample) initiated after three days of incubation. The relevant mineralization extent reached a plateau in correspondence of similar biodegradation values (37-42 %) after 47 days of incubation. However, a slight difference in the biodegradation profile was observed in the case of the blend film, whose mineralization started early, then slowed down after 10 days of incubation. A second exponential phase was observed after 25 days of incubation, thus closely resembling a typical dual phase growth curve (Fig 5).

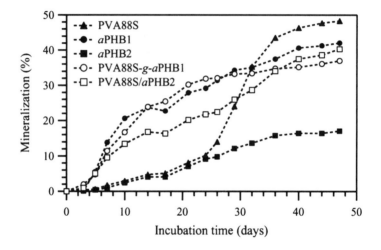

Figure 5. Biodegradation curves of atactic PHB and PVA samples and their graft copolymers and blends in an aqueous medium in the presence of PVA-degrading microorganisms.

Microbial attack of pure PVA having 88% hydrolysis degree, and of high molecular weight (10 kD) atactic PHB (aPHB2 sample) was delayed, displaying a sort of lag phase of 7 days. However, PVA reached about 50 % of mineralization in 48 days. By contrast, the biodegradation profile of high molecular weight synthetic PHB almost matched that of the vinyl polymer during the first 24 days of incubation, then slowed down to an almost constant value of about 15% mineralization (Fig 5).

The limited bioassimilation observed in the case of the aPHB2 sample can be explained by considering that very likely PVA-degrading microorganisms are able to use only the low molecular weight fractions of atactic PHB.

GPC analysis of the atactic PHB having the highest molecular weight (aPHB2 sample) in the original status and once retrieved from the degradation treatment, clearly evidenced that a small, but significant, microbial assimilation was restricted to the low molar mass fractions present in the sample (Fig. 6). This finding confirms the limited biodegradation recorded in the respirometric biodegradation test.

These results were partially in accordance with previous findings on the unusual biodegradation behavior of amorphous high molecular weight *a*PHB, whose attack by PHB-depolymerase seems to take place only in the presence of a crystalline counterpart[15,23-24]. Consequently PVA appear to stimulate the biodegradation of amorphous high molecular weight PHB, as recently reported for a microbial polyester sample[14], while the microbial attack of high molecular weight *a*PHB itself was restricted to low molar fractions. Moreover, atactic PHB oligomers can be assimilated by PHB-degrading bacteria, as well as by bacterial species, which are not able to release specific PHB depolymerase[16].

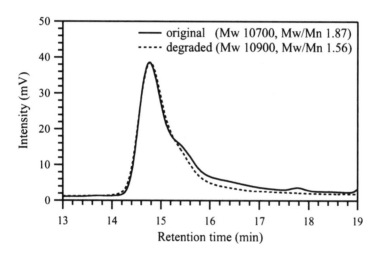

Figure 6. GPC chromatograms of *a*PHB2 sample before and after biodegradation in the presence of PVA-degrading microorganisms.

4. CONCLUSION

A biodegradation study was carried out on ternary blend systems represented by PVA containing 12% residual acetyl units, PHB of both natural and synthetic origin, and PVAc. Two basic issues were tentatively approached. In the first instance, the influence of hydrophobization of PVA-based materials, obtained by addition of bacterial PHB and PVAc, on their biodegradation in soil environment was investigated. It was ascertained that the biodegradation of PVA in soil is very limited, most likely due to the irreversible adsorption of the polymer on the clay component through the hydroxyl groups[25]..

Another basic issue to be considered was represented by the influence of PHB on the biodegradability of PVA in the presence of specific PVA-degrading microorganims, by considering that biodegradation of the polyester and of the polyhydroxylated polymer is strictly mediated by PHB-depolymerase[26] and PVA-oxidase[27] specific enzymes, respectively. In this connection, the biodegradability of both a graft copolymer and a cast blend was ascertained in the presence of a select bacterial culture able to utilize PVA.

The results so far obtained in a soil burial test evidenced that PVA-based hot pressed films containing PVAc and bacterial PHB experienced an appreciable degradation, that is still lower than that recorded for the biodegradable reference material represented by filter paper. The presence of both plasticizers and biodegradable bacterial PHB does not seem to affect significantly the biodegradation of the vinyl polymer by soil microorganisms. PVAc was not degraded at all, but rather surprisingly addition of this material apparently enhanced the PVA mineralization under soil burial condition, very likely due to a substantial reduction of PVA adsorption on soil particles.

PVA/atactic PHB graft copolymer and relevant blends were assimilated by selected PVA-degrading microorganisms. Their mineralization extents were comparable to that of pure PVA. Biodegradation extents recorded for the homopolymers, namely atactic PHB having 0.6 kD molecular weight, and PVA, account for a very reliable microbial attack of both the related fractions in the graft copolymers. The assimilation of *a*PHB also indicates that the PVA-degrading bacterial species, utilized in the present study, are able to hydrolyze the polyester to some extent, and may secrete enzymes that are not stereo-specific, in contrast to the majority of poly(β-hydroxy alkanoate) depolymerases. This is going to further substantiate previous findings on the bioassimilation of *a*PHB oligomers by the non-PHB-degrading bacterium *Ralstonia eutropha*[16]. In addition, the biodegradation of high molecular weight *a*PHB occurred in blends with PVA in the presence of non-PHB-degrading microorganisms. Up to now, biodegradation of amorphous high molecular weight PHB was observed only in blends and copolymers containing a second crystalline component[15,23-24].

REFERENCES

1. Marand, H., and Collins, M., 1990, Crystallization and morphology of poly(vynilidene fluoride)/poly(3-hydroxybutyrate) blends. *Polym. Prep.* **31**: 552-553.

2. Dubini Paglia, E., Beltrame, P.L., Canetti, M., Seves, A., Marcandalli, B., and Martuscelli, E., 1993, Crystallization and thermal behaviour of poly(D(-)-3-hydroxybutyrate)/poly(epichlohydrin) blends. *Polymer* **34**: 997-1001.

3. Greco, P., and Martuscelli, E., 1989, Crystallization and thermal behaviour of poly(D(-)-3-hydroxybutyrate)-based blends. *Polymer* **30**: 1475-1483.

4. Kumagai, Y., and Doi, Y., 1993, Biodegradable plastic. 1. The biodegradability of polymer blends. *Sumitomo Search* **52**: 155-162.

5. Kumagai, Y., and Doi, Y., 1992, Enzymic degradation of binary blends of microbial poly(3-hydroxybutyrate) with enzimically active polymers. *Polym. Degrad. Stab.* **37**: 253-256.

6. Kumagai, Y., and Doi, Y., 1992, Enzymic degradation and morphologies of binary blends of microbial poly(3-hydroxybutyrate) with poly(ε-caprolactone), poly(1,4- butylene adipate) and poly(vinyl acetate). *Polym. Degrad. Stab.* **36**: 241-248.

7. Lisuardi, A., Schoenberg, A., Gada, M., Gross, R.A., and McCarthy, S.P., 1992, Biodegradation of blends of poly(β-hydroxybutyrate) and poly(ε-caprolactone). *Polym. Mater. Sci. Eng.* **67**: 298-300.

8. Avella, M., and Martuscelli, E., 1988, Poly-D(-)(3-hydroxybutyrate)/polyrthylene oxide) blends: phase diagram, thermal and crystallization behaviour. *Polymer* **29**: 1731-737.

9. Avella, M., Martuscelli, E., and Greco, P., 1991, Crystallization behaviour of poly(ethylene oxide) from poly(3-hydroxybutyrate)/poly(ethylene oxide) blends: phase structuring, morphology and thermal behaviour. *Polymer* **32**: 1647-1653.

10. Azuma, Y., Yoshie, N., Sakurai, M., Inoue, Y., and Chûjô, R., 1992, Thermal behaviour and miscibility of poly(3-hydroxybutyrate)/poly(vinyl alcohol) blends. *Polymer* **33**: 4763-4767.

11. Kumagai, Y., and Doi, Y., 1992, Enzymic degradation of poly(3-hydroxybutyrate)-based films: poly(3-hydroxybutyrate)/poly(ethylene oxide) blend. *Polym. Degrad. Stab.* **35**: 87-93.

12. Yoshie, N., Azuma, Y., Sakurai, M., and Inoue, Y., 1995, Crystallization and compatibility of poly(vinyl alcohol)/poly(3-hydroxybutyrate) blends: influence of blend composition and tacticity of poly(vinyl alcohol). *Appl. Polym. Sci.* **56**: 17-24.

13. Ikejima, T., Yoshie, N., and Inoue, Y., 1996, Infrared analysis on blends of poly(3-hydroxybutyric acid) and stereoregular poly(vinyl alcohol): influence of tacticity of poly(vinyl alcohol) on crystallization of poly(3-hydroxybutyric acid). *Macromol. Chem. Phys.* **197**: 869-880.

14. Ikejima, T., Cao, A., Yoshie, N., and Inoue, Y., 1998, Surface composition and biodegradability of poly(3-hydroxybutyric acid)/poly(vinyl alcohol) blend films. *Polym. Degrad. Stab.* **62**: 463-469.

15. Scandola, M., Focarete, M.L., Adamus, G., Sikorska, W., Baranowska, I., Swierczek, S., Gnatowski, M., Kowalczuk, M., and Jedlinski, Z., 1997, Polymer blaends of natural poly(3-hydroxybutyrate-co-hydroxyvalerate) and a synthetic atactic poly(3-hydroxybutyrate). Characterization and biodegradation studies. *Macromolecules* **30**: 2568-2574.

16. Focarete, M.L., Scandola, M., Jendrossek, D., Adamus, G., Sikorska, W. and Kowalczuk, M., 1999, Bioassimilation of atactic poly[(R,S)-3-hydroxybutyrate] oligomers by selected bacterial strains. *Macromolecules* **32**: 4814-4818.

17. Shuai, X., Jedlinski, Z., Kowalczuk, M., Rydz, J., and Tan, H., 1999, Enzymatic synthesis of polyesters from hydroxyl acids. *Europ. Polym J.* **35**: 721-725.

18. Jedlinski, Z., Kowalczuk, M., Adamus, G., Sikorska, W., and Rydz, J., 1999, Novel synthesis of functionalised poly(3-hydroxybutanoic acid) and its copolymers. *Int. J. Biol Macromol.* **25**: 247-251.

19. Solaro, R., Corti, A., and Chiellini, E., 1998, A new respirometric test simulating soil burial conditions for the evaluation of polymer biodegradation. *J. Environ. Polym. Degr.* **6**: 203-208.

20. Chiellini, E., Corti, A., and Solaro, R., 1999, Biodegradation of poly(vinyl alcohol) based blown films under different environmental conditions. *Polym Degrad. Stab.* **64**: 305-312.

21. Krupp, L.R., and Jewell, W.J., 1992, Biodegradability of modified plastic films in controlled biological environments. *Environ. Sci. Technol.* **26**: 193-198.

22. Sawada, H., 1994, Field testing of biodegradable plastics. In *Biodegradable Plastics and Polymers* (Y. Doi and K. Fukuda, eds.), Elsevier publ., Amsterdam, pp. 298-312.

23. Scandola, M., Focarete, M.L., Gazzano, M., Matuszowicz, A., Sikorska, W., Adamus, G., Kurcok, P., Kowalczuk, M., and Jedlinski, Z., 1997, Crystallinity-induced biodegradation of novel [(R,S)-β-butyrolactone]-*b*-pivalolactone copolymers. *Macromolecules* **30**: 7743-7748.

24. Focarete, M.L., Ceccorulli, G., Scandola, M., and Kowalczuk, M., 1998, Further evidence of crystallinity-induced biodegradation of synthetic atactic poly(3-hydroxybutyrate) by PHB-depolymerase A from *Pseudomonas lemoignei*. Blends of atactic poly(3-hydroxybutyrate) with crystalline polyesters. *Macromolecules* **31**: 8485-8492.

25. Chiellini, E., Corti, A., Politi, B., and Solaro, R., 2001, Biodegradation of poly(vinyl alcohol) on solid substrates. *Polym and Environ.* In press.

26. Delafield, F.P., Doudoroff, M., Palleroni, N.J., Lusty, C.J., and Contopoulos, R., 1965, Decomposition of poly(b-hydroxybutyrate) by pseudomonads. *J. Bacteriol.* **90**: 1455-1466.

27. Suzuki, T., 1979, Degradation of poly(vinyl alcohol) by microorganisms. *J. Appl. Polym. Sci., Appl. Polym. Symp.* **35**: 431-437.

Structural Studies of Natural and Bio-Inspired Polyesters by Multistage Mass Spectrometry

MAREK KOWALCZUK and GRAZYNA ADAMUS
Polish Academy of Science, Centre of Polymer Chemistry, ul. Marii Curie Sklodowskiej 34, 41800 Zabrze, Poland

Abstract   Hyphenation of electrospray ionization with ion-trap multistage mass spectrometry (ESI-MS^n) provides characterisation of subtle molecular structure of both synthetic and natural aliphatic polyesters as well as oligomeric products of their biodegradation. Implementation of multistage MS technique for structural studies of the novel bio-inspired polymers including poly[(R,S)-3-hydroxybutyrate], a-PHB, telechelics and a-PHB conjugates with oligopeptides has been demonstrated. The novel results concerned with evaluation of the subtle structure of macroinitiators obtained by partial depolymerisation of selected natural poly(3-hydroxyalkanoate)s, PHA, are also presented.

1. INTRODUCTION

The creation of useful bio-inspired polymeric materials with defined molecular architecture and properties requires detailed and unambiguous information about their structure.[1] Chemical structures of polymer end groups play an important role in determining the polymerisation mechanism as well as functional properties of polymeric materials. Recently, electrospray ionisation multistage mass spectrometry (ESI-MS^n) has been successfully applied for characterisation of subtle molecular structure of aliphatic polyesters.[2,3] ESI permitted the generation of intact, charged high mass ions whereas MS provided the opportunity of their resolution with improved mass accuracy. With the aid of multistage mass spectrometry, polymeric ions produced by electrospray ionization were subjected to further fragmentation, and the induced fragment ions were found to be very informative for characterisation of polymer end group structure as well as distribution of co-monomers in the individual chains of copolymer macromolecules.[2-4]

The present work deals with novel issues related to the application of this method for evaluation of the subtle structure of bio-inspired synthetic homo- and copolymers of (R,S)-β-butyrolactone as well as macroinitiators obtained by partial depolymerisation of selected natural PHA.

2.   EXPERIMENTAL SECTION

The electrospray ionisation mass spectrometric analyses were performed using a Finnigan LCQ ion trap mass spectrometer. The samples were dissolved in methanol or chloroform methanol system (10:1 v/v) and such solutions were introduced to the ESI source by continuous infusion by means of the instrument syringe pump with the rate of 3 μL/min. The ESI source was operated at 4.25 kV and the capillary heater was set to 200°C. For ESI-MS^n experiments mass selected mono-isotopic parent ions were isolated in the trap and collisionally activated with 33% ejection RF–amplitude at standard He pressure. The experiments were performed in the positive and negative-ion mode.

3.   RESULTS AND DISCUSSION

Aliphatic polyesters are the most representative examples of biodegradable polymeric materials. Poly(3-hydroxyalkanoate)s, PHA, are well known biocompatible and biodegradable polyesters that are produced by various organisms. Poly [(R)-3-hydroxybutyrate], (R)-PHB, was the first discovered example of PHA and many attempts have been undertaken in order to synthesise (R)-PHB via biofermentation or ring-opening polymerization of β-butyrolactone.[5] Recently, a novel facile method of regioselective synthesis of biomimetic (R)-PHB has been developed.[6] It has been also found that among all possible synthetic stereoisomers of PHB the atactic poly[(R,S)-3-hydroxybutyrate], a-PHB, can significantly improve physical and biodegradation properties of polymeric materials obtained thereof.[7,8]

3.1   ESI-MS^n of Bio-inspired Homo- and Copolyesters

Ring opening polymerisation of (R,S) β-butyrolactone mediated by activated anionic initiators leads to a-PHB with well-defined chemical structure of the end groups as well as block, graft and random copolymers.[9] The ESI-MS^n technique allowed fast and reliable identification of a-PHB macromolecules. It was demonstrated that the biomimetic polymer studied contained various end groups, depending on the anionic initiator employed.[2] The obtained results supported previously proposed mechanism of the ring-opening polymerization of β-butyrolactone initiated by activated alkali metal alkoxides.[10] The utility of this technique was also illustrated for the evaluation of polymer end-capping reactions and characterisation of novel poly[(R,S)-3-hydroxybutyrate] telechelics containing primary hydroxyl groups at both polymer chain ends.[11] The telechelic a-PHB was synthesised by the ring-opening polymerisation of (R,S)-β-butyrolactone in the presence of 18-crown-6 complex of 4-hydroxybutanoic acid sodium salt as an initiator, followed by termination of polymerisation with corresponding bromoalcohol:

The positive ESI-MS spectra of a-PHB telechelics revealed that the polymer was terminated by primary hydroxyl groups. The corresponding set of sodium adduct ions of macromolecules containing 4-hydroxybutanoate and hydroxyethyl ester groups is presented in Fig 1. The most abundant sodium adduct ion was located at *m/z* 1031.6 and was assigned to the decamer (Fig 1a). The MS^2 fragmentation experiment on this mass-selected ion revealed fragmentation pathways from both ends of individual macromolecule (Fig 1b). The fragment ion at *m/z* 927 was formed by the loss of 4-hydroxybutanoic acid (104 Da). On the other hand, the fragment ion at *m/z* 901 was formed by the expulsion of 2-hydroxyethyl crotonate (130 Da).

The characterisation of novel a-PHB telechelics has been thus accomplished with the aid of ESI-MS^n analysis, and the convenience of this technique for evaluation of polymer end-capping reactions has been demonstrated.

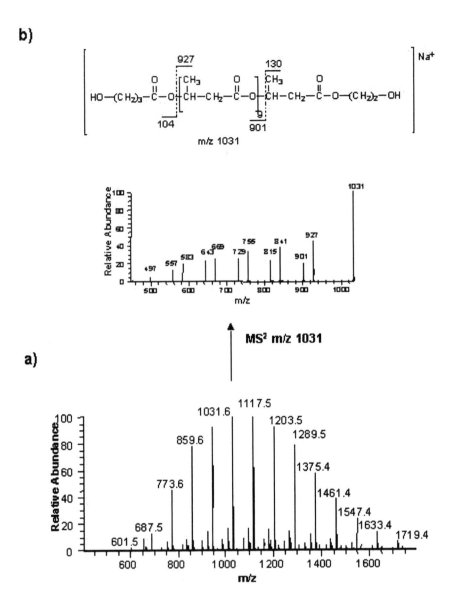

Figure 1. The ESI-MS spectrum (positive ion–mode) and sequential fragmentation spectrum (MS²) of a-PHB telechelic obtained in the polymerization of (R,S) β-butyrolactone initiated by 18-Crown-6 complex of 4-hydroxybutanoic acid sodium salt and terminated with bromoethanol.

The implementation of multistage mass spectrometry significantly improved analysis of novel bio-inspired materials that combined both oligopeptide and a-PHB structural units.[12] The ESI-MS spectrum (performed in negative ion mode) presented in Fig 2a confirmed that a copolymer obtained in the reaction of (R,S) β-butyrolactone with Ala-Ala-Ala oligopeptide possessed (Ala-Ala-Ala)-*block*-(a-PHB) structure.

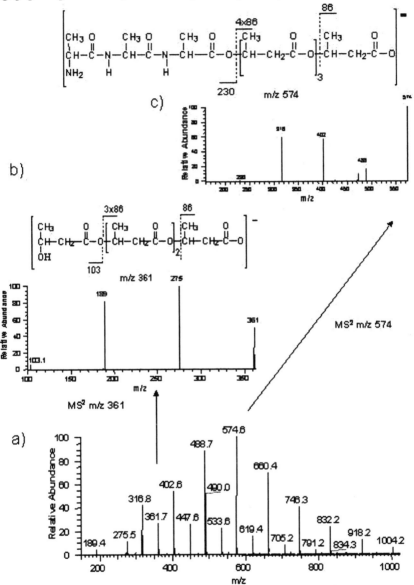

Figure 2. The ESI-MS spectrum (negative ion–mode) and sequential fragmentation spectra (MS²) of oligomers obtained in the reaction of Ala-Ala-Ala oligopeptide with (R,S) β-butyrolactone.

This spectrum exhibited the singly charged negative ions corresponding with the oligopeptide covalently bonded to different numbers of a-PHB units going from one (m/z 316) up to nine (m/z 1004). Moreover, this ESI-MS spectrum revealed the presence of minor series of ions corresponded to the oligomers containing 3-hydroxybutyrate and carboxylic acid residues. The most abundant ion of this series was located at m/z 361 and was assigned to the a-PHB tetramer (Fig 2a). The ESI-MS^2 experiments performed for the parent anions appearing at m/z 361 and m/z 574 (Fig 2b and 2c) confirmed the above assignments. Based on the fragmentation mechanism of a-PHB negative parent ions, that was reported previously,[13] it was assumed that for the parent anion at m/z 361 the respective fragment ions at m/z 275, 189 and 103 were formed because of successive loss of crotonic acid (86 Da) at the carboxylic side of a-PHB oligomer. The fragment anion at m/z 103 corresponded to the anion of 3-hydroxybutanoic acid. A similar mechanism was observed in the fragmentation experiment performed for the selected parent ion at m/z 574 (Fig 2c, MS^2). The fragment ions at m/z 488, 402, 316 and 230 were formed due to the successive loss of crotonic acid at the carboxylate side of the a-PHB block. The ion at m/z 230 corresponded to the Ala-Ala-Ala anion.

It has been recently demonstrated that the equimolar reaction of (R,S) β-butyrolactone with D,L-lactic acid, conducted in bulk at the temperature of 70^0 C, leads to the copolymer containing hydroxyl and carboxylic end groups.[14] The negative-ion ESI-MS spectrum of the copolymer obtained consisted with clusters of anions separated due to their different degree of oligomerization (from dimer up to 14-mer) and composition (Fig 3a).

The composition of co-monomer units in this copolymer was visualised by the fragmentation experiments (Fig 3b). The MS^2 fragmentation experiment performed for parent ion m/z 519 (Fig 3, HOB_5L anion selected from the hexamer cluster) induced sets of two fragment anions with the difference of 14 Da. The fragment anion at m/z 447 corresponded to HOB_5 oligomer and the fragment ion at m/z 433 corresponded to HOB_4L oligomer formed by the expulsion of crotonic acid (86 Da) from the carboxylic side of the HOB_5L parent anion. Such fragmentation pathway indicated clearly that the parent anion m/z 519 (Fig 3b) corresponds to the mixture of macromolecules of the general formula HOB_5L with the random location of one lactic unit along their chains.

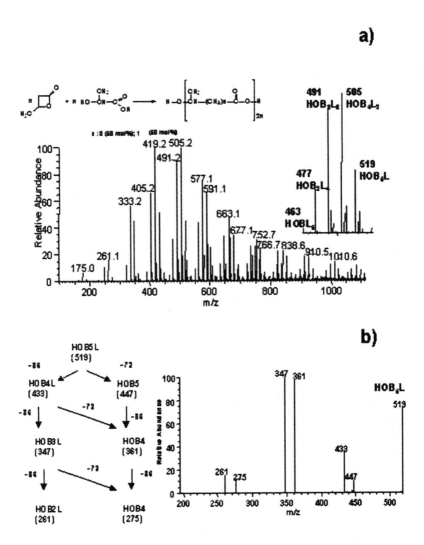

Figure 3. (a) The negative-ion ESI-MS spectrum of the copolymer obtained in the equimolar reaction of (R,S) β-butyrolactone with D,L-lactic acid, conducted in bulk at the temperature of 70^0 C, (b) the MS^2 fragmentation spectrum of selected HOB$_5$L parent ion m/z 519.

3.2   ESI-MS^n of Natural Homo- and Copolyesters

Sequence distribution and chemical structure of mass-selected macromolecules of macroinitiators derived from selected biopolyesters were accomplished recently with the aid of ESI-MS^n technique. The NMR and ESI-MS evaluation of the chemical structure of macroinitiators obtained by partial depolymerization of natural PHB, PHBV and PHO revealed that due to the elimination reaction they contain olefinic and carboxylic end groups.[4,15] Based on the ESI-MS^n studies of PHBV macroinitiator obtained by partial alkaline depolymerization of natural PHBV (containing 5 mole % of hydroxyvalerate units) the microstructure of this bacterial copolyester was assessed, starting from dimer up to the oligomer containing 22 repeat units.[4]

The PHO macroinitiator of the general structure shown in Figure 4 contained predominantly hydroxyoctanoate units and also a minor amount of hydroxyhexanoate as well as hydroxydecanoate units. The ESI-MS spectrum of this macroinitiator consisted of clusters of singly charged negative ions that are derived from this PHA after loss of the potassieted crown ether moiety (a). The anions were MS separated due to their different degree of oligomerization and composition (Fig 4a). The anion located at m/z 1419,1 (Fig 4a) was assigned to the PHO decamer. The MS^2 spectrum of this parent anion (Fig 4b) showed fragmentation at the carboxylic side of PHO. The set of fragment anions at m/z: 1277, 1135, 993, 851 and 709 is formed due to successive loss of 2-octenoic acid (142 Da). The MS^2 fragmentation experiment performed for selected parent anion of (HO)$_9$HD (m/z 1447, Fig 4c), contained one hydroxydecanoate unit, induced the set of clusters containing two fragment anions, with the difference of 28 Da. The fragment anion at m/z 1305 corresponds to (HO)$_8$HD oligomer, formed due to the loss of 2-octenoic acid (142 Da), and the fragment ion at m/z 1277 corresponds to (HO)$_9$ oligomer formed by the expulsion of 2-decenoic acid (170 Da) from the carboxylic side of the (HO)$_9$HD parent anion. The fragment ion m/z 1163 (HO)$_7$HD is formed due to expulsion of 2-octenoic acid from fragment anion m/z 1045, (HO)$_8$HD. However, the fragment anion m/z 1135 (HO)$_8$ may by formed from fragment anion m/z 1277 (HO)$_9$ due to the loss of 2-octenoic acid or from the fragment ion m/z 1305, (HO)$_8$HD, by the expulsion of 2-decenoic acid.

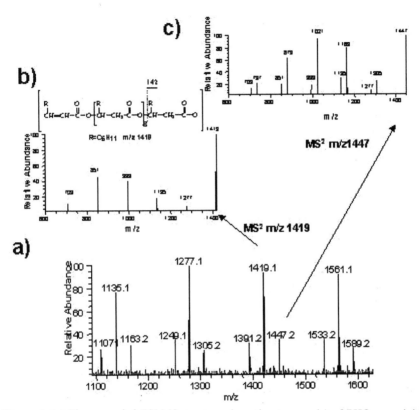

PHO: R = C_3H_7, C_5H_{11}, C_7H_{15}

Figure 4. **(a)** The expanded ESI-MS spectrum (negative ion–mode) of PHO macroinitiator, **(b)** the MS^2 fragmentation spectrum of PHO parent ion m/z 1419, **(c)** the MS^2 fragmentation spectrum of $(HO)_9HD$ parent ion m/z 1447.

Thus, the following fragmentation pathway may illustrate the result of the MS^2 experiment of parent ion m/z 1447$(HO)_9HD$:

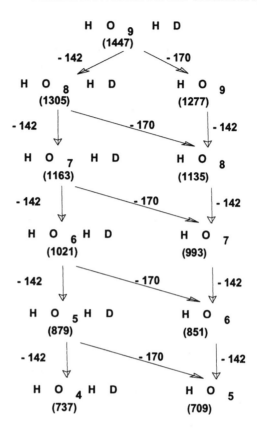

4.   CONCLUSIONS

The results of the present study indicate that the ESI-MS technique enables determination of the subtle structure of bio-inspired and natural polyesters. Moreover, the distribution of comonomers in the mass selected copolyester macromolecules can be shown by the ESI-MS^n fragmentation experiments, thus demonstrating the utility of this technique for the analysis of individual macromolecules including their end groups and composition.

ACKNOWLEDGEMENT

This work was partially supported by Polish State Committee for Scientific Research (Grant EUREKA E! 2004 "MICROPOL").

REFERENCES

1. Barron, A.E., Zuckermann, R.N., 1999, Bioinspired polymeric materials: in –between proteins and plastics, *Current Option in Chemical Biology* 3: 681.
2. Jedlinski, Z., Adamus, G., Kowalczuk, M., Schubert, R., Szewczuk, Z., Stefanowicz, P., 1998, Electrospray tandem mass spectrometry of poly(3-hydroxybutanoic acid) End groups analysis and fragmentation mechanism, *Rapid Commun. Mass Spectrom.* 12: 357.
3. Focarete, M.L., Scandola, M., Jendrossek, D., Adamus, G., Sikorska, W., Kowalczuk M., 1999, Bioassimilation of atactic poly[(R,S)-3-hydroxybutyrate] oligomers by selected bacterial strains, *Macromolecules* 32: 4814.
4. Adamus, G., Sikorska, W., Kowalczuk, M., Montaudo M., Scandola, M. 2000, Sequence distribution and fragmentation studies of bacterial copolyester macromolecules: characterization of PHBV macroinitiator by electrospray ion-trap multistage mass spectrometry, *Macromolecules.* 33: 5802.
5. Lenz, R.W., Jedlinski, Z., 1996, Anionic and coordination polymerization of 3-butyrolactone, *Macromol. Symp.* 107: 149.
6. Jedlinski, Z., Kurcok, P., Lenz, R.W., 1998, First facile synthesis of biomimetic poly-(R)-3-hydroxybutyrate via regioselective anionic polymerisation of (S)-β- butyrolactone, *Macromolecules* 31: 6718.
7. Scandola, M., Focarete, M. L., Adamus, G., Sikorska, W., Baranowska, I., Swierczek, S., Gnatowski, M., Kowalczuk, M., Jedlinski, Z., 1997, Polymer blends of natural poly(3-hydroxybutyrate-co-3-hydroxyvalerate) and a synthetic atactic poly(3-hydroxybutyrate). Characterization and biodegradation studies, *Macromolecules* 30: 2568.
8. Focarte, M. L., Ceccorulli, G., Scandola, M., Kowalczuk, M., 1998, Further evidence of crystallinity-induced biodegradation of synthetic atactic poly(3-hydroxybutyrate) by PHB-depolymerase A from *Pseudomonas lemoignei*. Blends of atactic poly[(R,S)-3-hydroxybutyrate] with crystalline polyesters, *Macromolecules* 31: 8485.
9. Jedlinski, Z., Kowalczuk, M., Adamus, G., Sikorska, W., Rydz, J., 1999, Novel synthesis of functionalized poly(3-hydroxybutanoic acid) and its copolymers, *Int. J. Biol. Macromol.* 25: 247.
10. Jedlinski, Z., 1999, Ring-opening reactions of β-lactones with activated anions, *Acta Chem. Scand.* 53: 157.
11. Arslan, H., Adamus, G., Hazer, B., Kowalczuk, M., 1999, Electrospray ionization tandem mass spectrometry of poly[(R, S)-3-hydroxybutanoic acid] telechelics containing primary hydroxy end groups, *Rapid Commun. Mass Spectrom.* 13: 2433.
12. Arkin, Z.G., Rydz, J., Adamus, G., Kowalczuk, M., in press, Water-soluble L-alanine and Related Oligopeptide Conjugates with Poly[(R,S)-3- hydroxybutanoic acid] Oligomers. Synthesis and Structural Studies by Means of Electrospray Ionization Multistage Mass Spectrometry, *J. Biomat. Sci, Polym, Ed.*
13. Adamus, G., Kowalczuk, M., 2000, Electrospray multistage ion trap mass spectrometry for the structural characterisation of poly[(R,S)-3-hydroxybutanoic acid] containing a β-lactam end group, *Rapid Commun. Mass Spectrom.* 14: 195.
14. Kowalczuk, M., Adamus, G., Sikorska, W., Rydz, J., 2000, Structural studies of biorelated polymers derived from natural PHA and their synthetic analogues with the aid of electrospray multistage mass spectrometry, *ACS Polym. Prep.* 41: 1626.
15. Kowalczuk, M., Adamus, G., Sikorska, W., Hazer, B., Borcakli, M., Arkin, A.H., 2000, Novel biodegradable packaging materials based upon atactic poly[(R,S)-3-hydroxybutyrate] and natural poly(hydroxyalkanoate)s, *Proc. International "Food Biopack Conference", Copenhagen*, p. 39.

Adsorption Studies of Humidity Presented by an Unbleached Kraft Woodpulp

JORGE M. B. FERNANDES DINIZ[1], MARIA H. GIL[2], and JOSÉ A. A. M. CASTRO[2]

[1] Escola Secundária de Jaime Cortesão, Coimbra, Portugal; [2]. Department of Chemical Engineering, University of Coimbra, Coimbra, Portugal

Abstract: A direct adsorption kinetic study of an unbleached Kraft woodpulp has exhibited a sequence of regimes of adsorption. The kinetics of adsorption of the first two stages may be described as typical of a first order process. The results obtained are perfectly consistent with qualitative descriptions of structure and fibre bonding in these materials, as abundantly reported in the literature. For the two first regimes, values for time constants and maximum mass of adsorbed water were evaluated. Clearly this technique reveals an outstanding potential for crucial information about surface area characterisation of paper woodpulps.

1.   INTRODUCTION

If the relation between one variable with time, undergoing a step input, may be described as an exponential rise to a maximum, then the kinetics of the process is typical of a first order linear system[1,2]. Its corresponding equation is as follows:

$$y = a\left(1 - e^{-bt}\right) \quad (1)$$

For a humidity adsorption process on a dried-up substrate, the constants in the above equation, may be related to specific physical quantities characteristic of the recorded physical changes:

Biorelated Polymers: Sustainable Polymer Science and Technology
Edited by Chiellini *et al.*, Kluwer Academic/Plenum Publishers, 2001

353

$$m(t) = m(\infty) \times \left(1 - e^{\frac{-t}{\tau}}\right)$$ (2)

where m(∞) is the asymptotic mass of adsorbed water, and τ, the process time constant, is the time at which the mass gain of the system attains 63.2% of the maximum (or asymptotic) value.

Experimental data may then be used to adjust, by non-linear regression, to equation (2), from which m(∞) and τ can be obtained.

2. EXPERIMENTAL METHOD

In order to maximise the adsorption of humidity by the woodpulp to be studied, a sample of an unbleached Kraft woodpulp was prepared in order to convert all of the carboxylic acid groups into carboxylate groups. For this purpose the unbleached woodpulp under study was submitted to a process of ion-exchange, by transferring the woodpulp into a 0.1 mol dm^{-3} aqueous solution of sodium chloride at a pH = 8.5 in a sequence of eight successive equilibria for 10 min of duration each. The choice of pH had the purpose of favouring the kinetics of the ion-exchange process, but avoiding lactone formation in the woodpulp[3]. The pulp was divided into three portions, and dried up until a constant weight was obtained.

The samples of the dried-up woodpulp were placed inside a closed chamber (adapted from a dessicator vessel) containing distilled water underneath to guarantee an atmosphere with a relative humidity of 100%. At time intervals in the scale of a 0.5-1 day, the samples were taken out of the chamber and their mass registered as a function of time for two minutes, in order to convey the mass of each pulp sample as a function of time inside the closed chamber. The mass of each sample of woodpulp to be taken into account was obtained graphically by extrapolation to zero time. A higher correlation factor was obtained at this stage with regression to a polynomial of second degree.

This research follows the same pattern of humidity adsorption study in other polymeric structures previously performed [4,5].

3. RESULTS

By plotting the mass of adsorbed water per unit mass of pulp as a function of time, the result is summarised in Fig 1.

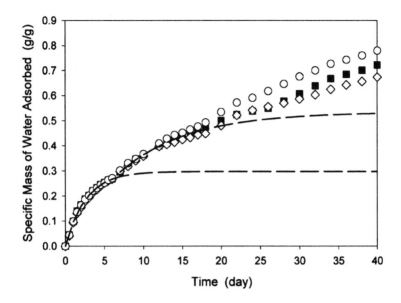

Figure 1. Global profile of the adsorption of humidity presented by three samples of non-bleached kraft woodpulp, showing the exponential regression curves in the two first regimes of adsorption.

This kinetic study of a woodpulp has exhibited a sequence of several regimes of adsorption, which may be identified by discontinuities in the first derivative of the curve. For the first two regimes, the degree of reproducibility in the three different samples is very high, quite unexpected for materials of this kind.

For the first regime, the exponential regression analysis indicated that $m_1(\infty) = 0.2974$ g g^{-1} and for the time constant $\tau_1 = 2.615$ day, with a correlation coefficient $r_1^2 = 0.998$.

For the second regime, the exponential regression analysis indicated that $m_2(\infty) = 0.2673$ g g^{-1} and for the time constant $\tau_2 = 8.865$ day, with a correlation coefficient $r_2^2 = 0.992$.

If one assumes that at the same level of relative humidity and degree of hydrophilic character of the substrate the number of layers of adsorbed water molecules is independent of the substrate, this number might correspond to seven[6,7]. Accepting as a good approximation that the molecular area of water molecules is 20×10^{-20} m^2 (20 Å^2)[8], then the initial surface area obtained for this unbleached woodpulp would be 2.5×10^2 m^2 g^{-1}.

4.   CONCLUSION

The results obtained in this study clearly confirm that the kinetics of adsorption for the first two stages may be described as typical of a first order process. It is also apparent that from the beginning of the third regime heterogeneity prevails over homogeneity in the three samples of woodpulp. Furthermore, the fact that $\tau_2 \gg \tau_1$ clearly suggests that the first stage of adsorption corresponds to a simple adsorption phenomenon of water molecules by the highly hydrophilic surface area initially available. As the adsorption proceeds, a second phenomenon occurs. The first layers of adsorbed water gradually destroy hydrogen bonds in the initial structure of the woodpulp fibres, thus increasing the available surface area for adsorption. A second regime of adsorption then proceeds at a slower rate as a new structure of micropores unveils. The fact that the amount of adsorbed water at the end of the first regime is nearly doubled at the end of the second regime clearly suggests an expansion of the available surface area, a fact always interpreted in terms of bulk swelling. These conclusions are perfectly consistent with descriptions of structure, fibre bonding and swelling in these materials, as abundantly reported in the literature[7,9,10]. Moreover, the technique adopted here to investigate the adsorption of water on woodpulp showed a remarkable reproducibility in the first two regimes of adsorption.

The value hereby derived for the initial surface area of the studied woodpulp is of the same order of magnitude of surface areas for unbleached woodpulps already published ($\approx 2.5 \times 10^2$ m^2 g^{-1}) [11], and based upon nitrogen adsorption studies. Although in need of a refinement in terms of experimental technique, this approach clearly reveals an outstanding potential for surface area determination of paper woodpulps.

REFERENCES

1. Stephanopoulos, G., 1984, Chemical Process Control: An Introduction to Theory and Practice, 1st. ed., Prentice-Hall Int. Ed., Englewood Cliffs, Chapters 7 and 10.
2. Ogunnaike, B. A., Ray, W. H., 1994, Process Dynamics, modelling and control, 1st. ed., Oxford University Press, New York, Chapter 5.
3. Fernandes Diniz, J. M. B., Pethybridge, A. D., 1995, Interfering Lactone Formation in Alkalimetric Studies of Paper Woodpulps, *Holzforschung* **49**: 81-83.
4. Pinto, P. V., Gil, M. H., Dias, A. M., Marques, A. T., Silva, M. A., 1989, Estudo de Algumas Características dos Adesivos de Epóxido, Materiais 89, Conference of the Portuguese Society of Materials, 351-360.
5. Piedade, A. P., Guthrie, J. T., Kazlauciunas, A., Gil, M. H., 1995, Characterisation of cellulose derivatives — relevance to sensor development, *Cellulose* **2**: 243-263.
6. Chatterjee, P. K., Nguyen, H. V., 1985, Mechanism of Liquid Flow and Structure Property Relationships. In *Absorbency* (P. K. Chatterjee, ed.), Elsevier, Amsterdam, 34.

7. Nissan, A. H., Walker, W. C., 1977, Lectures on Fiber Science, Joint Textbook Committee of the Paper Industry, Ed., Pulp and Paper Technology Series no. 4, Uppsala, pp 73-122.
8. Gregg, S. J., Sing, K. S. W., 1982, Adsorption, Surface Area and Porosity, 2nd. ed., Academic Press, 274.
9. Bristow, J. A., 1986, The Pore Structure and Sorption of Liquids. In *Paper Structure and Properties* (J. A. Bristow and P. Kolseth, eds.), Marcel Dekker, New York, pp 183-201.
10. Westman, L., Lindström, T., 1981, Swelling and Mechanical Properties of Cellulose Hydrogels. I. Preparation, Characterization, and Swelling Behavior, *J. Appl. Polymer Sci.* **26**: 2519-2532.
11. Herrington, T. M., 1985, The Surface Potential of Cellulose. In *Trans of the Eighth Fundamental Research Symposium* (V. Punton, ed.), Mechanical Engineering Publications Ltd, London, vol. I, pp 165-181.

Comparison of Quantification Methods for the Condensed Tannin Content of Extracts of *Pinus Pinaster* Bark

LINA PEPINO[1], PAULO BRITO[1], FERNANDO CALDEIRA JORGE[2], RUI PEREIRA DA COSTA[2], M. HELENA GIL[1] and ANTÓNIO PORTUGAL[1]
[1]*Departamento de Engenharia Química da Faculdade de Ciências e Tecnologia da Universidade de Coimbra, Pólo II – Pinhal de Marrocos, 3030 Coimbra – Portugal,* [2]*Bresfor, Indústria do Formol, S.A., Apartado 13, 3830 Gafanha da Nazaré – Portugal*

Abstract: Bark from Pinus Pinaster is an interesting source of polyphenolic natural compounds, that can be used successfully as total or partial replacement of conventional phenolic resins. These compounds, among other applications, are used as adhesives in the wood agglomerate industry. In this kind of application some problems remain to be solved in order to obtain a Pine extract of commercial value. It is necessary to optimise the extraction procedure and select a suitable method for the quantification of the tannin content of the bark. In order to study these problems, the tannin extraction from the Pine bark was tested with an alkaline solution (NaOH), and with a fractionation procedure based on a sequence of an organic (ethanol) and aqueous extraction. The phenolic content of each extract or fraction was evaluated by the Folin-Ciocalteu colorimetric assay for total phenols and two procedures using the Stiasny reaction: the gravimetric Stiasny method and the indirect colorimetric procedure that uses the Folin-Ciocalteu reagent to evaluate the total phenols present in the extract solution before and after it condenses with formaldehyde. The yield value when the alkaline extraction is used is substantially higher than the values obtained with organic or aqueous solutions. However, the selectivity of the process is low. In fact, it was found that the alkaline extract Formaldehyde Condensable Phenolic Material (FCPM) content represents 95-96 % of the total phenols content of the extract but this fraction is only ≈ 40 % of the total mass of extract. So, the alkaline extract is relatively poor in phenolic material, exhibiting a large variety of non-phenolic extractives. On the other end, ethanol provides a very rich phenolic extract, in which 96 % of total phenols are condensable with formaldehyde, but exhibits a relatively low extraction yield. The aqueous extract presents the lowest extraction yield with low content either in phenolic material as in FCPM, but, as most of the phenolics had already been extracted by the previous organic extraction, especially the low molecular weight fractions, this result was predictable.

1.   INTRODUCTION

In recent years, since the OPEC petroleum crisis in the early 70's, there has been an increasing interest on research for natural and cheap alternatives to synthetic phenolic petroleum-based adhesives for wood industry. Condensed tannins from wood barks that are widely used in the tanning industry, proved to be a viable phenolic source for these applications[6,7]. Tannins are natural polyphenols, usually classified as[2,6]: *Hydrolyzable* and *Condensed Tannins* or *Proanthocyanidins*. These groups involve structurally different chemical compounds. Their ability to bind and precipitate proteins is their main characteristic[2,9]. They are also able to condense with formaldehyde or undergo self-condensation without the need of any external reticulation agent. That characteristic confers adhesion properties to tannins[3,7]. Condensed tannins are complex polymers or oligomers of flavanoid units, namely flavan-3-ols and flavan-3,4-diols[6], that can be found in significant quantities in the bark of several species of trees, namely of the *Pinus* genre. *Pinus pinaster* is the most abundant forest specie in Portugal, and its bark is especially rich in condensed tannins (procyanidins and to a lesser extent prodelphinidins). Therefore, *Pinus pinaster* bark is an eligible source of polyphenolic natural compounds, that can be used as replacement of conventional phenolic resins (phenol-formaldehyde, urea-formaldehyde or melamine-formaldehyde) for the wood agglomerate industry. In order to obtain useful extracts, the following problems have to be addressed[6]: high reactivity of phloroglucinolic polymers (making the control of the polymerisation reaction difficult), high viscosities and low solubilities of the extract solutions (at least 40 % of solid contents). As a first approach, it is necessary to optimise the extraction procedure and select a suitable method for the quantification of the tannin content of the bark. In this paper, the results obtained for the tannin extraction from the Pine bark with an alkaline solution (NaOH) are presented. A fractionated extraction procedure was tested for tannin extraction of bark, using diethyl ether (for the elimination of lipophilic compounds) and a sequence of solvents with increasing polarity and decreasing specificity for tannins: ethanol and hot water[5]. The phenolic content of each extract or fraction was evaluated by the *Folin-Ciocalteu* colorimetric assay for total phenols[11] and two procedures using the *Stiasny* reaction for the quantification of the *Stiasny* phenols (phenols that condense with formaldehyde): ❶ the traditional gravimetric *Stiasny* method[12], ❷ the indirect colorimetric procedure using the *Folin-Ciocalteu* reagent to evaluate the total phenols present in the extract before and after the precipitation reaction with formaldehyde[10]. The results obtained in each extraction procedure in terms of extraction yields and phenolic content of the extracts, were compared in order to select the most suitable solvent and quantification method.

2. MATERIALS AND METHODS

2.1 Reagents

Gallic acid, catechin and *Folin-Ciocalteu* reagent were supplied by *Sigma Chemical Co.* Formaldehyde was graciously supplied by *BRESFOR-Indústria do Formol, S.A.*

2.2 Sample Preparation

Bark from 30 to 40 years old trees (*Pinus pinaster*) from the central region of Portugal was used. It was dried for 24 hours in an oven, with hot air flux at 100 °C. Dried bark was grinded by means of a hammer mill to less than 1 mm diameter particles.

2.3 Extraction Methods

Two procedures (Fig 1) for tannin extraction from the bark were used: alkaline extraction with a solution of NaOH, 1 % (w/w); fractionated extraction using a sequence of solvents with increasing polarity and decreasing specificity for tannins[5]: ethanol and hot water, which were utilised after a diethyl ether pre-treatment to remove the lipophilic components.

Alkaline Extraction: Tannins were extracted from samples (200 g) with 1000 ml NaOH, 1 % (w/w) (bark/solvent relation – 1:5) in a mechanically stirred reactor for 30 min at 90 °C. The suspension was separated by centrifugation and the clear extract was neutralised with HCl and conditioned in recipients at 4 °C. The extracted bark was dried to constant weight at 100 °C.

Fractionated Extraction: For the sequential extraction (Fig 1) a *Soxhlet* was used. Diethyl ether, ethanol and hot water was the solvent sequence selected. The extraction time for each procedure was defined by at least 50 cycles and the temperature was the boiling point of each solvent. The values used for the extraction were: 56 cycles (at 36 °C), 52 (at 78 °C) and 52 (at 98

°C) for each solvent, respectively. The extracts were dried by vacuum distillation. The extraction yields are defined as:

$$\eta = \frac{w_{dr.ext}}{w_{bark}} \cdot 100$$

(2.1)

where:

η- yield of the extraction;

$w_{dr.ext}$ - weight of the dried extract;

w_{bark} - initial weight of the bark in each extraction procedure.

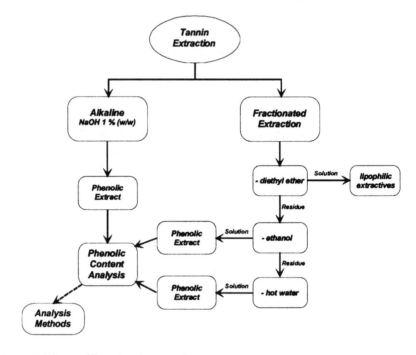

Figure 1. Scheme of the extraction procedures.

2.4   Methods of Analysis

2.4.1   Total Phenols:

Total phenols were determined by the *Folin-Ciocalteu* colorimetric method[11] (Fig 2). 2.5 ml of *Folin-Ciocalteu* reagent (diluted 10 times) and 2 ml of aqueous solution of sodium carbonate (75 g/l) were added to 0.5 ml of diluted extract, keeping the mixture 5 min at 50 °C. After cooling, absorbance was measured at 760 nm in a Jasco 7800 UV/VIS spectrophotometer. Aqueous solutions of catechin and gallic acid (8-40 µg/ml) were used as standards.

Calibration Curves Construction: calibration curves were constructed for both catechin and gallic acid using solutions of 8, 16, 24, 32 and 40 µg/ml and applying the procedure above described for the extract samples. Three replicates for each point were used.

2.4.2   Stiasny Polyphenols:

Stiasny polyphenols were determined by two different methods based on their ability to precipitate with formaldehyde (Fig 2):

Gravimetric Method[12]: 25 ml of extract were acidified with HCl to pH=1 and a molar excess of formaldehyde was added. This excess was calculated assuming that catechin and formaldehyde react in a 1:1 proportion. The suspension was refluxed during 30 min. The reaction products were filtrated, washed several times with hot water and dried to constant weight at 100 °C. The polyphenol content is expressed in absolute terms by the "*Stiasny* precipitation number with formaldehyde", η_{St}, defined by:

$$\eta_{st} = \frac{w_{pp}}{w_{ini}} \cdot 100$$

(2.2)

where:

η_{st} - *Stiasny* precipitation number;
w_{pp} - weight of precipitate that reacted with formaldehyde;
w_{ini} - weight of initial extract.

The *Stiasny* number is useful as an absolute measure for the condensable polyphenols in the extract but does not represent itself the content of this phenolic material. Therefore, it can not be used directly as a quantification method for this type of materials. Additionally, it is a laborious, slow and messy procedure that demands large quantities of extract solution.

Colorimetric Indirect Method: based on the typical reaction of polyphenols with formaldehyde, Singleton[10] proposed an assay for the quantification of phenolic material condensable with formaldehyde. He described a colorimetric procedure, instead of the gravimetric method, combining the *Folin-Ciocalteu* assay with the method of *Stiasny* (Fig 2). First, the total phenol content of the extract was measured before the precipitation reaction by the *Folin-Ciocalteu* method following the procedure described above. Then, 25 ml of extract solution was acidified to pH=1 and an excess of formalin 37.2 % was added to the solution. The suspension was refluxed during 30 min. The reaction products were removed by filtration and the total phenol content of the filtrate was measured again using the *Folin-Ciocalteu* method. The content of phenols that reacted was inferred by difference.

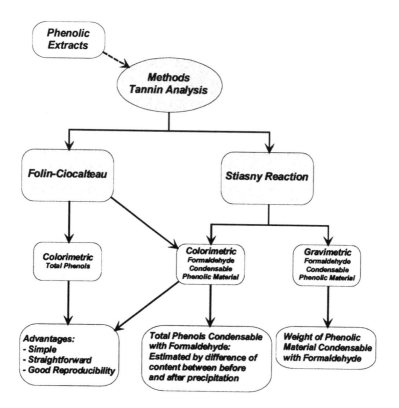

Figure 2. Methods used for the quantification of tannin content.

3.   RESULTS

3.1   Pine Bark Extraction

The yields of each extraction procedure and its comparison with the values reported in the literature[1] are presented on Table 1.

Table 1. Yield obtained for each extraction.

Fractionated Extraction		
Solvent	Yield (%)	Yield from the literature (%)[1]
Diethyl ether	6.7	-
Ethanol	8.9	10.3
Water	5.9	3.2
Alkaline Extraction		
NaOH 1% (w/w)	27.7	-

The extraction yields obtained are of the same order of the ones available in the literature and it is noticeable that the yield for the alkaline extraction is substancially higher than the typical yields obtained by organic or aqueous extraction. However, it is predictable that the rougher conditions provided by the alkaline extraction would promote the extraction of considerable quantities of non-phenolic extractives which can negatively affect the adhesion properties of the extract. Therefore, the comparatively higher value for the extraction yield of the alkaline extraction can be explained by the lower specificity of the solvent to phenolics and the effective composition of the alkaline extract (which present a much greater variety of extractives) has to be carefully analysed and controlled to prevent the degradation of the properties of the extract.

3.2 Quantification of Phenolics by the Method of Folin-Ciocalteu

To quantify the phenolic material content of each extract by the method of *Folin-Ciocalteu*, a calibration curve absorbance-concentration is necessary. A known standard that could be considered to infer the concentration of the phenolic material, namely the oligomeric molecules of condensed tannins has to be selected. In this work, gallic acid and catechin, that can be considered the basic units for the most common molecules of hydrolyzable and condensed tannins, were used. Both standards usually provide good linear correlations, but catechin presents a lower colour intensity[4]. Therefore, gallic acid is the most used standard for these applications.

The calibration curves are presented in Figures 3 and 4. We can conclude that the relation absorbance vs. concentration can be acceptably fitted by a linear model for both standards. A slightly better correlation for the gallic acid was obtained (Fig 3).

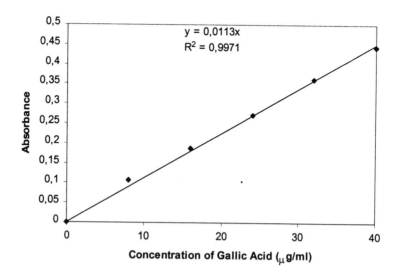

Figure 3. Calibration curve for the Gallic Acid.

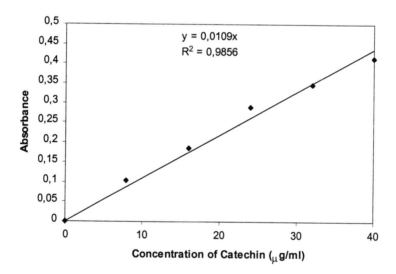

Figure 4. Calibration curve for the Catechin.

The results with gallic acid and catechin calibration curves are consistent (Table 2).

Table 2. Total phenols content in each extract by the *Folin-Ciocalteu* method.

	% of total phenols in the extract	
	Catechin equivalents	Gallic Acid equivalents
Fractionated Extraction		
Ethanol	92.9	92.9
Water	12.6	11.6
Alkaline Extraction		
NaOH 1% (w/w)	42.3	41.2

By analysis of Table 2 we observed that the richest extract in phenolic materials is the ethanol extract, as we would expect. The aqueous extract presented the lowest value. That can be explained by the fact that most of the phenolics had already been extracted by the previous organic extraction, especially the low molecular weight fractions, more soluble in water. Another possible justification for this low value could be the quantification method used that, as every redox method, is affected by the variation of the hydroxylation base and by the degree of polymerisation of the phenols[8]. The alkaline extracts exhibit an intermediate value that enforces the idea that the alkaline treatment provides a reasonable quantity of phenolic extractives but also significant quantities of other types of compounds.

3.3 Quantification of Phenolics Using the Stiasny Reaction

Comparing the results obtained for the two *Stiasny* procedures adopted, we can conclude that the results obtained are consistent (Table 3).

Table 3. Reaction of each extract with formaldehyde.

	% FTFti		% FTm extr		
	Catechin equivalents	Gallic Acid equivalents	Catechin equivalents	Gallic Acid equivalents	ηSt
Fractionated Extraction					
Ethanol	95.8	95.8	89.0	89.0	75.2
Water	36.9	34.2	4.6	4	-
Alkaline Extraction					
NaOH 1%	96.1	95.1	40.6	39.6	57.3

FT_{Fti} - Percentage of formaldehyde condensable material related to total phenols content in the extract.

FT_{mextr} - Percentage of formaldehyde condensable material related to total mass of initial extract.

η_{St}- *Stiasny* Number, defined by the relation between the mass of formaldehyde condensate and the mass of initial dry extract used in the reaction.

Considering the organic extraction, we verify that the Formaldehyde Condensed Phenolic Material (FCPM) content represents almost 96 % of the initial total phenol content of the extract and 89 % of the initial mass of the same extract. This result confirms the results presented in the above section. It is a clear indication of the abundance of phenolic material in the extract. Furthermore, it is noticeable that most of this phenolic material is condensable with formaldehyde. Therefore, the gravimetric *Stiasný* number is relatively high although not as high as we would expect.

The aqueous extract reveals low contents either in phenolic material as in FCPM. These values can be due to the fact that water dissolves preferentially low weight polyphenols. However, it is important to emphasise that this extract was concentrated in a rotating evaporator, where the removal of water was increasingly difficult as the extract became more concentrated and viscous. In the later stages of the operation the ebullition was violent making the polymerisation of the polyphenols possible, diminishing their reactivity to formaldehyde. Considering the very low value for the FT_{mextr} of the water, we can conclude that the aqueous extract contains a significant amount of non-phenolic compounds, typically simple sugars and polymeric carbohydrates[7]. It was not possible to apply the gravimetric method to the aqueous extract since the polyphenols-formaldehyde complexes formed were soluble and could not be separated from the solution by filtration.

Finally, the alkaline extract FCPM content represents 95-96 % of the total phenols content of the extract. However if this fraction is compared to the total mass of extract, we verify that this value is much lower: $\approx$ 40 %; η_{St} = 57.3, supporting the idea that the alkaline extract is relatively poor in phenolic material, presenting a number of non-phenolic extractives.

4.   CONCLUSION

The yield for the alkaline extraction of tannins from *Pinus pinaster* bark, is substancially higher than the typical yields obtained by organic or aqueous extraction. However, the specificity of the solvent to phenolics is low. Enforcing that conclusion we observed that the alkaline extract FCPM content represents 95-96 % of the total phenols content of the extract. However, when this fraction is compared to the total mass of extract, we verified that this value is much lower ($\approx$ 40 %; η_{St} = 57.3), which confirms the idea that the alkaline extract is relatively poor in phenolic material, presenting a variety of non-phenolic extractives. The most specific solvent is ethanol, that, in spite of revealing a relatively low extraction yield, provides a very rich phenolic extract, in which 96 % of total phenols are condensable with formaldehyde. The aqueous extract presents the lowest extraction yield, with a low content either in phenolic material and FCPM. However, that can be due to the fact that most of the phenolics had already been extracted by the previous organic extraction, especially the low molecular weight fractions.

REFERENCES

1. Anonymus (undated); Desenvolvimento de Adesivos com Incorporação de Taninos da Casca do Pinheiro para a Produção de Aglomerados de Madeira, Relatório Final – Projecto PBIC/C/AGR/2331/95, JNICT.
2. Cannas, A., 1999; "TANNINS: Fascinating but Sometimes Dangerous Molecules", http://www.ansci.cornell.edu/plants/toxicagents/tannin/index.html.
3. Jorge, F. C., Neto, C. P., Irle, M., Gil, H., Pedrosa de Jesus, J., 1998, *Wood Adhesives Based on Self-Condensation of Pine Bark Tannins*, Proc. 2nd European Panel Products Symposium, Llandudno, Wales, 21-22 Sept.
4. Julkunen-Tiitto, R., 1985, Phenolic constituents in the leaves of northern willows: methods for the analysis of certain phenolics, *J. Agric. Food Chem.*, **33**: 213-217.
5. Laks, P. E., 1991, "*Chemistry of Bark*", in Wood and Cellulosic Chemistry (Hon, D. N. S. and Shiraishi, N., eds.) Cap. 7, Marcel Dekker Inc., New York, pp 257-330.
6. Pizzi, A., 1983, "*Tannin-Based Wood Adhesives*", in Wood Adhesives: Chemistry and Technology (Pizzi A., ed.) Cap. 4, Marcel Dekker Inc., New York, p. 177-248.
7. Pizzi, A., 1994, "*Tannin-Based Wood Adhesives*", in Advanced Wood Adhesives Technology (Pizzi A., ed.) Cap. 5, Marcel Dekker Inc., New York, p. 149-217.
8. Price, M. L., Butler, L. G., 1977, "Rapid visual estimation and spectrophotometric determination of tannin content of sorghum grain", *J. Agric. Food Chem.*, **25 (6)**: 1268-1273.
9. Ribéreau-Gayon, P., 1972, *Plant Phenolics*, Cap. 7, Oliver & Boyd, Edinburgh, pp. 169-197.
10. Singleton, V. L., 1974, *Analytical fractionation of phenolic substances of grapes and wine and some practical uses of such analyses*, American Chemical Society, Adv. Chem. Ser., Washington, D.C., **137**: 184.
11. Singleton, V. L., Rossi Jr., J. A., 1965, Colorimetry of total phenolics with phosphomolybdic-phosphotungstic acid reagents, *Am. J. Enol. Viticult.*, **16**: 144-158.
12. Wissing, A., 1955, The Utilization of Bark II – Investigation of the Stiasny Reaction for the Precipitation of Polyphenols in Pine Bark Extractives, *Svensk Papp.*, **58 (20)**: 745-750.

The Role Of Life-Cycle-Assessment For Biodegradable Products: Bags And Loose Fills

[1]BEA SCHWARZWÄLDER, [1]RENÉ ESTERMANN and [2]LUIGI MARINI
[1]Composto+, Geheidweg 24, 4600 Olten, Switzerland; [2]Novamont SpA, via Fauser, 28100 Novara, Italy

Abstract: Life cycle assessments are of increasing importance for industrial development and for decision making in politics and at customers. The International Standard ISO 14040 guarantees high quality of LCA studies specifying the principles and requirements. The LCA study of Mater-Bi bags – incl. critical external review – performed the advantage of the biodegradable product compared to PE- and paper-bags and is used as selling argument for potential customers. The LCA of the Mater-Bi loose fills compared to EPS loose fills investigated different production variants and serves for process improvement.

1. INTRODUCTION

Biodegradable materials (BDM) were developed and sold as an ecological alternative to conventional plastics. The use of BDM instead of normal plastics is supposed to be less ecologically harmful, e.g. to reduce the impacts on climate changes (CO_2) and to save non renewable resources.

The economy developed a wide range of instruments to measure the economic success or failure of enterprises. Similar tools are requested to quantify the ecological impacts associated with manufactured and consumed products. One of the techniques being developed for this purpose is Life Cycle Assessment (LCA) a mean of environmental management systems[1]. Since 1997 the International Standard ISO 14040[2] has described principles and set frameworks for conducting and reporting LCA studies.

In practice there is a big number of LCA-software-tools available using different models for the impact assessment[3,4]. Several high quality LCA-

Biorelated Polymers: Sustainable Polymer Science and Technology
Edited by Chiellini et al., Kluwer Academic/Plenum Publishers, 2001

studies on BDM have been recently accomplished, so that some data and benchmarks of BDM are available now[5-13].

2.   PURPOSE OF LCA

LCA studies the environmental aspects and potential impacts throughout a product's life (i.e. from the cradle-to-grave) from raw material acquisition through production, use and disposal (Fig 1). The general categories of environmental impacts needing consideration include the exploitation of resources, human health, and ecological effects.

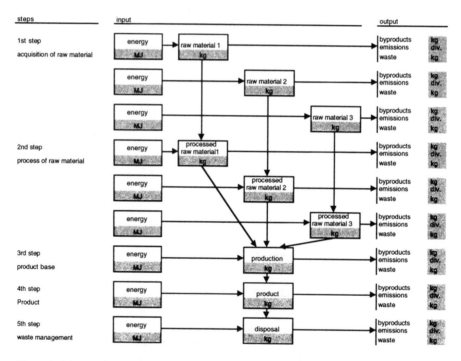

Figure 1. Scheme of a products life cycle

2.1   Internal use

LCA assists in identifying opportunities to improve the environmental aspects of products at various points of their life cycle. As it shows the environmental impacts separately for every production step in a processing plant, it is easy to find the most expensive step from the ecological (e.g. CO_2-emissions) and – in a restricted sense – from the economical point of view (e.g. energy demand) as well. Improving the most relevant steps is most advantageous for the environmental management system of an enterprise and might bring in money too, especially in the long term.

Cargill Dow Polymers (CDP) regards LCA to be an excellent process development tool to provide an insight about the polylactic acid (PLA) production, to set process improvements goals and to provide eco-profile data to be used by practitioners in performing their own life cycle studies[9]. On the other hand, CDP wants to make information available for the shareholders and the public due to the increasing demand of transparency. CDP started the first LCA in 1993 and updated it 3 times. In 2003, when the 140000 metric ton plant (MTPA) facility will be operative, new engineering data will be available for a next update.

As the biodegradable Mater-Bi loose fills can be obtained by using a conventional extruder, Novamont SpA evaluated their LCA to find the most ecologically favourable way for their production. The production of loose fills from granules directly at the customer's site, turned out to be most convenient as it minimized the transportation costs[8].

2.2   External Use

LCA can be used for decision making in industry, governmental and non-governmental organisations, e.g. for strategic planning and priority setting, and for marketing as an environmental claim, eco-labelling scheme or environmental product declaration.

The German agency of environment insists on LCA studies of BDM including comparative assertions as a base for the political strategy for waste management (disposal of BDM in the organic waste or not) and packaging taxes.

The LCA study of the biodegradable Mater-Bi bags serves for decision making i.e. in waste management federations to choose the bags, which are allowed in the organic waste collection and as a selling argument for potential customers. This study was the first LCA on bioplastics to be calculated and reported according to the International Standard ISO 14040 including an external critical review[7].

An increasing number of customer cares about the environmental engagement of the bidders and therefore a good documentation of the products' life cycle might be advantageous.

However, LCA studies proposed for external use – especially if results support comparative assertions – raise special concerns and require a critical review, since this application is likely to affect interested parties which were not involved in the calculating process.

3.   LCA ACCORDING TO INTERNATIONAL STANDARD ISO 14040

LCA studies accomplished according to the International Standard ISO 14040 guarantee high quality, because the International Standard specifies the general framework, principles and requirements for conducting and reporting them. However, no description of the LCA technique is reported in detail. Further International Standards ISO 14041, ISO 14042 and ISO 14043 provide additional details regarding methods concerning the various phases of LCA.

The International Standard ISO 14040[2] dictates that LCA include 4 phases: definition of goal and scope, inventory analysis, impact assessment and interpretation of results (Fig 2).

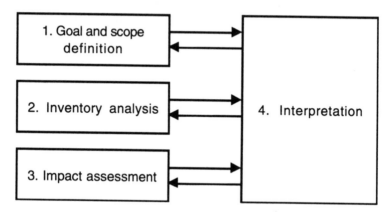

Figure 2. Phases of an LCA

3.1   Goal and Scope

The goal of the LCA study has to state unambiguously the intended application, the reasons for carrying out the study and the intended audience.

The most crucial points in defining the scope of an LCA study are the functional unit, the product system studied, the system boundaries, the types of impact and the methodology of impact assessment and subsequent interpretation to be used, the data requirements and the type of critical review, if any.

3.2   Inventory Analysis

Inventory analysis involves data collection and calculation procedures to quantify relevant inputs and outputs of a product system. The qualitative and quantitative data for inclusion in the inventory have to be collected for each unit process within the system boundaries. The data constitute the input to the life cycle assessment.

3.3   Impact Assessment

The impact assessment phase of LCA is aimed at evaluating the significance of potential environmental impacts by using the results of the life cycle inventory analysis. In general, this process involves association of inventory data with specific environmental impacts and attempting to understand those impacts. The phase of the impact assessment mostly includes elements such as: classification (assigning of inventory data to impact categories), characterisation (modelling of the inventory data within impact categories) and possibly evaluation (aggregating the results). Evaluation is a very sensitive process and should only be used in special cases provided that the data prior to evaluation remain available.

The methodological and scientific framework for impact assessment is still being developed. Models for impact categories are in different stages of development. There is subjectivity in the life cycle impact assessment phase such as the choice, modelling and evaluation of impact categories. Therefore, transparency is critical to impact assessment to ensure that the assumptions are clearly described and reported.

3.4   Interpretation

Interpretation is the phase of LCA in which the findings from the inventory analysis and the impact assessment are combined to reach conclusions and make recommendations.

The findings of this interpretation may take the form of conclusions and recommendations to decision-makers, consisting of the goal and scope of the study.

3.5   Report

The results of the LCA have to be reported fairly, completely and accurately to the intended audience. Which means that the results, data, methods, assumptions and limitations have to be transparent and presented in sufficient detail to allow the reader to understand the complexities and trade-offs inherent in the LCA study.

When the results of the LCA are to be communicated to a third party, a third party report has to be prepared, covering a critical review among other things.

3.6   Critical Review

The use of LCA results to support comparative assertions raises special concerns and requires critical review since this application is likely to affect interested parties that are external to the LCA study. A critical review may facilitate understanding and enhance the credibility of LCA studies, for example, by involving interested parties.

The critical review process has to ensure that

- the methods used to carry out the LCA are consistent with the International Standard ISO 14040.

- the methods used to carry out the LCA are scientifically and technically valid.

- the data used are appropriate and reasonable in relation to the goal of the study.

- the interpretations reflect the limitations identified and the goal of the study.

- the study report is transparent and consistent.

Dependent on the size and the intended audience, different kinds of critical review are possible: internal or external expert review and review by interested parties.

4.   KEY FEATURES OF LCA

LCA is always a snapshot that considers the actual state of the art. Therefore comparing results from LCA studies calculated 2 years before or later is "apple-to-pear-comparison".

LCA is never an exact calculation because most of the parameters can not be measured exactly and are more or less assumptions. Data from literature (e.g. for energy or transports) are average values. Some LCA-software tools consider this inaccuracy and allow a special error of calculation.

The scope, assumptions, description of data quality, methodologies and output of LCA studies have to be transparent. LCA studies should discuss and document the data sources and be clearly an appropriately communicated. The depth of detail and time frame of an LCA study may vary to a large extent, depending on the definition of goal and scope. Therefore results of different studies can hardly be compared to each other. The quality of the database determines the quality of the LCA.

5.   BIODEGRADABLE BAGS FOR ORGANIC WASTE COLLECTION

Life cycle assessments were applied in 1997/98 to analyse the degree of ecological damage caused by the production and disposal of Mater-Bi bags used in households to collect organic waste. Paper bags "haushalt kompost", which can be composted, and PE multipurpose bags, which cannot be composted, were used as points of reference.

The life cycles included raw material acquisition, the production and processing and/or disposal of the bags as well as routes of transport. Packaging, distribution, utilisation and collection as well as transport to the wholesalers could not be considered due to the dependency of these processes on the respective bulk buyers and retailers.

Life cycle profiles were drawn up using the modified impact-oriented model [3,5] and the impact categories of Eco-Indicator '95[4]. All in all the degree of ecological damage could be identified in thirteen different impact categories. The calculations were obtained by application of the life cycle assessment software EMIS (Environmental Management and Information System, Version 2.2). Most data were taken either from internationally recognised literature (energy supply[14], production and processing of paper, PE [polyethylene][15,16], disposal processes[17,18], transport[19]) or they were supplied by the manufacturers. In order to analyse sensitivity, new unit processes for the agricultural production of maize in France and for organic waste incineration were created. Assessments were carried out separately for each impact category because of previously specified boundaries of reliability.

The production and disposal of Mater-Bi bags (Table 1) causes less environmental damage than that of paper bags "haushalt kompost" in eleven out of thirteen impact categories. In the two remaining categories the Mater-Bi bag causes the same or greater degree of ecological damage.

The Mater-Bi bag and the multipurpose PE bag are equivalent in seven impact categories; the Mater-Bi bag achieves better results in four categories, but worse results in the two remaining categories. However, Mater-Bi bags generate less environmental damage than PE multipurpose bags in ten categories if one considers the waste adhering to the bags and being incinerated together with them. In two categories both bags obtain the same results, while in one category the production of Mater-Bi bags generates more environmental damage. It does not prove relevant for the overall results whether maize produced in Switzerland or respectively in France, is used. Mater-Bi bags made of French maize were selected for the overall assessment as maize on the European market is mainly produced in France.

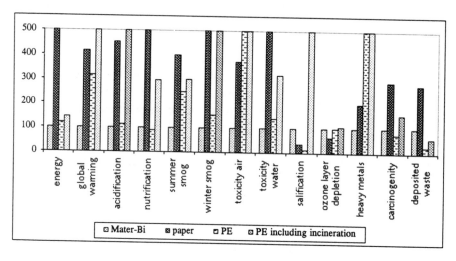

Figure 3. Assessment of the three products (Mater-Bi, paper and PE bag) ac. to the modified impact oriented model[3,5] and Eco-Indicator '95[4] standardised to the Mater-Bi bag.

Assessments of the Mater-Bi bag and the multipurpose PE bag show that both can be regarded as equivalent, as long as the focus remains on production and disposal (disregarding compostable waste incineration). If the compostable waste that is incinerated with the PE bags is taken into account, the Mater-Bi bag offers a better ecological value. The production and disposal of the paper bag is bound to cause considerably more damage to the environment than that of the Mater-Bi bag (Table 1).

Table 1. Mater-Bi bag in comparison with other products. Read: "In comparison with product A in 13 impact categories a Mater-Bi bag made of French maize is significantly better / better / comparable / worse / significantly worse."

Mater-Bi bag in comparison with	paper bag	PE bag	PE bag [a]
much better	5	2	6
better	6	2	4
comparable	1	7	2
worse	1	0	1
much worse	0	2	0
total result	**better**	**comparable**	**better**

a) Including organic waste incineration

There is no doubt that for the municipal collection of organic waste biodegradable bags should be recommended. From an ecological point of view the Mater-Bi bag has to be given preference over the other compostable bag (paper bag "haushalt kompost"). Short routes of transport and minimal use of packaging material should be weighty criteria as for the choice of product.

The study was conducted and reviewed according to International Standard ISO 14040 (ISO 14040, paragraph 7.3.2 ["external expert review"] by Dr. Gérard Gaillard from the Eidgenössische Forschungsanstalt für Agrarwirtschaft und Landtechnik, "Federal Research Centre for Agriculture and Cultivation Methods").

6. BIODEGRADABLE LOOSE FILLS

Life cycle assessments were applied to analyse the degree of ecological damage caused by the production and disposal of loose fills made out of Mater-Bi pellets in comparison to those made of expanded polystyrene.

The life cycle analysis includes raw material acquisition, production, processing and disposal of the loose fills as well as transport. Packaging, distribution, use and collection are not considered due to the dependency of these processes on the respective bulk buyers and retailers.

Life cycle profiles were drawn up using the modified impact-oriented model[3,5] and the impact categories of Eco-Indicator '95[4]. All in all the degree of ecological damage could be identified in thirteen different impact categories. The calculations were obtained by application of the life cycle assessment software EMIS (Environmental Management and Information System, Version 2.2). Most data were taken either from internationally recognised literature (energy supply[14], production and processing of Expanded Poly(styrene) [EPS][15,16], disposal processes[17,18], transport[19]) or the manufacturers supplied them. In order to analyse sensitivity new unit processes for the wastewater treatment of Mater-Bi loose fills were created. Assessments were carried out separately for each impact category because of previously specified boundaries of reliability.

The production and disposal of Mater-Bi loose fills causes less environmental damage than the one of EPS loose fills in eight of thirteen impact categories, but more in two categories. The emissions in the three remaining categories are comparable (Table 2).

Assessments of the Mater-Bi loose fills under different production and disposal management show small changes in the eco-profiles in comparison to the basic calculation.

- Disposal of Mater-Bi loose fills in a wastewater treatment instead of composting plant
- Production of Mater-Bi loose fills from pellets directly at the customer's site, avoiding transportation of the loose fills.
- Production of loose fills directly from starch, avoiding production and transportation of pellets

As Mater-Bi loose fills were not longer compared with an other product but with themselves considering different production variants, two new, closer limits were defined for the assumption of the production variants of Mater-Bi loose fills (Table 2).

Table 2. Mater-Bi loose fills in comparison with EPS loose fills and different production or disposal management. Read: " In comparison with product A in 13 impact categories Mater-Bi loose fills are significantly better, better, (slightly better), comparable, (slightly worse), worse, significantly worse."

Mater-Bi loose fills in comparison with	EPS loose fills	Mater-Bi loose fills in waste water treatment	Mater-Bi loose fills produced at the customer's site	Mater-Bi loose fills produced directly from starch
much better	3	0	0	0
better	5	2	0	0
slightly better	-	2	0	0
comparable	3	9	7	10
slightly worse	-	0	6	3
worse	2	0	0	0
much worse	0	0	0	0
total result	**better**	**slightly better**	**slightly worse**	**comparable**

The sensitivity analysis of the Mater-Bi loose fills under different production and disposal management doesn't show relevant changes to the basic calculation (see Fig 4 and Table 2).

In each two categories it is better or respectively slightly better than the waste water treatment of Mater-Bi loose fills. The production of Mater-Bi loose fills at the customer's site and directly from starch is slightly better than the normal production of Mater-Bi loose fills (basic calculation), in 6, respectively 3 categories.

From the ecological point of view the Mater-Bi loose fills have to be given preference over EPS loose fills. The disposal of Mater-Bi loose fills in wastewater treatment is possible but shouldn't be pushed due to the small ecological disadvantage. Producing Mater-Bi loose fills at the customer's site or directly from starch save economically and ecologically expensive transports.

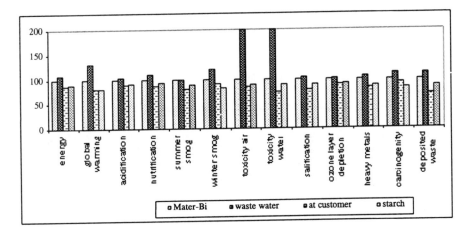

Figure 4. Assessment of the four variants of Mater-Bi loose fills (basic calculation, waste water treatment, production at the customer, production directly from starch) ac. to the modified impact oriented model[3,5] and Eco-Indicator '95[4] standardised to the Mater-Bi loose fills basic calculation.

7.   CONCLUSION

- LCA studies are of increasing importance for biodegradable products: to improve the production process, for external communication, for politics.
- International Standards such as ISO 14040, available database and special software tools guarantee a high quality and decreasing costs of the calculations.
- The change of agriculture to a more sustainable farming would amplify the ecological advantage of biopolymers made out of renewable resources. This aspect becomes more and more relevant regarding the increasing excess of farmland.
- LCA is a necessity for products that are intended to be sold using ecological arguments.

REFERENCES

1. Braunschweig, A. and Müller-Wenk, R., 1993, *Ökobilanzen für Unternehmungen – Eine Wegleitung für die Praxis*, Verlag Paul Haupt, Bern.
2. International Organisation of Standardisation, 1997, *Umweltmanagement - Produkt - Ökobilanz - Prinzipien und allgemeine Anforderungen. Schweizerische Normenvereinigung*, Zürich.

3. Heijungs, R., Guinee, J.B., Huppes, G., Lankreijer, R.M., Udo De Haes, H.A. & A. Wegener Sleeswijk 1992, *Environmental Live Cycle Assessment of Products, Guide and Backgrounds*, (R. Heijungs ed), CML Leiden.

4. Goedkoop, M. 1995, *The Eco-indicator 95*, Amersfoort 1995.

5. Buwal, 1996, Ökobilanz stärkehaltiger Kunststoffe. SRU 271, Bundesamt für Umwelt, Wald und Landschaft, Bern

6. Carbotech and FAT, 1995, *Bewertung nachwachsender Rohstoffe (Zwischenbericht). Bundesamt für Landwirtschaft, Bern.*

7. Composto, 1998, *Life Cycle Assessment of Mater-Bi bags for the collection of compostabl waste. Study acc. to ISO 14040 incl. critical review.* Novamont SpA, Novara (Italia)

8. Composto, 2000, *Life Cycle Assessment of Mater-Bi and EPS Loose Fills. Study acc. to ISO 14040.* Novamont SpA. Novara (Italia)

9. Eichen Conn, R., 2000, Life Cycle Assessment of PLA Polymers. 7th Conference Biodegradablee Polymers 2000, Würzburg (D)

10. FAT and Carbotech, 1997, *Beurteilung nachwachsender Rohstoffe in der Schweiz in den Jahren 1993 - 1996.* Bundesamt für Landwirtschaft, Bern.

11. IFEU, in prep: Nachhaltige Nutzung von nachwachsenden Rohstoffen zur Förderung regionaler Stoffkreisläufe – Beurteilung der Hemmnisse und Möglichkeiten auf dem Gebiet des Bauwesens. Institut für Energie- und Umweltforschung Heidelberg GmbH, Heidelberg (D)

12. IFEU, NOVA, IAF 1996, Hanfprodukte im Vergleich zu konventionellen Produkten. Übersichts-Ökobilanz im Rahmen des Projektes "Erarbeitung von Produktlinien auf Basis von einheimischem Hanf – aus technischer, ökonomischer und ökologischer Sicht". Deutsche Bundesstiftung Umwelt, Osnabrück (D)

13. Würdinger, E., Wegener, A., Roth, U., Reinhardt, G.A., Detzel, A. 2000: Kunststoffe aus nachwachsenden Rohstoffen: Vergleichende Ökobilanz für Loose-fill-Packmittel aus Stärke und Polystyrol. BifA, Augsburg (D), IFEU, Heidelberg (D).

14. ETH, 1996: Ökoinventare von Energiesystemen, ENET Bern.

15. Buwal, 1995, Vergleichende ökologische Bewertung von Anstrichstoffen im Baubereich, Band 2. SRU 232, Bundesamt für Umwelt, Wald und Landschaft, Bern

16. Buwal, 1996, Ökoinventare für Verpackungen. SRU Nr. 250 (2 Bände), Bundesamt für Umwelt, Wald und Landschaft, Bern

17. ETH, 1996: Ökoinventare von Entsorgungsmodulen, ENET Bern.

18. Aebersold, A., Eichenberger, S., Künzli Hauenstein, M., Schmid, H. And Schmidweber, A., 1993, Vergären oder Kompostieren? Entscheidungshilfen für die Systemwahl. NDS Umweltlehre 1993, Universität Zürich, Zürich.

19. SSP Umwelt, 1995, Ökoinventar Transportarten, Modul 5, Verlag Infras.

Index